Chinese Academy of Sciences · Ministry of Education of the PRC · Chinese Academy of Social Sciences · Chinese Academy of Engineering · National Data Administration · China Association for Science and Technology · National Natural Science Foundation of China · Chinese Academy of Agricultural Sciences
Editors

China's e-Science Blue Book 2025

中国工信出版集团 電子工業出版社 PUBLISHING HOUSE OF ELECTRONICS INDUSTRY http://www.phei.com.cn
Springer

Editors
Chinese Academy of Sciences
Beijing, China

Ministry of Education of the PRC
Beijing, China

Chinese Academy of Social Sciences
Beijing, China

Chinese Academy of Engineering
Beijing, China

National Data Administration
Beijing, China

China Association for Science and Technology
Beijing, China

National Natural Science Foundation of China
Beijing, China

Chinese Academy of Agricultural Sciences
Beijing, China

ISBN 978-981-96-8251-5 ISBN 978-981-96-8252-2 (eBook)
https://doi.org/10.1007/978-981-96-8252-2

Jointly published with Publishing House of Electronics Industry
The print edition is not for sale in China (Mainland). Customers from China (Mainland) please order the print book from: Publishing House of Electronics Industry.

This Springer imprint is published by the registered company Springer Nature Singapore Pte Ltd.
The registered company address is: 152 Beach Road, #21-01/04 Gateway East, Singapore 189721, Singapore

Preface

In recent years, the rapid development of emerging information technologies such as big data and artificial intelligence (AI) has facilitated e-Science that serves as an increasingly critical driving force in scientific and technological innovation. Research activities are evolving towards greater digitization and intelligence. Notably, in the 2024 Nobel Prizes for Natural Sciences, two of the three major awards were related to AI, solidifying "AI for Science" as a cutting-edge focus in the global research. In addition, in the era of digital economy, data has become a new type of production factor, a core element of scientific research and innovation, and a fundamental pillar for AI application. The synergy between data and AI results fuels the rapid development of scientific research paradigms and technological innovation.

Over the past 2 years, the integration of big data and artificial intelligence ("Big Data + AI") has led to significant breakthroughs across various research fields, including protein structure prediction, automated processing of astronomical data, extreme weather forecasting and climate trend analysis, the discovery and optimization of new materials, pharmaceutical R&D and disease diagnosis, as well as ecological and environmental monitoring. "Big Data + AI" is driving a new wave of paradigm revolution in scientific research, accelerating advancements in science, technology, and industry. To support these transformations, e-Science is evolving to meet the demands of this paradigm shift by developing a new digital, intelligent, and networked research informatization infrastructure platform. This new platform will facilitate the efficient supply of data, computing power, and network resources while enabling the intelligent coordination of various scientific and technological resources. It is not only an upgrade to the traditional e-Science platforms but also a forward-looking, comprehensive system designed to meet the evolving needs of future research, providing a safe, efficient, and flexible digital environment for scientific research.

To summarize and demonstrate the progress and practical applications of China's e-Science infrastructure development over the past 2 years, and to further explore the research paradigm shifts driven by "Big Data + AI", as well as advancements in the new research informatization infrastructure platform, the Chinese Academy of Sciences, in collaboration with the Ministry of Education of the PRC, the Chinese

Academy of Social Sciences, the Chinese Academy of Engineering, the National Data Administration, the China Association for Science and Technology, the National Natural Science Foundation of China, and the Chinese Academy of Agricultural Sciences, has jointly compiled and published the *China's e-Science Blue Book 2025*. This book was compiled from 17 articles by leading experts in the fields of e-Science in China, focusing on the latest developments, fundamental capabilities, and practical applications of research informatization. Rich in content and detailed case studies, this book serves as a valuable reference for policymakers, researchers, and professionals in government departments, scientific research institutions, colleges and universities, and relevant enterprises and institutions engaged in the field of informatization.

Finally, due to the limited preparation time and incomplete access to information, this book may not fully capture all aspects of China's e-Science development. We sincerely apologize for any omissions. We warmly welcome feedback and suggestions from readers across various fields. Your valuable insights will help us continuously refine and enhance our work. We remain dedicated to pursuing even greater excellence in our future endeavors.

Beijing, China
March 2025

Working Group on *China's e-Science Blue Book 2025*

Contents

Forefront Trends

Artificial Intelligence Large Models for Scientific Research: Developments, Challenges and Paths

Wen Gao, Shu Liu, Jiankun Yang, Wei Zhang, Juntong Yan, and Ge Li

Abstract Artificial intelligence is becoming the core driving force of the new generation of technological and industrial revolutions. And the large models have marked the beginning of a new path towards artificial general intelligence. Meanwhile the rapid development of artificial intelligence has brought significant changes and opportunities to scientific research. AI for science as a new paradigm, has more potential to accelerate the pace of innovation than ever before, and has become a frontier hotspot with remarkable achievements. In this article, it is elaborated in the first chapter that the development history of artificial intelligence and the technological evolution of large models, as well as the transformation of scientific paradigms driven by "big data + AI". Then the state of the art in large models for science is systematically presented, including mathematics, biomedicine, chemistry, materials, and earth science. Furthermore, the challenges faced by artificial intelligence large models for science is analyzed. Finally, the recommended development paths of artificial intelligence large models for science are proposed.

Keywords AI for science · Large models · The fifth paradigm

In recent years, artificial intelligence (AI) has experienced accelerated development, particularly with AI large models empowering various industries, initiating a new phase toward artificial general intelligence (AGI). "AI for Science" (AI4S) has emerged as a global frontier in AI research. Comprehensively understanding AI's developmental trajectory and the opportunities and challenges involved by AI in the field of scientific research, plays an important role for scientific and technological innovation advancement.

W. Gao · S. Liu · J. Yang (✉) · W. Zhang · J. Yan
Pengcheng Laboratory, Shenzhen, China
e-mail: yangjk@pcl.ac.cn

W. Gao · G. Li
Peking University, Beijing, China

China's e-Science Blue Book 2025, https://doi.org/10.1007/978-981-96-8252-2_1

1 AI Drives a Paradigm Shift in Scientific Research

The aim of scientific research is to uncover fundamental principles and solve practical problems. In recent years, with the advancements in big data, massive computing power, and large models, AI has provided new tools and novel approaches for scientific research, which is driving paradigm shifts in scientific methodologies and accelerating the processes of scientific innovation.

1.1 AI Development Enters a New Phase

AI is a technological science that focuses on studying and developing theories, methods, technologies, and application systems to simulate, extend, and expand human intelligence. Since its formal proposal at the Dartmouth Conference in 1956, the concept of AI has undergone three major development waves: the symbolic logic reasoning phase, the expert systems with manual rules phase, and the current data-driven deep learning phase. In 2006, Professor Geoffrey E. Hinton from the University of Toronto proposed the deep learning algorithm, laying the theoretical and methodological foundation for the new wave of AI development. In 2012, the deep learning neural network model AlexNet defeated other non-neural network models by a large margin in the ImageNet image recognition challenge, marking the rise of deep learning. Data-driven deep learning effectively integrates data, algorithms, and computing power, shifting AI research focus from creating intelligence to acquiring intelligence. Deep learning models such as Convolutional Neural Networks (CNN), Recurrent Neural Networks (RNN), and Graph Neural Networks (GNN) have been widely adopted. In 2022, AI large models represented by ChatGPT demonstrated human-like capabilities in performing diverse tasks, ushering in the new phase toward AGI.

Large models, also known as large-scale pre-trained models or foundational models, are deep learning models trained on massive datasets with substantial parameter scales and generalization capabilities. The Transformer architecture proposed by Google in 2017 laid the algorithmic foundation for mainstream large models. In 2018, Google and OpenAI released BERT and GPT-1 (Generative Pre-trained Transformer) respectively, that established large models as the main approach in natural language processing. In 2020, OpenAI introduced GPT-3 as the largest language model with 175 billion parameters, which achieved significant performance improvements in zero-shot learning tasks. This breakthrough was listed among MIT Technology Review's "10 Breakthrough Technologies of 2021" and sparked a global research hotspot in large models. In November 2022, ChatGPT powered by GPT-3.5 was released, amassing 100 million active users within two months to become the fastest-growing application in history. It demonstrated exceptional capabilities in natural language generation, code generation, and logical reasoning. This milestone marked the beginning of accelerated development in AI large models. These models rapidly expanded from natural language processing to other general intelligence domains like vision and hearing, evolving toward multimodal capabilities. Models such as Gemini, GPT-4, and GPT-4o can process text, images, audio, and other data types, achieving qualitative leaps in real-time conversation and multimodal analysis. The "pre-training + fine-tuning" paradigm of large models can well adapt to different downstream tasks, demonstrating strong versatility. During the pre-training phase, self-supervised learning based on large-scale data is used to obtain an "understanding" of the data. In the fine-tuning stage, the model is optimized according to the labeled data of specific tasks to meet the requirements of specific tasks. Through alignment, the behavior of the large model is made to conform to human intentions and values (Fig. 1).

In China, AI large models have entered a phase of vigorous development, with a series of prominent models emerging successively. These include BAAI's "WuDao·TianYing", Baidu's "ERNIE", Huawei's "Pangu", Pengcheng Laboratory's "Pengcheng Mind", and Alibaba's "Tongyi". These models have not only achieved significant improvements in intelligence levels but also expanded application scenarios continuously. In 2023, general-purpose large models experienced rapid development, while 2024 witnesses accelerated practical implementation of industry-specific large models. The empowerment effect of large models across various sectors is unleashing new economic growth momentum, serving as a crucial engine driving the development of new quality productive forces.

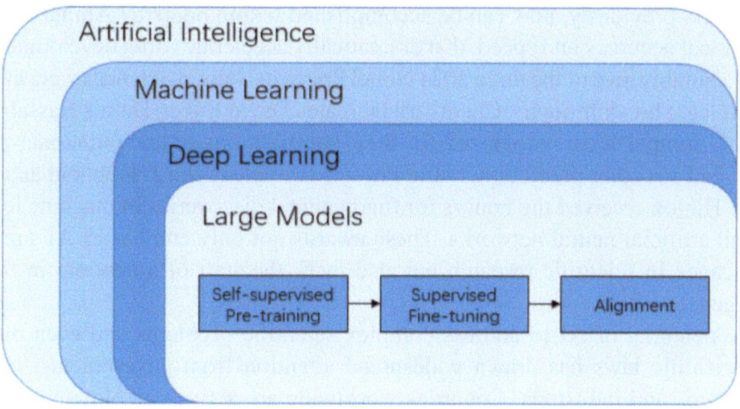

Fig. 1 Evolution of AI technology and typical training process of large models

1.2 The AI-Driven Paradigm Shift in Scientific Research

The breakthroughs in AI have caused paradigm shifts across multiple scientific fields. Turing Award laureate Jim Gray categorized the historical evolution of scientific discovery into four paradigms: empirical science, theoretical science, computational science, and data-intensive science [1]. Since 2018, numerous researchers have proposed a fifth paradigm, named as AI-driven scientific research or Science Intelligence [2], representing a new phase where AI becomes fully integrated into scientific, technological, and engineering investigations. This fifth paradigm integrates the strengths of previous four paradigms, combining empirical and theoretical approaches and merging AI with computational science. Through the integration of "scientific big data + massive computing power + high-performance models", it provides essential resources and tools for scientific exploration. Data serves as a critical component in this new paradigm, where AI models acquire machine intelligence through data "learning", subsequently employing this intelligence as a tool to facilitate scientific research and discovery. A landmark example is DeepMind's AlphaFold2 for protein structure prediction, recognized as revolutionary progress in AI applications for scientific research and honored by both Nature and Science as one of 2021's most significant scientific breakthroughs [3]. AlphaFold2 integrates protein structures and biological information into deep learning algorithms through Multiple Sequence Alignment (MSA), and enables prediction of three-dimensional protein structures from amino acid sequences. What required years of effort for researchers previously, now can be accomplished within hours by AlphaFold2 with exceptional accuracy and speed, that dramatically accelerates drug development efficiency. Notably, two of the three 2024 Nobel Prizes in Natural Sciences were awarded to AI-related breakthroughs. Chemistry laureates David Baker, Demis Hassabis, and John M. Jumper were recognized for their contributions to computational protein design and structure prediction, while Physics laureates John J. Hopfield and Geoffrey E. Hinton received the honors for fundamental discoveries in machine learning through artificial neural networks. These awards not only emphasize AI's growing significance in scientific research but also mark the start of a new era in Science Intelligence.

The potential of AI to address complex scientific problems and even discover new scientific laws has drawn widespread attention from governments, research institutions, and industries. Countries worldwide are actively deploying AI-driven scientific research as a crucial direction for future technological innovation. The U.S. National Science Foundation has funded universities to establish AI research institutes since 2021; while the U.S. Department of Energy released a policy report in May of 2023, that outlined frontier directions and funding agendas for AI in science. Global tech leader enterprises like Google, Microsoft, Huawei, and Baidu are actively advancing scientific AI development. Dedicated research teams such as Microsoft Research AI4Science and the Beijing Academy of Artificial Intelligence have emerged. Leading Chinese AI institutes like Pengcheng Laboratory, Pujiang Laboratory, and Beijing Academy of Artificial Intelligence continue to innovate in this field. The Bulletin of the Chinese Academy of Sciences in the January 2024 issue focused on "Advancing Transformations in Research Paradigms" to discuss AI-driven scientific research. As an emerging and rapidly growing field [4], AI-driven scientific research has promoted increasingly substantial outcomes, with remarkable achievements published in top journals like Nature and Science. The number of papers and patents in this field has grown significantly in global scale, with China ranking first in paper publications.

2 Development of Scientific Research Empowered by AI Large Models

The newborn paradigm of AI-driven scientific research has been accelerating innovative scientific research and the commercialization of scientific and technological achievements in an unprecedented manner. Over the past decade, AI technologies such as machine learning and deep learning have been increasingly adopted in various applications in scientific research, ranging from reducing repetitive work for scientists, addressing the "curve of dimensionality", assisting scientific research to stimulating scientific discoveries. We have witnessed a growing number of impactful AI-driven scientific achievements since 2023 [5, 6]. The rapid development of AI large models offers a new path for scientific research, bringing about countless notable advances in scientific domains including but not limited to mathematics, biomedicine, chemistry and materials science, and geoscience. Via comprehending scientific data, AI large models have demonstrated their potentials to leverage approaches of generating and planning to accelerate scientific discoveries.

2.1 Mathematics

The mathematical problems include arithmetic, word problems, geometry, automatic theorem proving, and mathematics in visual content, among others. For solving mathematical problems, it's required to improve the mathematical reasoning ability of AI large models, which is based on the abilities of language understanding, image interpretation, table analysis, symbolic manipulation, logical operations, and the understanding the principles of mathematical knowledge.

With the high-quality specialized corpora and the improved fine-tuning strategies, the ability of AI large models has been significantly enhanced in mathematical tasks. In 2023, the Mathematical Computing Model "Abel", achieved the first place in authoritative evaluation rankings in mathematical reasoning, such as GSM8K and MATH, with an accuracy rate of 83.6%. It was developed by the Generative AI Research Group of Shanghai Jiao Tong University, and fine-tuned with a carefully designed corpus pattern. In 2024, the Alibaba Qianwen large model team released

Qwen2-Math, which was specifically for solving mathematical problems. The was pre-trained with a specialized mathematical corpus, which contained large-scale high-quality mathematical network texts, books, codes, exam questions, and some mathematical pre-training data generated by the Qwen2 model. The mathematical capabilities of Qwen2-Math were the best among similar large models.

Supported by knowledge bases and external tools, AI large models can effectively carry out formal theorem proving. The theorem prover architecture has evolved from classical machine learning algorithms (such as KNN, GNN) to Transformer-based large models, who treat expressions as strings. The open-source theorem prover LeanDojo [7], developed by Caltech and NVIDIA, builds datasets by extracting theorems, proofs, strategies, and premises from Lean's Mathlib database. It trains a retrieval-augmented large model that enables the model to learn how to effectively select and apply premises from mathematical libraries, generate the next proof strategy based on the current proof state, and interact programmatically with Lean to complete theorem proofs.

AI large models are able to collaborate with mathematicians and improve reasoning abilities through prompt engineering, chain of thought, and other methods, effectively assisting in solving complex mathematical problems. Researchers from Microsoft Research Asia, Peking University, and Beihang University proposed a universal problem-solving framework called "Socratic" reasoning. This framework guides large language models recursively to discover, solve, and integrate problems through continuous dialogue, promoting self-assessment and refinement. It encourages the large model to coordinate various sub-problems and establish high-level reasoning paths when solving highly complex tasks, in order to find answers in a vast solution space. Through 97 rounds of strict reasoning, GPT-4 reached the conclusion that "P≠NP" [8], demonstrating the ability of large model to collaborate with human in exploring complex problems.

AI large models show tremendous potential in proposing mathematical solutions and generating new mathematical discoveries. In 2023, the journal **Nature** published a mathematical discovery using large model programs by the DeepMind team. The study developed FunSearch, which pairs the multimodal large model PaLM2 with an automated "evaluator." The large model generates creative solutions in the form of computer code, while the evaluator checks whether the solution outperforms correct solutions and provides feedback to the large model. The iterative process evolves the initial solution into new knowledge. FunSearch successfully found novel and better-performing solutions to classic mathematical problems such as the Cap-Set and Bin-Packing, and can also provide the solution process. This demonstrated the feasibility of using large models for mathematical solutions and new knowledge discovery [9].

2.2 Biomedicine

Biomedicine is an interdisciplinary frontier science that integrates life sciences, biology, medicine, pharmacy, and other fields. Omics data from life sciences such as genomics and proteomics, along with medical imaging and literature data, provide a rich corpus for large-scale AI models. The empowerment of AI large models in the field of biomedicine is significant, which has been enabling the establishment of cross-modal representation models for text, images, proteins, molecules, and genomes. This enhances the understanding of biomedical knowledge and facilitates the design of small molecule drugs and proteins by analyzing different data patterns, including images and sequences.

AI large models are effectively applied in medical Q&A and clinical diagnostics. The GPT-4-Medprompt developed by Microsoft's research team employs a combination of dynamic few-shot selection and automatic generation of thought chains. The GPT-4-Medprompt has achieved an accuracy rate of over 90% on the MedQA dataset (US Medical Licensing Examination questions), surpassed the famous Med-PaLM2 model [10]. Meanwhile, the MediTron-70B was developed by the Swiss Federal Institute of Technology Lausanne, and based on the open-source large model LLaMA2. It is pre-trained with articles from PubMed and internationally recognized medical guidelines. Its performance on medical knowledge understanding and reasoning was enhanced through supervised fine-tuning via medical reasoning benchmark datasets. In collaboration with Waterwood Molecular, Tsinghua University has constructed a foundational model BioMedGPT [11] for the multimodal biomedical field. This model integrates multi-source heterogeneous data such as genes, molecules, cells, proteins, literature, patents, and knowledge bases, and it creates a dataset that aligns natural language with biomolecular representations. Fine-tuned on a large-scale biomedical corpus based on LLaMA2, it merges text knowledge embedded in molecular language and graph knowledge summarized through wet lab experiments. It enhances the model's generalization ability and interpretation ability, and then supports cross-modal Q&A across natural language, molecules, and proteins.

AI large models have made breakthrough progress in drug development and gene prediction, and have effectively enabled protein structure prediction, drug target identification, drug-target binding prediction, and synthetic pathway planning. The Pengcheng Laboratory has launched the Pengcheng-Brain-Shennong large AI model in the biomedical field. Based on the Pengcheng-Cloud-Brain-II supercomputer and the MindSpore AI framework, it leverages big biomedical data, computational biology theories and technologies, AI algorithms, and computing clusters to achieve novel drug screening and creation, virus evolution prediction, protein structure prediction, small molecule generation, target-small molecule interaction prediction, and new antimicrobial peptide design and efficacy evaluation. Taking antimicrobial peptides as an example, utilizing the powerful amino acid sequence generation capability of Pengcheng-Brain-Shennong model, combining with existing antimicrobial peptide datasets, thousands of candidate peptide amino acid sequences can be generated in a short time. Subsequent biological experiments and clinical validation can significantly accelerate the discovery progress of new antimicrobial peptides. A research project completed by the Pengcheng Laboratory in collaboration with Peking University and Shandong University on the Pengcheng-Brain-Shennong platform, titled as "Leading the Evolution of Viruses—Predicting Future High-Risk SARS-CoV-2 Variants through AI Simulation," successfully was selected as the finalist of the 2022 Gordon Bell Special Prize for COVID-19. Furthermore, Waterwood Molecular has released the ChatDD-FM, which is a multimodal biomedical dialogue model with hundreds of billions of parameters. The ChatDD-FM model can integrate and understand multimodal data such as small molecules, macromolecules, DNA, single-cell sequencing data, and natural language text. Enhanced with biomedical expertise, fine-tuned with professional domain instructions, and reinforced with human feedback, it can interact naturally with researchers and collaborate with humans, to improve drug development efficiency. In addition, Huawei, in collaboration with the Shanghai Institute of Materia Medica, has trained the PanGu Drug Molecule Model, which enables AI-assisted drug design for the entire process of small molecule drugs.

2.3 Chemistry and Material Science

Chemistry and Material Science is also a key area of AI large model, and the training data usually comes from literature, tool books, and professional databases. Chemistry and material AI large models are being applied to many areas such as experimental automation, scientific research assistants, new material discovery, and so on.

Trained with specific corpora, this kind of large models show excellent performance on the chemistry-related tasks. The open source AI large model ChemLLM, developed by Shanghai AI Lab, undergoes two-stage command fine-tuning based on InternLM2.0. The first stage uses open source general datasets for training, and the second stage uses general datasets added with the chemistry field datasets for training. The chemistry field datasets include various chemical command data synthesized by some templates and chemistry-related data retrieved from internet (including PubChem, USPTO database, chemistry books and so on). This AI large model outperforms GPT-3.5 in tasks like chemical names transforming, title analysis, and chemical reaction conditions prediction. Moreover, AISpeech Shanghai Jiao Tong University Human-Computer Interaction Joint Lab and Suzhou Lab jointly released a ten-billions-scale professional chemistry large model with the name of ChemDFM [12]. It is based on the open source large model LLaMa, and uses almost four million literature of chemistry, thousands of chemistry books and some other databases to form a corpora with 34 billion words for pre-training. To fully learn and master the chemical technical terms and expression, the ChemDFM has collected numerous data from PubChem molecule database and USPTO chemical reaction database, and has builted 1.7 million items of data for fine-tuning. Both the domain pre-training corpora and command fine-tuning corpora are constructed with equivalently number and corresponding format. It has enhanced the nature language ability of ChemDFM. And finally the ChemDFM model with 13 billion parameters comes out, and outperforms GPT-4 in most chemistry-related tasks. On the other hand, Tsinghua University has proposed ChatMol large model, which helps researchers to realize molecular design in conversation manner by describing naturally and editing target molecule. The ChatMol model has combined experimental property information, molecular space knowledge and the relationship between natural language and chemistry language [13].

AI has many great achievements on material discovery, material generation, material property prediction, material synthesis path optimization and so on. For instance, GNoME from DeepMind has discovered more than 2.2 million potential chemical compounds [14]. MatterGen from Microsoft can generate the materials with special chemical composition, symmetry, or scalar properties (like magnetic density) [15]. AI large model shows great potential in the field of material science. AI for Science Institute Beijing and DPTechnology applied the AI large model Uni-Mol to predict the properties of organic luminescent molecules, to realize high efficiency sift of the 2nd generation OLED molecule lr(III) compounds, to find potential patent molecule with good optical properties, and to discover new OLED materials [16]. Chinese Academy of Sciences extracted 35675 solid-state reaction synthesis of inorganic materials from over 4 million papers, and leaded to 13878 items high confidence synthetic pathway descriptions. What's more, they fine-tuned the open source large model LLaMA2-7B, and developed the large model—MatChatt [17], focusing on inorganic material synthesis pathway prediction tasks. This large model owns material synthesis fields' knowledge generation and inference ability, which presents the potential innovation and application of large models in the field of material science.

2.4 Geoscience

Earth science studies the properties of rocks, minerals, and landforms, as well as phenomena and principles related to Earth's climate, oceans, atmosphere, ecosystems, and other aspects. Earth science data includes scientific literature data, geological data, meteorological data, remote sensing data, geographic knowledge bases, news articles, and points of interest, and more.

Specialized large models trained on geoscience datasets can provide new perspectives and tools for earth science research, and enhance research efficiency. In May 2023, Shanghai Jiao Tong University launched K2, a 7-billion-parameter geoscience large model [18], and in July 2023, they introduced GeoGalactica, a 30-billion-parameter geoscience large model. K2 was built on the original LLaMA-7B model and underwent secondary training using over 1 million geoscience research papers and geoscience-related Wikipedia articles. It was further fine-tuned with GeoSignal, a specialized geoscience dataset containing article content, categories, references, and domain-specific entities. It can finish tasks like answering geoscience questions, entity recognition, and hierarchical concept analysis of geoscience concepts, and more. The GeoGalactica large model extends this foundation of K2. It corporates 6 million geoscience research papers and leverages the enhanced GeoSignal dataset from K2, aiming to align with domain expertise and support advanced geoscience research applications better. To evaluate performance in understanding and applying geoscience knowledge, the team also established a benchmark for assessing geoscience language models.

Remote sensing large model can process multispectral remote sensing data, hyperspectral remote sensing data, and radar remote sensing data. It usually integrates geographic knowledge, expert experience, and domain knowledge together, to effectively enhance generalization capabilities and promote innovative applications of remote sensing technology. In May 2023, Pengcheng Laboratory, the National Remote Sensing Center, the Department of Science and Technology of Guangdong Province, Guangzhou University, and The University of Hong Kong jointly released the Pengcheng Star Cloud Data Service Platform and the corresponding datasets. These initiatives aim to advance China's active participation in international big science programs and projects in the fields of remote sensing and astronomy. The Pengcheng Star Cloud Data Service Platform, as a scientific research platform for remote sensing and astronomy, is built upon the large scientific infrastructure Pengcheng-Cloud-Brain-II with E-level AI computing power. Based on the Pengcheng-Mind large model, the platform offers three major services: intelligent multi-domain and multi-modal data production services, multi-domain AI+ disciplinary algorithm research services, and interdisciplinary research exchange services. Base on this platform, the dataset is analysis ready data and can be directly applied to scientific research. Since the launch of the platform, there have been nearly 6 thousand independent registered users, who are from 74 countries and regions. The platform has cumulatively responded to over 300,000 times of data download requests, with more than 700 thousand data entries downloaded, totally nearly 100 TB. Users are widely from international top-tier institutions such as the University of Cambridge, University of Oxford, Tsinghua University, and Wuhan University, and from research organizations like the National Development and Reform Commission, China Geological Survey, and Chinese Academy of Agricultural Sciences, and from international institutions such as the United States Geological Survey (USGS), Google, and the UK Centre for Ecology & Hydrology, and so on. This platform will leverage the advantages of AI technology in data-driven scientific research applications, and aim to break through key technologies in water cycle, carbon cycle, and urban sustainable development, as well as in advanced research areas like the origin of the universe and galaxy evolution. At the same time, these initiatives will promote interdisciplinary integration and advance scientific intelligence. They hold significant importance for supporting China's anticipation of international big science programs and engineering projects, and for expanding the frontiers of knowledge, exploring the unknown world, and addressing major global issues.

AI large models have also been developing rapidly in the field of weather forecasting, especially showing remarkable results in short-and-medium term weather forecast. The GraphCast weather forecasting model, released by Google DeepMind, contains 36.7 million parameters and can predict global weather for the next 10 days at a high resolution of 0.25° longitude/latitude (28 km × 28 km) [19]. In April 2023, the Shanghai AI Laboratory released the AI large model *Fengwu.* This model achieved effective 10-day forecasts of core atmospheric variables at high resolution, reportedly reducing the error rate by 10.87% compared to Google's AI model. It can generate global high-precision forecasts for 10 days in just 30 s. In March 2024, the upgraded Fengwu-GHR model was released. It integrated big data and high-performance computing with meteorological prediction to establish an intelligent cross-domain, multi-scale, and accurate meteorological system. Its forecast resolution was improved to 0.09° (9 km × 9 km), expanding the precise prediction range seven times compared to the first-generation Fengwu. Another important approach is Huawei's Pangu Weather Model, which endeavors to provide global weather forecasts in seconds. Its predictions include meteorological elements such as geopotential, humidity, wind speed, temperature, and sea-level pressure, and it can be directly applied to various specialized meteorological research scenarios. In 2024, the Pangu Weather Model was upgraded and achieved a leap in accuracy from global 25 km resolution model to more difficult regional forecasts with resolutions of 1, 3, and 5 km, covering meteorological elements such as temperature, precipitation, and wind speed. In June 2024, the China Meteorological Administration launched three major AI-driven systems: the global short-and-medium term forecasting system Fengqing, the AI-based real-time casting system Fenglei, and the global subseasonal-seasonal prediction system Fengshun. The application scope of meteorological models is expanding into industry services, such as pollutant prediction and agricultural production guidance. They are becoming important tools to support decision-making across various sectors.

3 Challenges Faced by AI large models for Scientific Research

Although AI has achieved numerous breakthroughs in scientific research and has been credited by government, academia and industry, there are still many challenges regarding data aquisition, data quality, key model technologies, massive computing power supplies, interdisciplinary collaborations, and more. The following four aspects are getting more attention.

3.1 Data: Aggregation and Sharing of Multimodal Scientific Data

Data is the key to the high-quality development of AI large models. The data volume, quality and diversity are the critical factors that influence the capabilities of large models. As revealed by OpenAI, the amount of text data for training GPT-3.5 reached 45 TB. On top of this, multimodal datasets were further included in GPT-4. It must be admitted that data has become one of the most conspicuous technological barriers in developing AI large models. Big scientific data not only provides a wealth of data sources for training AI large models for scientific research, but also enables the models to strengthen generalization capabilities and extend their potentials for better data exploitation and data mining.

Many challenges exist in the acquisition, processing, applications and management of big scientific data. In terms of data acquisition, datasets from public aspects are the primary training data for current AI large models. Most existing public datasets comprise English texts alone, while the shortage of high-quality Chinese corpus is becoming more and more prominent. Furthermore, although vast amounts of knowledge, experimental data, observational data, survey data and simulation data have been aggregated by enterprises, universities and research institutes through long-term scientific research, these valuable resources still remain underutilized and have not been fully converted into high-quality training datasets for AI large models. Other high-quality datasets such as those from books, journals or the internet are usually fragmented, isolated and not easy to access. In some cases, scientific research would rely on data resources that are challenging and costly to acquire. This is by all means a severe bottleneck that hinders the iteration and optimization of AI technologies.

In terms of data processing and applications, scientific research may involve datasets that differ in volume and type, and require different processing compared to those for general use. There are unique traits associated with scientific data from different domains. For example, remote sensing images, obtained through aircrafts, satellites, or other remote sensing instruments, have covered a large number of spectral bands, so that they can provide detailed information about surface features. These images are also large in size and contain objects with a wide range of scales. Similarly, genomics data is usually highly multi-dimensional, encompassing both temporal and spatial dimensions. In short, a critical challenge lies in how to effectively store, process and analyze such domain-specific data from various scientific domains for one particular science objective.

In terms of data management, the use of scientific data concerns ownership and access issues. Moreover, the standards, rules and regulations for utilizing data require further improvement. Appropriate technical and management measures are also necessary to promote data sharing and ensure data security, in order to allow the use of data to comply with laws, regulations, and ethical standards. In this regard, China has been making tremendous efforts in building scientific data platforms. For example, the National Basic Science Data Center systematically integrates and provides open access to data from scientific domains such as physics, chemistry, astronomy, geospace and biology. The overall data volume reaches 431 TB in total.

While certain achievements have been made in the construction of big data platforms, the development of AI would require even larger amounts of high-quality data, more diverse knowledge sources, and more open and accessible data and information exchange mechanisms. It is essential to promote interdisciplinary and cross-domain collaborative innovation, leveraging data-driven approaches to discover new scientific laws, generate new knowledge, and accelerate new advancements and reforms in scientific research paradigms.

3.2 Computing Power: Flexible Availability of High-Performance and AI-Driven Computing Resources

The rapid advancement and ongoing evolution of large AI models have led to a dramatic surge in the demand for computing power. For instance, training ChatGPT at a rate of 10 trillion operations per second would take 3,640 days, equivalent to the computing time required for 64 NVIDIA A100 GPUs running continuously for one year. Scientific research data processing typically requires large-scale, high-precision computations, which in turn requires the seamless integration of high-performance computing resources and AI-driven computing resources.

The computing resources needed by large model developments primarily come from intelligent computing centers, supercomputing centers, and cloud computing centers. The computing power scale is limited by the supply of AI chips like GPUs, so there is no enough AI computing resources in many countries. And then, the scale of intelligent computing power at intelligent computing centers and supercomputing centers is not sufficient to meet the broad demands of scientific research. On the other hand, for the specialized requirement of scientific task, AI computing and high-performance computing abilities are sometime needed to solve one problem together. Therefore, it is necessary to coordinate general computing power, intelligent computing power, and high-performance computing power to achieve collaborative computing abilities. This will address the issues of providing flexible and accessible computing power, efficient sharing and transmission of massive data, and the green, low-carbon nature of facilities. In February 2022, China officially launched the East Data West Computing Initiative to promote the construction of a nationwide integrated computing network. The first phase of the China Computing Network (C2Net), led by Pengcheng Laboratory, has been deployed. However, the rapid development of China's computing network still faces several challenges [20]. On the technical side, it is crucial to address issues such as the insufficient compatibility and interconnection of various heterogeneous resources, high network transmission latency between computing nodes, and limited data exchange bandwidth. Organizationally, there is an urgent need to strengthen national coordination, optimize resource allocation, and provide robust computing power support to empower scientific research, particularly for large AI models.

3.3 Algorithm: Applicability of Artificial Intelligence Models

From classical machine learning to deep learning and now to foundation models, AI has made significant advancements in natural language processing, computer vision, and multimodal learning. These fields typically have well-defined datasets, standardized information encoding methods, and clear optimization objectives. However, scientific research presents far greater complexity. Scientific tasks often involve intricate and domain-specific challenges that require a deep understanding of underlying principles, dynamic data structures, and evolving methodologies. Effectively integrating AI into scientific research requires more than just adapting existing models, while it demands a fundamental alignment between AI capabilities and the nature of scientific inquiry. Researchers must bridge the gap between AI-driven pattern recognition and the rigorous hypothesis-driven processes of scientific discovery. To achieve this integration presents significant challenges, from optimizing AI architectures for scientific reasoning to ensuring research objectives with transparency, reliability, and alignment.

From an information representation perspective, scientific knowledge—such as the structures and properties of proteins, genes, and molecules—must be encoded in a structured scientific language to be processed by computers and effectively learned by AI models. Most large language models rely on the Transformer framework for pre-training, learning semantic relevance in language, and adapting well to downstream tasks through fine-tuning. However, these methods may not be optimal for scientific research, since they require specialized optimization tailored to the unique characteristics of scientific tasks. Ethical and security concerns also pose challenges for applying large AI models in scientific research. Like general-purpose models, data-driven scientific models can inherit biases from their training data, generate misleading hallucinations, or produce contents that contradict established scientific knowledge, potentially leading to errors in research. Additionally, these models face risks such as adversarial attacks and data breaches particularly when handling sensitive biological data, that necessitates strict security and governance measures. Model evaluation presents another significant challenge. Scientific knowledge and tasks are highly specialized and complex, yet conventional AI evaluation benchmarks often lack the depth to assess scientific reasoning and accuracy effectively. There is an urgent need to develop domain-specific benchmarks for rigorous evaluations to AI models on complex scientific tasks. It is also important to recognize that the powerful intelligence exhibited by foundation models today primarily results from that the model parameters reach a certain scale. However, these models remain largely black boxes, lacking transparency in their decision-making processes. Moreover, key cognitive functions such as complex causal reasoning and logical thinking are still underdeveloped and not good enough. As a result, research in alternative AI development pathway continues in pursuit of more explainable, reliable, and scientifically capable models.

3.4 Ecosystem: Advancing Interdisciplinary Research and Engineering Collaboration

AI-driven scientific research is fundamentally reshaping disciplines and knowledge systems. It demands not only the integration of fields such as computer science, data science, material science, chemistry, and biology but also the deeper theoretical and algorithmic contributions from foundational disciplines like mathematics and physics. Beyond technological advancements, AI-driven research is also transforming the organization of scientific activities. In the era of foundation models, where data, computing power, and model development serve as critical infrastructure, collaboration among research institutions, enterprises, and other stakeholders becomes essential. The convergence of cutting-edge research across multiple disciplines fosters new ideas and methodologies, which are then validated and translated into practical applications through engineering implementation. Such interdisciplinary and cross-sector collaboration requires robust mechanisms and technical frameworks. To ensure sustainable development, it is crucial to establish new research institutions, cultivate interdisciplinary talent, and strengthen collaboration between academia, industry, and research organizations. Additionally, in the context of AI-driven scientific research, new challenges arise in research integrities, skills, and ethical considerations, which are based on the development of corresponding guidelines and governance frameworks.

4 Development Paths for AI Models in Scientific Research Fields

To promote the healthy development of AI foundation models for scientific research, it is essential to strengthen the construction of infrastructure such as data and computing power, to build multimodal scientific AI foundation models and platforms, to boost an interdisciplinary innovation ecosystem, and to integrate scientific research with AI innovation, thereby advancing the role of AI in supporting scientific research and scientific discoveries.

4.1 Strengthen Scientific and Intelligent Infrastructures

The construction of AI infrastructure for scientific research is a crucial foundation for driving AI empowerment. In recent years, the development of large AI models has confirmed the Scaling Law, which asserts that increases in model size, training data, and computational resources directly enhance model performance. So that, data and computing power are essential prerequisites for the advancement of large-scale scientific models.

In terms of data resources, it is essential to advance the development of high-quality corpora and scientific databases to provide robust support for both general and scientific large models. First, efforts should focus on strengthening the construction of data resource systems, expanding the volume, content diversity, and quality of corpora, and promoting openness and sharing. In 2022, China's data output reached 8.1 ZB, and by the end of that year, the data storage volume had reached 724.5 EB, reflecting a 21.1% year-on-year increase and accounting for 14.4% of global data storage [21]. Institutions are continuously advancing the sharing of high-quality datasets. For example, in 2023, Beijing released over 600 TB of high-quality datasets for large AI models, including the People's Daily corpus dataset and the national laws and regulations corpus dataset. In May 2024, the China Meteorological Administration released the Artificial Intelligence Meteorological Large Model Training Thematic Data Catalog for the first time, which includes six types of meteorological data and products across three major categories. Additionally, there is a need to build and integrate scientific databases and knowledge databases. Internationally recognized scientific databases, including research-generated data, experimental data from technology transfer processes, and cross-disciplinary knowledge, should be compiled into high-quality scientific databases and knowledge databases with gradually promoted sharing through effective mechanisms and technologies. Synthetic data is also a valuable resource for scientific research. In natural science, data can be generated based on scientific laws. For example, the microscopic properties of electronic structures and molecular systems can be derived by solving the Schrödinger equation. Furthermore, it's critical to foster multi-level data sharing and application through the development of scientific big data platforms. For instance, the sharing of earth observation data has evolved from providing raw satellite data to offering analysis-ready data, data products, and customized data analysis. In terms of data management, in 2022, the Central Committee of the Communist Party of China and the State Council issued the Opinions on Building a Basic Data System to Better Leverage the Role of Data Elements, which accelerates the development of a fundamental data system, focusing on data ownership, circulation, trading, and security governance.

In terms of computing power, the advancement of high-performance AI chips provides critical computational support for large models. Additionally, more software and hardware solutions tailored to large model training and inference are being developed. Intelligent computing platforms, built on large-scale heterogeneous computing power, will become a key enabler of scientific research. Led by Pengcheng Laboratory, the China Computing Network (C2Net) are serving as a critical infrastructure to support the high-quality development of the digital economy. The network connects multi-source heterogeneous and massive, ubiquitous computing centers, enabling efficient resource scheduling, green and low-carbon facilities, flexible computing power supply, and intelligent on-demand services. The first phase of the C2Net was officially launched in June 2022. Using Pengcheng Cloud Brain as the hub node, it has successfully managed more than 20 heterogeneous computing centers across domains and now aggregates over 5 EFlops of computing power. A national intelligent computing interconnection system and an open-source AI platform have been established, making computing power and AI open-source services available to users nationwide. In December 2023, the National Development and Reform Commission, along with other departments, issued the Implementation Opinions on Deeply Implementing the East Data West Computing Project and Accelerating the Construction of a National Integrated Computing Power Network. This aims to continue advancing the development of both national and urban computing networks. Technology companies, in collaboration with universities and research institutions, are also jointly building dedicated computing platforms to foster scientific intelligence research. For example, Alibaba Group and Fudan University have partnered to create a cloud-based scientific research intelligent computing platform. This platform features an intelligent computing cluster for multidisciplinary integration and innovation, called QWen, as well as a high-performance computing cluster dedicated to large-scale tasks, known as Jin Si. Through a public cloud model, it provides intelligent computing with over one thousand processing units, supporting the training of large models with more than 100 billion parameters.

4.2 Develop Multimodal Scientific Foundation Models and Platforms

AI models have transferred from specialized small models to pre-trained general large models, which enable AI models to evolve from single modal ones to multimodal ones. The multimodal models have the ability to learn from vast amounts of unlabeled data. It is commonly believed that combining large model training methods with industry-specific big data can produce high-quality, industry-specific models. However, an overemphasis on knowledge from a single domain might limit the ability of large models to develop new capabilities. Therefore, during pre-training, large models learn a broad range of knowledge from extensive datasets to maintain their adaptability and potential for innovation.

Currently, the research and application of scientific large models are still in early stages. We can draw inspiration from the development of general large models in other fields and create multimodal foundation models in scientific domain. These models can learn scientific knowledge and reasoning abilities from diverse scientific documents, integrate scientific languages across multiple disciplines, break down barriers between different fields, and shift scientific research from isolated breakthroughs to platform-based approaches. The Microsoft Research team has proposed building a unified scientific base model to solve a wide range of scientific challenges with a single model. China has launched general large language models such as Pengcheng Mind, Huawei Pangu, and Baidu Wenxin Yiyan, gradually establishing several scientific large model implementations which are internationally influential.

These efforts have laid a solid foundation for the construction of multimodal scientific foundation models and platforms. Scientific large models must be tailored to the specific needs of scientific research, integrating both data-driven and physics-based approaches. They should effectively incorporate scientific laws and expert knowledge into model construction and training process. By combining other AI technologies or specialized tools, these models can design effective strategies and evaluation criteria for addressing scientific challenges, creating a research framework that integrates multiple scientific paradigms and fosters human-computer collaboration. Additionally, large models must deepen their understanding of ethics and social norms, align with human values, and address more complex ethical issues.

4.3 Promote Theories and Practices of Scientific Intelligence

The iterative updates of AI algorithms continuously provide new tools for scientific research. Deep learning is currently widely applied in various scientific fields. Although foundation models, as a milestone technology in the development of AI, have demonstrated potential in accelerating scientific research and been driving scientific discoveries, their key technological advancements and applications in scientific fields are not as good as those in general fields. In the predictable future, large models are surely the optimal path toward general AI, the ongoing development of AI technology will undoubtedly continue to empower scientific research.

AI can be integrated into every aspect of scientific research. In 2023, a paper published in *Nature* explored the comprehensive empowerment of AI across the scientific research process, including hypothesis generation, experimental design, and data analysis. Technologies such as deep learning, foundation models, and generative AI enable researchers to mine high-dimensional scientific data, and uncover the scientific laws hidden within multimodal data. By generating synthetic data, researchers can overcome the limitations of experimental observations and the theoretical constraints of numerical simulations, and expand the scope for scientific hypotheses. Foundation models also make it easier to interact with humans, bridging natural language and scientific language, allowing machines to better understand human intentions and perform scientific operations. Fully exploring the potential of large AI models in assisting the entire scientific process—from proposing hypotheses to computational simulation and experimental verification—would fundamentally transform human capabilities into scientific discovery. However, the intelligence of foundation model has not yet reached the level of human's thinking. AI-driven scientific research does not mean that machines will complete all complex tasks independently. Instead, it is crucial to combine human intelligence, machine intelligence, and other computational tools. At this stage, human-machine collaboration is a key feature of scientific intelligence. Foundation models will continue to expand their application scenarios in the scientific field.

AI models designed for scientific research must be optimized for specific disciplines and application areas to better capture the patterns and characteristics inherent to those fields. The Transformer architecture, commonly used in foundation models, provides flexibility through tokenization, position embedding, and attention mechanisms. This flexibility allows for the joint modeling of various modalities and tasks. Researchers are continually exploring advanced solutions for scientific challenges, focusing on scientific language representation, such as proteins, genes, molecules, and chemical expressions. They are also investigating novel model architectures, like multimodal fusion architecture, in order to enhance the collaboration between foundation models and external knowledge, with professional tools and intelligent agents. For instance, in the field of chemistry, AI has been employed to improve the efficiency of chemical experiments and manufacturing, aiding new material creation and drug discovery. Furthermore, foundation models have shown significant advantages in analyzing and synthesizing complex scientific data across diverse disciplines, including biology, drug discovery, computational chemistry, and material design. Additionally, the iterative process of AI models proposing designs, researchers testing these designs in the lab, and feeding the resulting data back into the AI for continuous learning, fosters innovation through data enrichment. This cycle is expected to become even more powerful with the emergence of agent AI, which facilitates conversational interaction and knowledge exchange between researchers and AI-driven agents. AI technologies are evolving rapidly. They can leverage techniques and insights from fields such as natural language processing, computer vision, and large multimodal models to address scientific challenges. For instance, in the field of protein prediction, before the rise of deep learning, scientists used machine learning methods like Markov random fields and classical neural networks to infer structures based on co-evolutionary information. AlphaFold introduced a residual convolutional network similar to ResNet from computer vision, while AlphaFold2 adopted the Transformer architecture and introduced novel innovations like the Equivariant Attention mechanism. Many of the network structures in AI incorporate biological considerations, greatly enhancing the accuracy of protein predictions. AlphaFold3, proposed in 2024, takes a step further by incorporating innovative modules such as diffusion models, expanding the prediction scope to include more biomolecular structures, such as proteins, DNA, RNA, and a variety of ligands and ions [22]. The rapid advancing AI technologies offer new methods and ideas for scientific research. Large model technologies in scientific fields, when being tailored to the specific characteristics of data and tasks across disciplines, present significant opportunities for further exploration and optimization. However, it is important to recognize that AI models, particularly large ones, are still often regarded as black boxes. As such, foundation models for scientific research should be carefully designed, tested, and validated to ensure the objectivity and accuracy of the scientific outcomes.

4.4 Improve Scientific Intelligent Innovation Ecosystem

AI-driven scientific research includes not only AI technologies but also the resolution of scientific problems and the application of research results. This is an incredibly challenging and complex interdisciplinary field. While computing power, models, and data are essential for the development of AI-driven scientific research, the construction of an innovative ecosystem is the key factor to succeed.

First, it is crucial to leverage the strengths of enterprises, research institutes, and universities to build a collaborative innovation system that integrates industry, academia, and research, linking upstream, midstream, and downstream industries. New research institutions should be established in key areas, interdisciplinary platforms should be developed, and efforts should be made to strengthen the formation of interdisciplinary teams to foster the cultivation of interdisciplinary talents. Second, investment in AI-driven scientific research should be increased, and an appropriate science and technology financial system should be established to enhance the transformation of scientific research outcomes and their industrial application, promoting the integrated development of technological and industrial innovation. In 2023, China proposed the development of New Quality Productive Forces to reshape both productivity and production relations. AI-driven scientific research is an important support to this proposal. Addressing key scientific issues in basic science is essential to solving industrial problems, which in turn will drive breakthroughs in core industrial technologies and their applications. International cooperation and the establishment of an open-source ecosystem are also critical. Open-source initiatives play a vital role in accelerating AI innovation, and open-source foundation models have already made significant contributions to scientific research across various disciplines. It is essential to apply the principles of open science to make meaningful AI-driven scientific contributions, to create opportunities for resource sharing and collaboration, to promote open cooperation in AI-driven research, to conduct joint studies on global cutting-edge AI scientific challenges, and to accelerate the scientific research process by sharing knowledge, technology, models, and data.

5 Conclusion

The world is witnessing rapid advancements in science and technology. AI-driven scientific research has evolved from concept to reality. From solving the Schrödinger equation to accelerating molecular simulations, from predicting protein structures to enabling drug design, a large number of outstanding applications have been emerging at an accelerated pace. It has become a crucial direction for future scientific and technological innovation and will reshape the process from scientific research to industrial development. China has elevated AI-driven scientific research to an important national strategic position. Since 2022, continuous breakthroughs in large AI models have propelled artificial intelligence into a new phase of rapid development. The large AI models have been shifting from perception to cognition, from analysis to generation, and from specialized to general applications. Application demands in scientific fields such as drug discovery, materials science, and earth remote sensing are driving continuous innovation breakthroughs in scientific large models. However, the potential of large model in scientific research has not to be fully realized, with challenges persisting in data resources, computing power, algorithms, and ecosystem development. Sustained efforts are required in infrastructure development, model and platform construction, and innovation ecosystem improvement. By innovatively applying artificial intelligence who can understand scientific principles deeply, human will advance scientific discovery and technological transformation further. In the near future, human will surely witness more innovations in artificial intelligence technologies, which will contribute to more influential achievements in scientific research.

References

1. HEY T, TANSLEY S, TOLLE K. The Fourth Paradigm: Data-Intensive Scientific Discovery[M]. Washington: Microsoft Research, 2009.
2. SUN J L, HONG X H, WANG Y, et al. Scientific Research Informatization: A Key Driving Force in Facilitating the Transformation of Scientific Research Paradigms[M]. Beijing: Publishing House of Electronics Industry, 2022.
3. JUMPER J, EVANS R, PRITZEL A, et al. Highly Accurate Protein Structure Prediction with AlphaFold [J]. Nature, 2021, 596: 583–589.
4. World Economic Forum. Top 10 Emerging Technologies in 2024[R]. 2024.
5. Stanford Institute for Human-Centered Artificial Intelligence. AI Index Report [R]. 2024.
6. WANG H, FU T, DU Y, et al. Scientific Discovery in the Age of Artificial Intelligence[J]. Nature, 2023, 620: 47–60.
7. YANG K, SWOPE A M, GU A, et al. LeanDojo: Theorem Proving with Retrieval-Augmented Language Models[C]. In Proceedings of the 37th NeurIPS. New York: Curran Associates Inc. 2023.
8. DONG Q, DONG L, XU K, et al. Large Language Model for Science: A Study on P vs. NP[J]. arXiv:2309.05689, 2023.
9. ROMERA-PAREDES B, BAREKATAIN M, NOVIKOV A, et al. Mathematical Discoveries from Program Search with Large language Models[J]. Nature, 2024, 625: 468–475.
10. NORI H, LEE Y T, ZHANG S, et al. Can Generalist Foundation Models Outcompete Special-Purpose Tuning? Case Study in Medicine[J]. arXiv:2306.05064, 2023.
11. LUO Y, ZHANG J, FAN S, et al. BioMedGPT: Open Multimodal Generative Pre-trained Transformer for BioMedicine[J]. IEEE Journal of Biomedical and Health Informatics, 2024:1–12.
12. ZHAO Z, MA D, ChEN L, et al. ChemDFM: A Large Language Foundation Model for Chemistry[J]. arXiv:2401.14818, 2024
13. ZENG Z, YIN B, WANG S, et al. ChatMol: Interactive Molecular Discovery with Natural Language[J]. Bioinformatics, 2024, 40(9): btae534.
14. MERCHANT A, BATZNER S, SCHOENHOLZ S S, et al. Scaling Deep Learning for Materials Discovery[J]. Nature, 2023, 624:80–85.
15. ZENI C, PINSLER R, ZUGNER D, et al. MatterGen: A Generative Model for Inorganic Materials Design[J]. arXiv:2312.03687, 2023.
16. CHENG Z, LIU J, JIANG T, et al. Automatic Screen-out of Ir(III) Complex Emitters by Combined Machine Learning and Computational Analysis[J]. Advanced Optical Materials, 2023, 11(18):1–21.
17. CHEN Z, XIE F, WAN M, et al. MatChat: A Large Language Model and Application Service Platform for Materials Science[J].Chinese Physics B, 2023, 32:118104.
18. DENG C, ZHANG T, HE Z, et al. K2: A Foundation Language Model for Geoscience Knowledge Understanding and Utilization[C]. In Proceedings of the 17th ACM International Conference on Web Search and Data Mining. New York: Association for Computing Machinery, 2024.

19. LAM R, SANCHEZ-GONZALEZ A, WILLSON M, et al. Learning Skillful Medium-Range Global Weather Forecasting[J]. Science, 2023, 382: 416–1421.
20. GAO W. Promoting Industrial Innovation through Construction of China Computing NET [N]. People's Daily, 2024-04-12, p. 9.
21. Cyberspace Administration of China. Digital China Development Report (2022) [R]. 2023.
22. ABRAMSON J, ADLET J, DUNGER J, et al. Accurate Structure Prediction of Biomolecular Interactions with AlphaFold3[J]. Nature, 2024, 630: 493–500.

Wen Gao an Academician of the Chinese Academy of Engineering, serves as the director of Pengcheng Laboratory and the dean of the School of Information and Engineering Sciences at Peking University, where he holds the Boya Chair Professorship. He is a fellow of both the Institute of Electrical and Electronics Engineers (IEEE) and the Association for Computing Machinery (ACM). In the fields of artificial intelligence, multimedia technology, computer vision, pattern recognition, image processing, and virtual reality, he has authored over 300 academic papers in international journals.

Shu Liu a member of the Strategic Research Expert Group in the Research Department of Pengcheng Laboratory. She received the Ph.D. degree in computer architecture from Peking University in 2010. Her research focuses on artificial intelligence and system software.

Jiankun Yang a researcher in the Pengcheng Laboratory and a part-time professor in the Xi'an Jiaotong University. He received the Ph.D. degree in rehabilitation engineering from Tsinghua University in 2006. His current research interests include rehabilitation engineering, service-robot fundamental theories, and human-machine interaction.

Wei Zhang the director of the Institute of Data Intelligence in the Department of Networked Intelligence at Pengcheng Laboratory. He received the Ph.D. degree in Computer Science and Technology from the Harbin Institute of Technology in 2025. He holds an M.S. degree (2000) in Safety Science and Engineering from Anhui University of Science and Technology. His research focuses on cross-modal representation learning and data compression. He has authored over 10 publications and holds more than 10 patents and software copyrights.

Juntong Yan an assistant researcher in Pengcheng Laboratory. She received the Master degree from Nanjing University in 2017. Her current research fields include scientific research management, technological strategy research and technological ethics.

Ge Li a professor at the School of Electronic and Computer Engineering, Peking University, Shenzhen, China. His research interests include image, video and 3D signal processing and analysis, machine learning and artificial intelligence.

The Transformation of Network Architecture in the Era of Artificial Intelligence

Yunjie Liu, Zhengjie Sun, Shuo Wang, and Tao Huang

Abstract Driven by the tide of digitalization, the rise of artificial intelligence (AI) technology is reshaping the future of network architecture, leading a wave of transformation centered on intelligence. The rapid development of AI technology has driven a swift expansion in the density of terminal connections and the scale of networks. Traditional network architectures are unable to keep pace with this development and urgently need to leverage AI technology to achieve large-scale network automation management. Therefore, network architectures are undergoing transformation with automation, intelligence, and efficiency as core objectives. This transformation not only supports existing network services but also adapts to the needs of future emerging technologies. This paper systematically reviews the evolution of network architecture, summarizes the key characteristics of service-oriented networks, deterministic networks, intelligent computing networks, and self-intelligent networks, and analyzes the design of network architecture and key technologies. It examines the future direction of network development—service-generating networks, from the perspectives of national policy and development needs, introduces their architectural design and application use cases, and further assesses the development trends and challenges. The research indicates that service-generating networks are the cornerstone of future network development and will be combined with other emerging technologies such as 6G, the Internet of Things, and edge computing to jointly build an intelligent digital society and a smart interconnected world.

Keywords Network architecture transformation · Service customized network · Deterministic network · Intelligent computing power network · Self-intelligent network · Service generation network

Y. Liu · S. Wang (✉) · T. Huang
State Key Laboratory of Networking and Switching Technology, Beijing University of Posts and Telecommunications, Beijing, China
e-mail: shuowang@bupt.edu.cn

Y. Liu · Z. Sun · S. Wang · T. Huang
Purple Mountain Laboratories, Nanjing, China

China's e-Science Blue Book 2025, https://doi.org/10.1007/978-981-96-8252-2_2

1 Introduction

In the contemporary digital information age, applications and services based on Artificial Intelligence (AI) are thriving. For instance, machine learning and deep learning have achieved state-of-the-art performance in various fields such as facial recognition [1], natural language processing [2], computer vision [3], traffic forecasting [4], and anomaly detection [5]. With the in-depth development of AI technology, information society is rapidly advancing towards digitalization and intelligence, leading to exponential growth in terminal connection density and traffic density. The increasing complexity of networks, driven by the continuous addition of diverse network protocols, topologies, and access methods, has rendered traditional static, single-channel network architectures incapable of coping with the expanding network scale. There is an urgent need to achieve large-scale network automation to adapt to frequent network dynamics [6]. Meanwhile, in addressing diverse service requirements, heterogeneous infrastructure, and high network complexity, AI demonstrates significant advantages over traditional manual mathematical models [7]. Therefore, AI technology has become an important driving force for the transformation of network architecture.

The development of network architecture mainly includes two modes: the "evolutionary" development that applies AI technology to existing networks and the "revolutionary" development that redesigns entirely new network architectures. The core of this development lies in designing an intelligent and efficient universal network platform that can support the integration of heterogeneous network technologies, meet the stringent requirements for computing and data processing in the AI era, and provide new impetus for the development of the communications industry [8]. The design paradigm of the new network architecture is evolving towards being dynamic, intelligent, self-learning, and adaptive, gradually transitioning from a single data processing center to an intelligent decision-making center [9]. This transformation will break down the boundaries of each network layer, forming an integrated communication design that encompasses sensing, computing, capability, security, AI, and other aspects. It will be compatible with various access modes such as fixed, mobile, and satellite, manage different types of networks including public, industrial, and digital networks, and utilize the collaboration of different network domains such as access networks, transport networks, and core networks. This will not only improve the operational efficiency of the network but also enhance its security, reliability, and quality of service, providing users with more personalized and intelligent service experience.

In this paper, the development process of network architecture in the age of artificial intelligence is first elaborated, including the key characteristics of service-oriented networks, edge intelligent networks, deterministic networks, self-intelligent networks, and intelligent computing networks, as well as the analysis of network architecture design and key technologies. Furthermore, a detailed study is conducted on the design requirements and challenges of the future network architecture development focus—service generation networks. The transformation of network architecture in the age of artificial intelligence is redefining the way humans interact with the digital world and providing impetus for the digital transformation of various industries. Future network architectures are rapidly evolving towards openness, white-boxing, cloudification, and programmability, paving the way for the development of global communication networks in the future.

2 The Development Requirements of the Intelligent Internet

(1) The support and promotion of national policies

In 2021, the Central Cyberspace Affairs Commission issued the "14th Five-Year Plan for National Informatization", emphasizing the application of new-generation information technologies to digitally and intelligently transform traditional infrastructure. The plan highlights the need to accelerate the construction of ubiquitous and intelligent digital infrastructure. It indicates that building an intelligent Internet to provide users with faster, higher-quality, more reliable, broader, and smarter information connectivity is an important support for enhancing network supply and service capabilities.

In 2023, the Central Committee of the Communist Party of China and the State Council issued the "Overall Layout Plan for the Construction of Digital China", proposing to accelerate the coordinated construction of 5G networks and gigabit optical networks, deeply advance the large-scale deployment and application of IPv6, promote the comprehensive development of mobile Internet of Things, and vigorously advance the large-scale application of Beidou. The plan also focuses on optimizing the layout of computing power infrastructure, promoting efficient complementarity and collaborative linkage of computing power between the eastern and western regions, and guiding the rational and tiered layout of general data centers, supercomputing centers, intelligent computing centers, and edge data centers. This indicates that the development of the intelligent Internet has become an important research direction across various industrial fields nationwide.

In 2024, the Ministry of Industry and Information Technology, together with six other departments, issued the "Implementation Opinions on Promoting the Innovative Development of Future Industries". The document indicates support for the innovative development of new network architectures, GPU chip cluster interconnection networks, and large-scale new intelligent computing centers. It emphasizes the importance of breaking through key core technologies to promote technological innovation and the transformation of achievements in future industries. Providing technical support for intelligent interconnection networks is identified as an important development goal.

(2) The Construction of an Intelligent Network Governance Framework

China will continue to participate in global artificial intelligence cooperation, incorporating it as a key scientific field within the framework of the national legal system. This will promote the development of a widely accepted governance framework and standards for artificial intelligence, drive technological innovation and the expansion of applications, and provide a sound governance system support for the development of the intelligent Internet [10].

(3) Support for Technological Convergence

The deep integration of AI technology with the Internet, the Internet of Things (IoT), and computing power networks involves the collaborative operation of data, algorithms, and hardware and software for computing power. It widely connects people, machines, and objects, and enables the transmission and exchange of intelligent resources, providing robust technological support and intelligent decision-making support for the development of the intelligent Internet [11].

(4) The rapid development of intelligent terminals

Driven by the development of technologies such as deep learning and computer vision, the integration of Internet terminals with large AI models is accelerating. For example, domestic smartphone manufacturers like Huawei and Vivo have successively launched mobile products equipped with large AI models. Lenovo and other manufacturers are also actively developing and launching personal computers equipped with dedicated AI processors and AI software. The initial commercialization of AI smartphones and AI personal computers has been realized [12]. The rapid development of intelligent terminals not only promotes the progress of intelligent internet technologies but also facilitates the maturation of the entire industrial chain and the formation of new business models. This has laid a solid foundation for the future development of the intelligent internet.

(5) The accelerated implementation of intelligent vertical applications

With the implementation of intelligent vertical applications such as smart office, smart healthcare, and smart education, enterprises in various industrial fields are actively deploying and developing these applications. This trend drives the construction of a new application industry model supported by AI, which not only promotes the development of the intelligent Internet, but also fosters innovation and transformation in related industries. As a result, the role of the intelligent Internet in promoting economic and social development is further enhanced [13].

3 Advances in Hot Network Technologies

The transformation of network architecture is shifting from traditional hardware dependency to software-based and programmable approaches to meet the growing demands of traffic and services. In the era of artificial intelligence, the development of network architecture has witnessed a series of innovations and transformations. Service-oriented networking can provide personalized network services based on different business needs, meeting the specific network performance requirements of particular scenarios. Deterministic networking, by precisely controlling the forwarding behavior of network data, offers predictable latency, jitter, and packet loss rates for critical services. This is essential for applications with extremely high network performance requirements, such as industrial automation and autonomous driving. Intelligent computing networks integrate computing and network resources deeply, enabling optimized resource allocation and efficient utilization. They support extensive computing needs from the cloud to the edge and provide robust support for emerging compute-intensive applications. Self-intelligent networks leverage artificial intelligence technologies to understand the intent behind network configurations and automatically execute operations to meet these intents. This simplifies network management and enhances efficiency. The evolution of these network architectures indicates that future networks will be smarter, more flexible, and more efficient. They will be able to meet increasingly complex business demands and growing data traffic, while also delivering richer and more personalized network service experiences for users. The following sections will provide a detailed analysis of the key technologies and architectural designs of each network architecture.

3.1 Service Customized Networking Architecture

The traditional network service model of "one-size-fits-all" and "best-effort" is difficult to ensure the quality of service for diverse businesses. Service Customized Networking (SCN) aims primarily at differentiated quality of service and has become one of the mainstream integrated network frameworks for the next generation [14]. SCN relies on advanced technologies such as the decoupling of data and control planes in Software-Defined Networking (SDN), Network Function Virtualization (NFV), and network slicing [15]. By decoupling functions from the underlying network, SCN provides differentiated network capabilities and on-demand services for large-scale businesses. Through innovative network architecture, the ability to provide differentiated and deterministic services has broken through the limitations of traditional microservice-based network operating systems. It has also solved the challenge of transforming wide-area networks from "manageable but not controllable" to "fully controllable across the network," providing strong technical support for major scientific infrastructure, supercomputing centers, and scientific data centers.

The design of SCN network architecture is service-oriented and content-centric, providing customized services for heterogeneous networks, using big data analysis and artificial intelligence technology, optimizing information scheduling and network management to cope with the explosive growth of data traffic, and providing flexible and customizable services for each user to satisfy the user's personalized demand for services. The SCN architecture consists of the infrastructure plane, control plane, management plane, Distribution Plane and Sensing Plane, as shown in Fig. 1. The infrastructure plane consists of computing resources that support functions such as network function virtualization, storage resources that store information transmitted in the network, and network resources. The control plane, located in the middle of the architecture, consists of the SDN controller, the NFV management module, and the cache resource manager, which meets the QoS guarantee requirements of differentiated users by constructing different network slices for different users. The two sides of the architecture are the distribution plane and the sensing plane. The distribution plane sends commands to the control plane to accomplish intelligent control, scheduling and distribution of information resources. The sensing plane analyzes the parameters collected in the network using big data technology and provides timely feedback to the distribution plane and control plane. The top management plane acts as the manager of the entire network by calling the northbound interfaces opened by other planes for custom development. The development of SCN provides strong support for the upgrading of network technology and the innovation of industrial applications, promotes the improvement of network performance and the optimization of resource utilization, and lays a solid foundation for the transformation and upgrading of the digital economy and the digitalization of industries. SCN lays a solid foundation for the digital economy and the digitalization of industries by providing limited latency, minimum jitter, zero packet loss and other deterministic guarantees, it accomplishes timely interaction of services in an isolated, flexible and highly reliable manner, and reduces performance loss due to resource competition.

3.2 *Deterministic Network Technology*

Driven by policies in the new generation of information technology industry, the rapid development of artificial intelligence (AI) technology and the continuous expansion of application scenarios have made an interconnected and secure system architecture an important driving force for network transformation [16]. The Deterministic Network (DetNet), with its "on-time and accurate" data transmission services, offers extremely low packet loss rates and limited end-to-end delivery latency, becoming one of the important development directions of the new generation of network architecture to meet the needs of future digital transformation [17]. DetNet supports long-distance and large-scale communication, providing various levels of Ultra-Reliable Low-Latency Communication (URLLC) services to achieve advanced communication with higher connectivity, security, and reliability. In the process of building

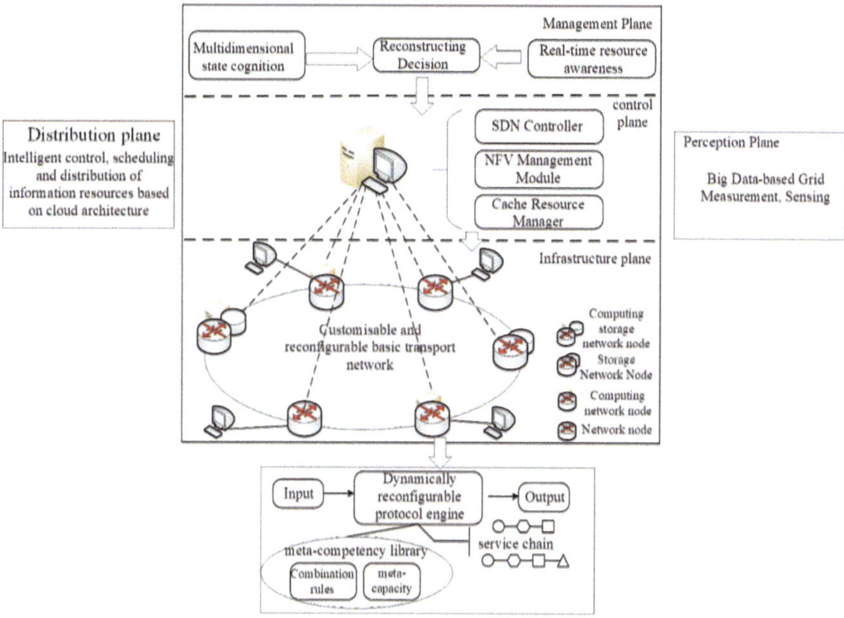

Fig. 1 SCN network architecture

major network infrastructure, deterministic networks play an important role, offering new opportunities to occupy the high ground in information technology development. As a development direction for the new generation of network communication systems, deterministic networks empower industrial upgrading comprehensively and become a key driving force for building strong nations in network, industry, agriculture, and services. This forms a "Deterministic Network+" technological and industrial pattern, which is of great significance for promoting the high-quality development transformation of various industries towards digitalization, networking, and intelligence.

The architecture of the DetNet is shown in the Fig. 2. The entire deterministic network architecture consists of terminal systems, edge nodes, relay nodes, and transport nodes. Terminal systems act as the source or destination of deterministic network flows and provide encapsulation services for deterministic network flows. Edge nodes serve as the source or destination nodes in the service sublayer, with the capability to initiate or terminate resource allocation in the forwarding sublayer. They also perform the mapping between application flows and deterministic network flows, providing service protection for one or more terminal systems (e.g., adding or removing packet sequencing information). A deterministic network flow is a series of packets containing a unique flow identifier. End-to-end services provide computational support for these packets. The forwarding sublayer selectively provides explicit routing and resource reservation and allocation functions for deterministic network flows on the underlying network paths. Relay nodes interconnect

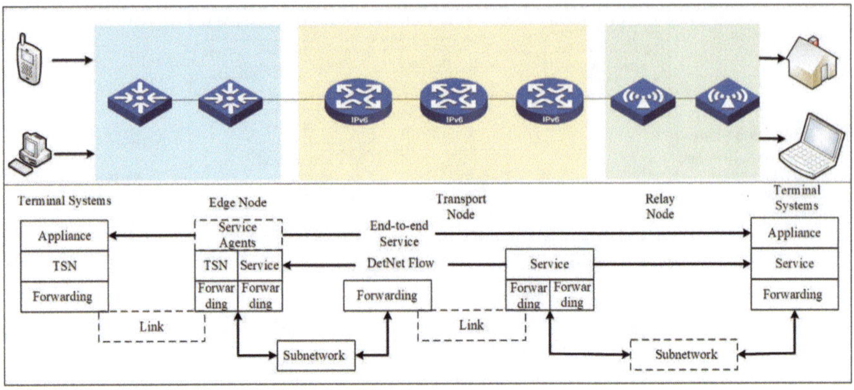

Fig. 2 Deterministic network grouping model architecture

different forwarding sublayer paths to provide service protection, mitigating or eliminating packet loss caused by device failures (including random media and memory failures) by distributing data across multiple disjoint forwarding paths. Links, as connections between two nodes, provide appropriate traffic delivery for deterministic network flows. Transport nodes use the link layer to enable switching across multiple link subnets, providing paths for resource allocation for service sublayer functions. The deterministic network employs Flexible Ethernet (FlexE) technology, which adds a FlexE shim layer directly between the link layer and the physical layer. The FlexE shim layer ensures low-latency data transmission through a precise time-slot switching mechanism and effectively decouples the transmission rate of service data from the rate of the physical channel. The network also introduces Time-Sensitive Networking (TSN) standards, requiring all devices in the network to achieve precise time synchronization. Advanced scheduling and traffic shaping gating techniques are used to enable efficient traffic scheduling. The implementation of deterministic networks involves key technologies such as traffic engineering, resource reservation, and queue scheduling. With the further integration of artificial intelligence technology and network architecture, deterministic networks will be more widely applied and practiced globally. They will drive the development of new business models and applications and become a key driving force for digital transformation.

3.3 Intelligent Computing Power Network Technology

Driven by the "East Data, West Computing" initiative and the national strategy for new infrastructure construction, accelerating the synergistic, service-oriented, and intelligent development of information network infrastructure has become an important direction for the new generation of information networks. In the process of construction, a large number of new applications with high concurrency and high-precision computing demands have emerged, especially the deep integration of AI services with 6G and edge computing. The computational volume and parameter scale of AI applications are growing exponentially, reaching hundreds of billions or even trillions [18]. To meet the development needs of digital infrastructure, the intelligent computing power network, which integrates "computing and networking," provides support and guarantee for enhancing the overall informatization and intelligence of the country [19]. Computing power, as a vital productive force in the digital economy era, has transcended the traditional narrow definition of merely computational capability. Instead, it has evolved into a multi-dimensional comprehensive system that includes computing, storage, input/output (I/O), software, platforms, and algorithms. This new type of intelligent, efficient, and on-demand computing power service system is actively promoting the development of new business models such as integrated space-ground networks and digital twins. It lays a solid technical foundation for the "East Data, West Computing" initiative and drives continuous innovation in the digital economy era.

The intelligent computing power network is a network architecture that allocates and flexibly schedules computing resources on-demand among the cloud, edge, and endpoints based on the requirements of computing services. The computing power network has inspired innovations in network addressing protocols, expanding from traditional network index-based addressing to include computing power indexing, namely "data + computing resources + algorithms". To meet the diverse future service demands, the traditional "network-centric" supply model is transformed into a "network-based" model. The providers of computing power resources are no longer limited to centralized cloud platforms but are extended to heterogeneous and distributed computing power nodes connected through the network, such as edge computing nodes close to users, supercomputing centers, and data centers. This enables the construction of a new type of integrated computing and

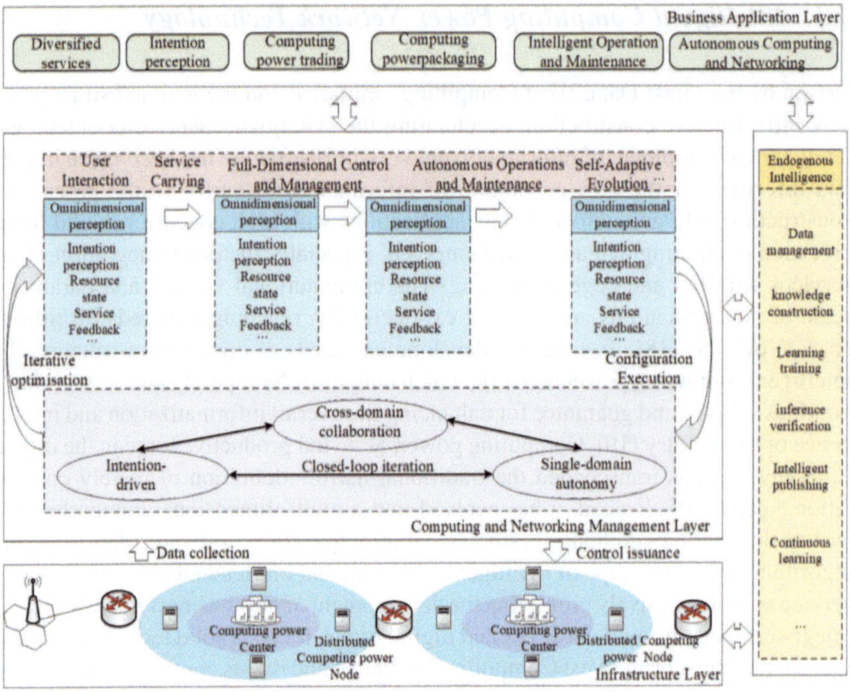

Fig. 3 Computing power network architecture

networking infrastructure. The entire intelligent computing power network architecture is shown in Fig. 3, consisting of three layers from bottom to top: the infrastructure layer, the computing and networking management layer, and the service application layer, which interact in real-time with the intrinsic intelligence module. The infrastructure layer includes various hardware devices, software applications, and data resources. It is responsible for the centralized monitoring and management of computing, networking, and storage resources across the entire network. It also aggregates the total computing power resources and their utilization within a region, ensuring efficient scheduling and resource management based on service demands. Through interaction with intrinsic intelligence, this layer enhances its self-awareness, including resource status, performance metrics, and potential faults, providing reliable data support for the upper computing and networking management layer. On the other hand, it enables real-time analysis and decision-making at the data source, allowing the network to achieve rapid service response, intelligent adjustment of devices, and detection and repair of computing and networking faults within milliseconds. This enhances the system's capabilities for self-response, self-repair, and self-optimization. The computing and networking management layer, as the core of the entire network architecture, is responsible for executing specific operations and controls of the network system. It receives status data and service demands from the infrastructure layer through north–south interfaces, performs comprehensive analysis and decision-making, and controls the network. This includes real-time monitoring of resource status, intelligent scheduling of resources, fine-grained management of computing power resources, orchestration of service processes, in-depth analysis of faults, and automated repair mechanisms. It ensures a closed-loop management process from information perception and analysis to decision-making and execution. This layer deeply integrates artificial intelligence technologies, applying them across various levels of the computing power network to enhance the intelligent learning capabilities and adaptability of each functional module to different scenarios. This deeply embedded AI technology supports the continuous development of intelligent services, ensuring that the network can meet current service demands and adapt to future emerging services. The service application layer is the user-facing service interface, integrating the abstract implementation of various service functions. The system can intelligently allocate and schedule service applications to the most suitable execution nodes based on users' specific needs and intentions. This ensures maximum resource utilization efficiency and provides users with efficient and personalized services. The intrinsic intelligence module constructs a complete intelligent closed loop, covering functions such as data management, learning and training, intelligent distribution, and continuous learning. It provides intelligent technical support for each layer, meeting service demands in different scenarios and covering the entire lifecycle from design, training, inference validation, deployment, and iterative optimization. The intelligent computing power network, as a unified resource-sharing platform, promotes the new operation model of integrated scheduling of computing and networking. It effectively facilitates the development of the AI industry and the AI-ization of various industries, serving as an essential foundation for digital infrastructure construction. The future direction of network development is gradually shifting from basic resource construction to scenario-based applications, accelerating the construction of new types of data centers and the "East Data, West Computing" initiative, and driving the collaborative progress of computing and networking capabilities.

3.4 Self-intelligence Networking Technology

With the deep integration of AI and networking, self-intelligent networks enhance the efficiency and quality of network services through automation and intelligence, addressing issues such as the increasing complexity of network structures and the challenges in operation and management brought about by the rapid development of AI [20]. By deeply integrating AI computing and communication technologies, self-intelligent networks strengthen the intelligent learning capabilities and scenario adaptation abilities across various layers, including the physical and network layers. They incorporate intelligence and data planes into the network architecture design, achieving intelligent network elements and interfaces, and thereby providing efficient and precise scenario-based services [21]. As the target of the digital transformation of communication network operations and maintenance, self-intelligent networks offer innovative network services to users with digital experiences such as "zero wait," "zero touch," and "zero failure." They build communication networks with digital operation and maintenance capabilities, including self-configuration, self-healing, and self-optimization. As a strategic technology leading a new round of technological revolution and industrial transformation, self-intelligent networks promote the mutual integration and enhancement of 5G and AI, further constructing an infrastructure system for industrial upgrading, integration, and innovation.

The self-intelligent network architecture consists of the application layer, intent layer, network management and orchestration layer, and infrastructure layer, as shown in Fig. 4. The application layer is composed of various vertical use cases and their users, exposing underlying network infrastructure and services to users through the Intent Northbound Interface (INBI). The intent layer expresses the desired outcomes through service providers, adhering to the declarative design intent of the INBI. It provides the necessary interfaces for intent controllers and orchestrators. Intent mapping enables providers to interpret intents based on system-specific models. The intent controller translates the user's intent by defining its perspective and establishing mappings with service and operator catalogs as well as service and network specifications. It translates intents into lookup tables and directories, including available resources in the network, and ensures the consistency of the service model by performing policy evaluations on the generated services. The network management and orchestration layer comprise modules such as network orchestration, configuration controllers, slice orchestrators, and monitoring and assessment of network status. Network orchestration consists of different domain controllers and orchestrators, including network function virtualization management and orchestration, SDN, and slice controllers interfaced with intent controllers via intent APIs. Configuration controllers receive service requirements in the form of application topology and orchestration specifications, generating models based on YANG (Yet Another Next Generation), Extensible Markup Language, or non-markup languages for various network devices. They expose resource status through open application programming interfaces. The monitoring and assessment module is responsible for interaction with controllers and coordinators, ensuring that service deployment requests and intents are followed at a higher level. It interacts with the infrastructure regarding resource status through the Southbound Interface (SBI) and network coordinators, obtaining state information required to indicate the status to the intent controller. The network infrastructure provides the necessary physical computing, storage, and communication resources, which are shared through the SBI. Resources are distributed across different types of services provided by various providers, including access, core, and transport segments of the network, based on domain-specific definitions. Future self-intelligent networks will evolve towards self-learning and self-evolution, optimizing network operation and maintenance efficiency. Moreover, the deep penetration of intelligence will enable the network system to possess a high degree of intelligence at all levels, providing users with seamless, efficient, and intelligent network experiences. At the same time, it will offer network operators and service providers broader service capabilities.

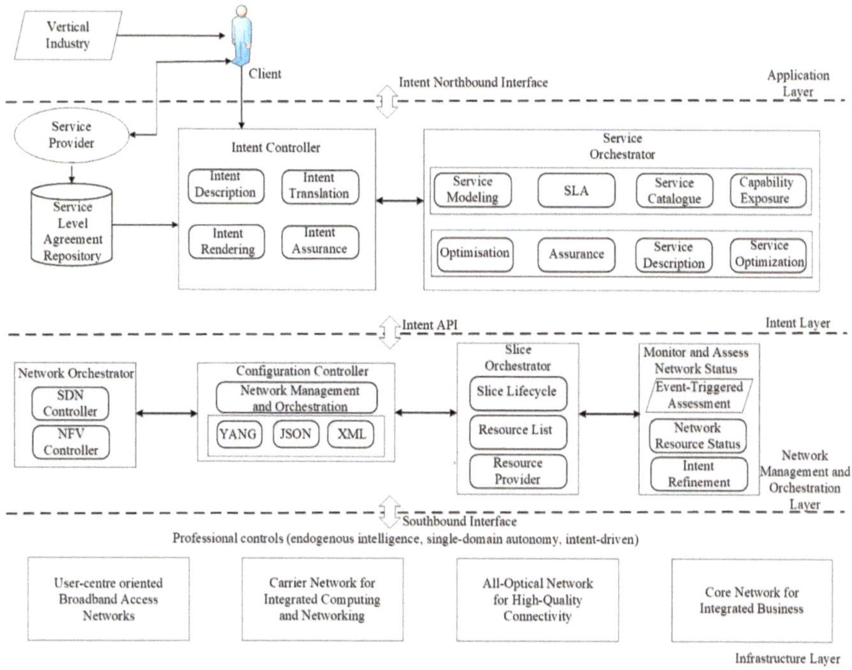

Fig. 4 Self-intelligent network architecture

In summary, SCN, DetNet, intelligent computing power networks, and self-intelligent networks have become important development directions in the field of network technology due to their unique advantages. However, the drawbacks in the evolution of network architecture cannot be ignored. Service-oriented networks, while providing personalized services, increase operational complexity and costs. Deterministic networks, which rely on mechanisms such as time synchronization, resource reservation, and queue control to offer high-bandwidth, low-latency, and highly reliable services, need to comprehensively consider demands and performance factors across different scenarios when widely applied. Computing power networks face challenges in application encapsulation technology, data migration, security assurance, and flexibility during their development. Self-intelligent networks, which deeply integrate AI with network operations and maintenance, confront challenges related to standardization and interoperability, as well as network stability and reliability. The future development trends of networks need to address challenges such as automated configuration and scalability in complex network environments. By integrating advanced technologies like artificial intelligence, networks can provide a smarter, more flexible, efficient, and secure environment to adapt to evolving business needs, promote technological innovation, and drive the sustainable development of network resources.

4 Service Generation Network Architecture for the Intelligent Internet

Inspired by the direction of the new round of technological revolution and industrial transformation, the future trend of network architecture development is to utilize advanced artificial intelligence technologies to build a smarter, more efficient, and secure network architecture for the digital age. This trend has led to the proposal of design requirements and challenges for network architecture development from four aspects: high-intensity data centers, large-scale network edges, high-complexity devices, and intelligent orchestration. These challenges have spurred the emergence of future highly intelligent network architectures [22]. The Service Generation Network (SGN) is an innovative AI-based network system that integrates natural language processing, deep learning, and multimodal integration. It consolidates capabilities such as intent-driven networking, automated network configuration, digital twin networking, programmable networking, and network optimization. The goal of SGN is to enable the automatic generation and customization of intelligent network services, making it a key research direction for future network architecture development. In the future, the Service Generation Network will be combined with other emerging technologies such as 6G, the Internet of Things (IoT), and edge computing to jointly build an intelligent digital society. This integration will promote technological innovation and the process of informatization.

4.1 Service Generation Network Architecture

The core of the Service Generation Network (SGN) lies in accurately understanding user requirements and generating corresponding intelligent network services. It involves key technologies such as intent-driven networking, automated network configuration, and full programmability of networks, providing users with personalized and intelligent solutions. The SGN encompasses three main layers: the Generative Artificial Intelligence (GAI)-empowered Intent-Driven Layer, the Digital Twin Network Mapping Layer, and the Deeply Programmable Network Layer, as shown in Fig. 5. SGN is capable of understanding user intents, automatically configuring networks, leveraging digital twin networks for real-time monitoring, operating programmable networks, and optimizing service delivery. It offers intelligent service solutions for various network environments.

The primary function of the Intent-Driven Layer is to interpret user intentions, including requirements for network design, planning, configuration, and operation and maintenance, as well as to uncover user reliability needs. Through user intent input, a Large Decision-Making Model (LDM) is trained. Based on network status and intent feedback, pre-training, retraining, and fine-tuning of labeled tasks are conducted to generate intelligent decision-making models. The layer also utilizes

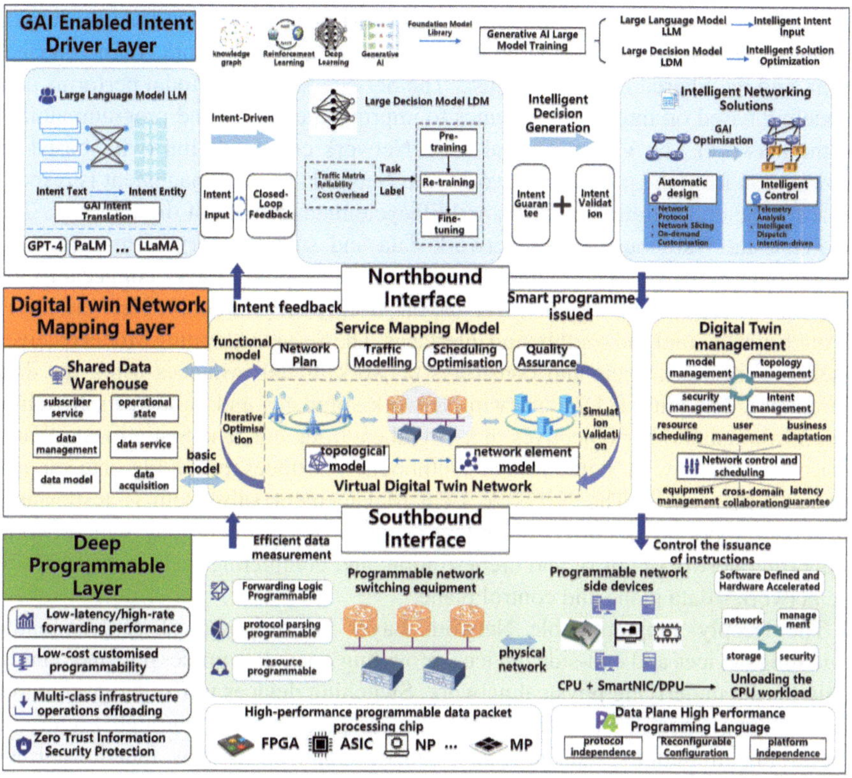

Fig. 5 Service generation network architecture

closed-loop feedback from the mapping layer to optimize GAI for intent assurance and validation of decision-making models, such as supporting the automatic design and telemetry analysis of network protocols and network slices in intelligent management and control. The generated intelligent decision-making plans can be adjusted, optimized, and verified within the mapping layer, thereby enabling the digital twin network to achieve real-time control, feedback, and optimization of the actual physical network. This ultimately realizes the closed loop of network self-learning, self-validation, and self-evolution [23].

The Digital Twin Mapping Layer provides a real-time and accurate representation of network status, which is beneficial for network monitoring, analysis, and prediction. By enabling real-time updates to the network's data plane and control plane, it facilitates efficient operation and configuration of the physical network, achieving low-cost trial-and-error validation of the physical network. Additionally, the Digital Twin Mapping Layer can simulate different network services and application scenarios based on specific user requirements, customizing personalized network services to enhance the efficiency and quality of network planning and operation.

The Digital Twin Network Mapping Layer consists of four subsystems: the Service Mapping Model, the Virtual Digital Twin Network, the Network Twin Management, and the Shared Data Warehouse. The Service Mapping Model performs data modeling based on intelligent schemes to improve the agility and programmability of the network. The Virtual Digital Twin Network creates digital twins based on information from the physical network. The Network Twin Management is responsible for the orchestration, control, and lifecycle management of digital twins. The Shared Data Warehouse stores network data and supports service mapping and digital twin networks. The Digital Twin Network Mapping Layer connects to the GAI-empowered Intent-Driven Layer and the Deeply Programmable Network Layer through northbound and southbound interfaces. The Service Mapping Model receives intelligent network plans and constructs network model instances using the data warehouse. The Virtual Digital Twin Network simulates and validates scheduling optimization and other objectives in a virtual environment. The Shared Data Warehouse collects network data through southbound interfaces, providing data support for network models. The Network Twin Management subsystem is responsible for model, topology, security, and intent management of the digital twin network, supporting network control and orchestration, and completing real-time updates to the network's data plane and control plane.

The Deeply Programmable Network Layer is composed of programmable switching devices and end-side devices, providing open data processing and control logic for the underlying physical network. Switching devices utilize programmable chips to implement customized packet processing, breaking through traditional forwarding limitations and enabling rapid development of new functions. End-side devices offload CPU tasks through hardware-software collaboration, such as networking, storage, and security functions, thereby enhancing processing capabilities. This includes the use of Smart NICs (Network Interface Cards) and DPUs (Data Processing Units), which reduce the burden on the CPU, support rapid networking and data processing, and achieve low-latency, high-speed forwarding performance. Additionally, this layer provides low-cost, customized programming capabilities, as well as infrastructure operation offloading and zero-trust security protection.

4.2 Key Technologies of Service Generation Network

(1) Generative Artificial Intelligence

Generative Artificial Intelligence (GAI), as an important subfield of AI, is composed of multi-layer interconnected nodes. It utilizes neural networks and deep learning technologies to learn new content based on the statistical patterns of existing input data. The working principle of GAI involves training models to learn and predict the probability distribution of data, thereby generating new data. This process is specifically divided into three steps: data measurement, data preprocessing, model pre-training and fine-tuning, and content inference and generation. Real-time measurement and transmission data are collected, cleaned, transformed, and standardized into a processable format. These processed data are then used to train deep learning models through extensive iterative training. Once trained, the models generate high-quality content that meets user needs based on seed data inputs, such as new image and video synthesis, voice synthesis, and language generation. During the content inference stage, the key is to utilize large AI models to perform inference and completion based on the generated content and user feedback. Through model integration and fusion techniques, the models further understand contextual information and infer the optimal output results. Finally, the inference results, including user preferences, behaviors, and feedback, are received and stored by the end-device users for subsequent training and improvement of the AI models.

From traditional programming to neural networks and then to generative models, Generative Artificial Intelligence (GAI) has emerged as a disruptive machine learning algorithm that progressively enables the generation of entirely new content, solutions, or concepts from limited amounts of information. Currently, supported by broader and more diverse datasets, improved algorithms, and more powerful computing hardware, the progress of GAI is rapid and continuous, significantly expanding the application prospects of the Service Generation Network (SGN) across various fields.

(2) Intent-Driven Network Technology

Intent-Driven Network (IDN) technology focuses on optimizing network services and behaviors by parsing user intentions and requirements. It combines Natural Language Processing (NLP) and machine learning to deliver more intelligent and customized service experiences for users. This includes key components such as intent recognition, intent understanding, service generation, personalized experiences, and real-time responsiveness. Using NLP techniques, IDN analyzes user requirements expressed in natural language, extracting and mapping key information to complete the topological identification of entities, actions, and relationships. It retrieves user requirements from a policy library, invokes functional blocks to perform specific operations, and generates services that align with user intentions. Intent-Driven Network technology provides each user with exclusive, immediate services that quickly respond to requests based on individual needs and historical behavior, ensuring an efficient and convenient service experience. Additionally, knowledge graphs offer strong technical support for the development of IDN. As a semantically rich and structurally compact database, knowledge graphs store and organize the information and knowledge involved in service generation. They can connect diverse data sources and integrate information from multiple modalities to establish mappings of entities and relationships. By fusing knowledge from different sources, they enable unified management and effective utilization of information.

Intent-Driven Network technology provides customized services based on user personalization needs and behaviors, learns and adapts to user preferences, and responds to user requests in real-time. Through IDN technology, the Service Generation Network (SGN) meets users' ever-changing and diverse needs while improving user experience and network efficiency.

(3) Network Automation Configuration

Network automation configuration optimizes the management of network resources and services through programmable logic, enabling network operations to quickly configure, scale, secure, and integrate network infrastructure and application services. It eliminates the cumbersome steps of manually logging into devices to make configuration changes and overcomes the drawbacks of relying on script programming or automation software to execute tasks on the operating system command-line interface (CLI). The Service Generation Network (SGN) leverages Large Language Models (LLMs) to perform command-line automation and automate software tasks. It begins by configuring YANG model parameters and status information, defining data models to write automation scripts or using automation tools to execute network configuration tasks. The Network Configuration Protocol (NETCONF) server processes configuration requests, translating them into corresponding YANG configuration operations for network devices, and completes the formulation of automation policies, including change management, error handling, and rollback mechanisms. Network automation removes the manual steps required to manage networks, such as configuring routers, switches, load balancers, and firewalls. It consolidates network tasks into pre-packaged programs that users can select, schedule, and execute from the application's front end. This enables automated operations for network configuration, security protection, orchestration, and provisioning across service providers.

The Service Generation Network (SGN) enhances the scalability and consistency of networks through network automation configuration technology. It reduces network operating costs while improving network reliability and security. By supporting industry digital transformation, SGN increases the flexibility of network services and promotes the development of intelligent networks, providing users with more efficient and reliable network service experience.

(4) Digital Twin Network Technology

Digital Twin Network technology creates a virtual model composed of data, sensor inputs, and other key parameters by mapping physical entities or systems from the real world into a virtual environment. This virtual model is capable of real-time monitoring, simulation, and optimization of the operation of the physical entities or systems. In the creation of Digital Twin Networks, data from various sensors and data sources are typically integrated. Data fusion techniques consolidate information from different sources to enhance the consistency and accuracy of the Digital Twin Network data, while reducing the likelihood of data redundancy and conflicts. During the construction of Digital Twin Networks, machine learning algorithms are applied to extract patterns of behavior and performance of physical entities or systems from vast amounts of data, thereby enhancing the precision and generalizability of the Digital Twin Networks. Model parameters are optimized and selected to best match real-world data, and appropriate model architectures are chosen to ensure the stability and performance of the Digital Twin Networks. Moreover, Digital Twin Network technology is crucial for predicting, optimizing service systems, and enabling predictive maintenance.

Digital Twin technology can provide the Service Generation Network (SGN) with real-time and accurate representations of network status, facilitating real-time monitoring, analysis, and prediction of network behavior. This supports data-driven optimization and decision-making by identifying optimal resource allocation and operational strategies to enhance system performance and efficiency. On the other hand, Digital Twin enables remote monitoring and maintenance. By simulating and analyzing data from network devices, SGN can predict equipment failures and maintenance needs in advance, allowing for remote preventive maintenance. This reduces production downtime and improves the stability and reliability of the network. By simulating network operations in a virtual environment, SGN can test and adjust configurations without directly affecting the actual network. This approach reduces costs and improves the efficiency of network deployment. In the future, Digital Twin networks will extend beyond individual entities or systems to build vast industrial ecosystems. Through Digital Twin networks, different enterprises, organizations, and individuals can collaborate and share data to achieve more efficient and intelligent service generation networks.

(5) Fully Programmable Network Infrastructure

The Service Generation Network (SGN) integrates Programming Protocol-Independent Packet Processors (P4) technology and other programmable methods, supporting the definition and processing of multiple network protocols. This provides a fully programmable solution with high customization and flexibility for networks. P4 allows network administrators to customize packet processing procedures and operations, supporting the definition and processing of various network protocols without relying on vendor-predefined functions. Instead, administrators can tailor packet forwarding and processing logic according to their needs. This enables SGN to rapidly iterate and innovate in response to changing demands and emerging network technologies, offering more advanced and flexible services. The programmability of P4 allows network devices to update and modify processing logic through software. Leveraging this capability, SGN dynamically sets packet processing methods based on user requests and service types, enabling rapid iteration and innovation of customized services.

Fully Programmable Network Infrastructure allows network administrators to customize network behavior and services according to their needs, enabling the generation of personalized end-to-end services. This programmability enables the Service Generation Network (SGN) to dynamically adjust service generation strategies in real-time based on network status monitoring, thereby providing smarter and more adaptive services. It is applicable across multiple fields, including communications, the Internet of Things, healthcare, and finance. This capability also drives the formation of a programmable application template library, which can be repeatedly invoked by different fields according to their specific needs.

(6) Deep Reinforcement Learning Network Optimization

Deep Reinforcement Learning (DRL) is an advanced machine learning technique that integrates Deep Learning (DL) and Reinforcement Learning (RL). Its core lies in training an agent to learn how to make optimal decisions in complex environments through continuous interaction with the environment, thereby maximizing cumulative rewards. Network optimization aims to find the best configuration in computer networks to enhance performance, resource utilization, and reduce latency and costs. DRL processes high-dimensional state and action spaces through neural networks, enabling the agent to learn the nonlinear relationships between complex network states and optimization objectives. It is a key tool for implementing network optimization in the Service Generation Network (SGN). DRL learns to select optimal node layouts and link connections in complex network environments and autonomously adjusts routing strategies based on real-time network conditions. This adaptability to network and traffic changes makes it suitable for networks of different scales. The network optimization solutions output by DRL after iterative training achieve comprehensive optimization of network resources and performance.

The integration of Deep Reinforcement Learning (DRL) and Large Language Models (LLMs) in network optimization enables the Service Generation Network (SGN) to enhance network performance and user experience while reducing resource waste and human intervention costs. This fusion optimizes various aspects, including automated network configuration, load balancing, network topology optimization, path optimization, cybersecurity optimization, and dynamic service quality optimization.

4.3 Application Scenarios of Service Generation Network

The application scenarios and use cases of the Service Generation Network (SGN) are extensive. This chapter only provides a detailed analysis of three scenarios: intelligent network planning and optimization, intelligent network deployment and intelligent network management and control.

(1) Intelligent Network Planning and Optimization

In network planning and optimization, the Service Generation Network (SGN) translates user requirements into intelligent solutions by analyzing demands and real-time status to determine the optimal network layout and resource allocation, thereby enhancing network performance and efficiency. It supports a variety of planning and optimization tasks, including wireless planning, performance evaluation, and traffic generation. In practice, planners need to conduct a comprehensive analysis of environmental information, service requirements, and resource conditions to identify potential bottlenecks and performance limitations within the network. Based on the analysis results, planners can enhance network coverage and capacity by adjusting the network topology, optimizing configuration parameters, and adding network facilities. Additionally, continuous monitoring enables the optimization of path routing to improve data transmission efficiency and reduce latency, thereby enhancing user experience.

The Service Generation Network (SGN) leverages key technologies of Generative Artificial Intelligence (GAI) to assist in the full lifecycle of air-ground collaborative wireless network planning. This includes accurately extracting service requirements through reasoning and decision-making capabilities, providing optimal recommendations that align with user intentions. Additionally, SGN trains large models using user behavior data to deliver more precise predictions of service requirements. This capability will become an essential tool for planning air-ground collaborative wireless networks.

In highly dynamic and complex topology air-ground collaborative networks, predicting communication needs is crucial for planning solutions. The Service Generation Network (SGN) leverages business requirements, wireless environment, and user behavior data to train large models, providing more accurate predictions of service demands. This enables network planners to design or evaluate plans based on predictive results, increasing design time and basis. After completing the planning and design, the plan needs to be converted into configuration information and deployed to physical devices. SGN provides configuration information for devices such as base stations, satellites, and wireless gateways, simplifying the entire configuration process, reducing costs, and saving human resources.

(2) Intelligent Network Deployment

Network deployment and construction is a comprehensive project involving the full lifecycle management of project initiation, design, and acceptance. It aims to deliver, configure, and accept specific performance indicators. In each phase of network deployment and construction, the Service Generation Network (SGN) can intelligently deploy and configure the network to meet service demands, automatically select and configure devices, and achieve rapid network expansion. This process simplifies the configuration of network devices, reduces complexity, and supports various service deployments, including semantic encoding and decoding, heterogeneous access, and digital twin technologies. Leveraging its large language models and domain expertise, SGN can accelerate the deployment of 5G private networks, cloud-network services, and edge computing networks, thereby enhancing the competitiveness of operators. Additionally, the application of multimodal technologies can facilitate automated quality inspection after network construction, saving human resources. The use of automation tools and large language models will improve the efficiency and speed of network construction, promoting the development and operational efficiency of networks.

SGN supports digital twin networks, which are virtual replicas of physical networks or their components. These digital twins can reflect the network state in real-time and with high precision, facilitating the monitoring, analysis, and prediction of network behavior. Based on this information, SGN can generate customized services tailored to the actual network conditions, including semantic encoding and decoding, as well as access for heterogeneous devices. The application examples of SGN demonstrate its potential in network deployment, not only providing highly personalized and intelligent solutions but also enhancing the efficiency and reliability of the network.

(3) Intelligent Network Management and Control

Intelligent Network Management and Control leverages advanced technologies and intelligent algorithms to achieve network operations and management, including network monitoring, fault diagnosis, resource management, traffic optimization, and security protection. By utilizing big data analytics, machine learning, and artificial intelligence, Intelligent Network Management and Control can monitor network status in real-time, identify potential issues, and automatically take corresponding measures for optimization and repair. Additionally, by integrating large language models, Intelligent Network Management and Control can perform real-time monitoring and analysis of network models to ensure effective assurance of network performance and service quality. Through model training and algorithm optimization, large language models can quickly identify anomalies and potential problems and automatically trigger corresponding response measures.

SGN monitors network conditions and predicts potential failures, enabling intelligent fault handling and automatic recovery functions. Its intelligent management and control unit can customize network slicing services, achieve self-healing of faults, and monitor anomalies. SGN's intelligent slicing services based on LLM (Large Language Model) meet the strict and diverse service requirements in multi-tenant environments and support resource management for multiple infrastructure providers. SGN supports cooperation in multi-tenant environments and among multiple infrastructure providers. Users can input customized slicing requirements through a large language model. After model training, real data is obtained via digital twins to generate management strategies, meeting the service requirements in dynamic multi-tenant environments and effectively managing the multidimensional resources of multiple infrastructure network providers. The novel SGN network architecture enhances network flexibility and manageability by decoupling hardware and software and separating the data plane from the control plane, allowing network devices to more flexibly adapt to service requirements. In the SGN architecture, real-time data is obtained through digital twins to train network state perception and failure prediction models, enabling targeted network monitoring and maintenance. This approach helps to defend against abnormal traffic attacks and enhances network security and stability.

5 Conclusions and Future Prospects

In the age of artificial intelligence, network architectures are undergoing profound transformations to adapt to the surging data traffic, diversified service requirements, and continuously evolving technological innovations. The integration of AI technologies has steered the development of network architectures towards intelligence, especially in the innovative network architectures designed for AI computing clusters, which enable high-performance and highly stable network interconnections. To promote interoperability and innovation, the transformation of network architectures also includes support for open interfaces and standardization, facilitating the integration of different systems and components. In summary, the transformation of network architectures in the AI era has enhanced network performance and efficiency, providing a solid foundation for the development of emerging applications and services. Looking ahead, future service-oriented network architectures are moving towards the direction of "intelligence + collaboration." By leveraging intelligent technologies, they achieve unified scheduling and orchestration of computing power, enabling efficient collaboration between the cloud, network, edge, terminals, and industries. Actively exploring new technologies and architectures will help adapt to the future trend of multi-operator co-construction and sharing of information infrastructure. Meanwhile, in response to the diverse business needs of future intelligent applications, more enriched deployment scenarios will be developed.

References

1. Jia S Y, Luckie M, Huffaker B, et al. Tracking the deployment of IPv6: Topology, routing and performance[J]. Computer Networks, 2019, 165: 1–15.
2. Mansour M, Gamal A, Ahmed A I, et al. Internet of things: A comprehensive overview on protocols, architectures, technologies, simulation tools, and future directions[J]. Energies, 2023, 16(8): 3465.
3. Liu Y J, Huang T, Wang S. Thoughts on innovation of future network architecture [J]. Proceedings of the Chinese Academy of Sciences, 2022, 37(1): 38–45.
4. Yuan X Z, Peng L, Zhang L F. Requirement and challenge of holographic-type communication to the future network [J]. Telecommunications Science, 2020, 36(12): 59–64.
5. Chen J, Yi C, Okegbile S D, et al. Networking architecture and key supporting technologies for human digital twin in personalized healthcare: a comprehensive survey[J]. IEEE Communications Surveys & Tutorials, 2023, 26(1): 706–746.
6. Simani S, Lam Y P, Farsoni S, et al. Dynamic neural network architecture design for predicting remaining useful life of dynamic processes[J]. Journal of Data Science and Intelligent Systems, 2024, 2(3): 141–152.

7. Kumar K S, Anand V. Underwater Acoustic Sensor Network: Architecture, Challenges, Future Possibilities In the Perspective of IoT[C]. 2023 Second International Conference on Electrical, Electronics, Information and Communication Technologies (ICEEICT), Trichirappalli: IEEE, 2023.
8. Kazmi S H A, Qamar F, Hassan R, et al. Survey on joint paradigm of 5G and SDN emerging mobile technologies: Architecture, security, challenges and research directions[J]. Wireless Personal Communications, 2023, 130(4): 2753–2800.
9. Lee H, Park S, Baek H, et al. AI-Native Network Algorithms and Architectures[J]. Fundamentals of 6G Communications and Networking, 2023: 573–584.
10. Tsinghua University Institute for AI International Governance. Framework and Implementation Pathways for AI Governance [R]. Beijing: Tsinghua University, 2021.
11. China Mobile Research Institute. White Paper on Computing Awareness Network (CAN) Technology [R]. Beijing: China Mobile, 2021.
12. People's Daily Online Research Institute. Report on the Development of China's Intelligent Internet (2024) [M]. Beijing: Social Sciences Academic Press, 2024.
13. Feng C, Sun L, et al. Accelerating Industry Intelligence White Paper [R]. Beijing: Huawei, 2023.
14. Wei L, Shuai J, Liu Y, et al. Service customized space-air-ground integrated network for immersive media: Architecture, key technologies, and prospects[J]. China Communications, 2022, 19(1): 1–13.
15. Huang T, Tan S, Tang Q, et al. NT-SCDC: Realizing Service Customized Networking in Distributed Clouds with NaaS Ticket[J]. IEEE Network, 2024, 38(3): 236–243.
16. Varga B, Farkas J, Fejes F, et al. Robustness and reliability provided by deterministic packet networks (TSN and DetNet)[J]. IEEE Transactions on Network and Service Management, 2023, 20(3): 2309–2318.
17. Li Z, Peng C, Yu G, et al. Detnet: Design backbone for object detection[C]. The European conference on computer vision (ECCV), Munich: Springer, 2018.
18. Tang X, Cao C, Wang Y, et al. Computing power network: The architecture of convergence of computing and networking towards 6G requirement[J]. China communications, 2021, 18(2): 175–185.
19. Ji M, Su C, Yan Y, et al. Adaptive provisioning in-band network telemetry at computing power network[C], 2023 IEEE/ACM 31st International Symposium on Quality of Service (IWQoS), Orlando: IEEE, 2023.
20. Mehmood K, Kralevska K, Palma D. Intent-driven autonomous network and service management in future cellular networks: A structured literature review[J]. Computer Networks, 2023, 220: 109477.
21. Foley M, Hicks C, Highnam K, et al. Autonomous network defence using reinforcement learning[C], 2022 ACM on Asia Conference on Computer and Communications Security, New York: ACM, 2022.
22. Tao H, Jiang L, Shuo W, et al. A Review of Future Network Technologies and Development Trends [J]. Journal on Communications, 2021, 42(1): 130–150.
23. Mihai S, Yaqoob M, Hung D V, et al. Digital twins: a survey on enabling technologies, challenges, trends and future prospects[J]. IEEE Communications Surveys & Tutorials, 2022, 24(4): 2255–2291.

Yunjie Liu academician of the Chinese Academy of Engineering, expert in communication and information system, has made pioneering work in data network, Internet and network convergence. His main research directions are future network architecture, network convergence and evolution, etc. He has presided over many national key scientific research projects in the field of data communication and achieved many important results. He has presided over the design, construction and operation of the national public data network, computer Internet and high-speed broadband network, which laid an important foundation for the development of China's informatisation; and presided over the design, construction and operation of China Unicom's 'Multi-service Unified Network Platform', which solved the problems of uncontrollable, unmanageable and unguaranteed quality of service of IP network.

Zhengjie Sun, Ph.D., Assistant Researcher of Purple Mountain Laboratory, is mainly engaged in the research of future network, deterministic network, edge computing, exploring and designing the next-generation network architecture, the core key technologies of large-scale end-to-end deterministic network, the deterministic network layered architecture, access mechanism and scheduling mechanism, adapting to the future needs of digital transformation, and actively constructing a smarter, more efficient and more reliable network environment.

Shuo Wang, Ph.D., Associate Professor of Beijing University of Posts and Telecommunications, selected for the 6th China Association for Science and Technology Young Talent Support Project, senior member of China Communications Association, mainly engaged in the research of future networks, programmable networks, deterministic networks, etc., and has presided over more than 10 projects of the National Natural Science Foundation of China, Beijing Municipal Natural Science Foundation, China Postdoctoral Science Foundation, and sub-projects of the National Key Research and Development Program. In the past five years, he has published more than 20 SCI/EI search papers, submitted more than 10 national technical invention patents, and won the special prize of the China Communications Association and the Innovation Team Award of the Chinese Institute of Electronics.

Tao Huang, Ph.D., Professor of Beijing University of Posts and Telecommunications, Chief Engineer of Future Network Experimental Facility of National Big Science Engineering Project, mainly engaged in scientific research on future network architecture, software-defined network, network 2030, etc. He has presided over more than 20 national scientific research projects, has published more than 100 papers in important journals and conferences, has authorized more than 30 patents, and has published 6 domestic monographs, and the related technologies have been deployed and applied in the departments of Relevant technologies have been deployed and applied in telecommunication, radio and television, education and other sectors, and his core achievements have won the first prize of the Technical Invention Award of the Communications Society of China and the Second Youth Science and Technology Award of the Communications Society of China.

Data Factor Empowering New Quality Productivity

Ke Tang and Gang Chen

Abstract The rapid growth of the digital economy has positioned data factors as essential drivers of new quality productivity. This paper examines the theoretical foundations and underlying mechanisms through which data factors contribute to new quality productivity and offers relevant policy recommendations. Specifically, we define the conceptual boundaries of data factors and distinguish them from traditional factors of production. Then, we explore the theoretical significance of new quality productivity and delves into the relationship between data factors and this emerging form of productivity, outlining the conditions necessary for effective empowerment. In analyzing the mechanisms of empowerment, we show how data factors drive technological breakthroughs, enable innovative configurations of production factors, and facilitate deep industrial transformation and upgrading. Additionally, we identify key challenges in the current stage of data factor marketization and utilization, including difficulties in valuation and pricing, slow development of application scenarios, strong resistance to digital economic transformation, and potential resource and environmental concerns. Finally, the paper proposes policy recommendations to enhance the role of data factors in fostering new quality productivity, providing theoretical insights and practical guidance for economic development in the digital age.

Keywords Data factor · New quality productivity · Factor synergy · Multiplier effect · Innovation · Technological breakthrough · Innovative configuration of production factors · Deep industrial transformation and upgrading

---

Portions of this text are translated with assistance from AI such as DeepSeek, o3-mini and Doubao, followed by review and editing by the authors.

K. Tang (✉) · G. Chen (✉)
Institute of Economics, School of Social Sciences, Tsinghua University, Beijing, China
e-mail: ketang@tsinghua.edu.cn

G. Chen
e-mail: chengang311@tsinghua.edu.cn

K. Tang
Zhishan College, Tsinghua University, Beijing, China

China's e-Science Blue Book 2025, https://doi.org/10.1007/978-981-96-8252-2_3

1 Introduction

As the global economy transitions toward digitalization, traditional productivity models are undergoing profound changes. This transformation underscores the growing need for higher efficiency, superior quality, and increased innovation in the digital era. In response, during an inspection tour in Heilongjiang in September 2023, the General Secretary introduced the concept of "new quality productivity," emphasizing the integration of technological innovation resources, the development of strategic emerging and future industries, and the acceleration of new quality productivity formation [1]. At the Central Economic Work Conference in December 2023, the General Secretary stressed the need for industrial innovation to be powered by technological progress, particularly through groundbreaking and emerging technologies that drive the creation of new industries, business models, and economic growth. The rise in new quality productivity has become a key force behind sustainable and stable economic expansion. As digital technologies continue to evolve, data has taken on a crucial role as a production factor in the digital economy, acting as a vital catalyst for advancing new quality productivity.

Although the role of data as a key driver of new quality productivity is widely acknowledged in academia [2], its underlying mechanisms remain underexplored. Advancing theoretical research to fully understand how data, as a new type of production factor, influences and enables the formation and growth of new quality productivity is crucial. Analyzing the economic logic and operational dynamics of data-driven productivity will not only uncover new patterns of economic development in the digital age but also contribute to the refinement and expansion of existing economic theories. In practice, despite the deep integration of digital and physical economies and the increasing prominence of data in driving industrial transformation, fostering new business models, and expanding economic opportunities, several challenges persist. These include limited data availability due to concerns over security, reluctance, and technical barriers, information asymmetry and paradoxes, valuation and pricing difficulties stemming from uncertain value attribution, issues of illegal usage and traceability, and regulatory challenges related to privacy and security. These constraints collectively hinder the full potential of data in fostering new quality productivity. Addressing these practical challenges through in-depth research can provide a theoretical foundation and empirical guidance for policymakers and business leaders navigating data-driven economic development. Strengthening research in this area is not only of significant theoretical importance but also has urgent practical implications.

2 Theoretical Foundation

2.1 *Definition and Characteristics of Data Factor*

2.1.1 Definition of Data Factor

A clear delineation of the concept of data factors is the essential prerequisite for exploring how data factors empower new quality productivity. In practice, there is often conceptual confusion among "data," "raw data," "data resources," "data assets," "data products," and "data factor," with data sometimes being directly equated with data factor. This conflation can lead to misunderstandings regarding the actual role of data factor in empowering new quality productivity within economic activities. From an economic perspective, these concepts are not synonymous; clear distinctions must be drawn to accurately define how data contributes to productivity growth.

As shown in Fig. 1, the term "data" commonly refers to the initial stage of the data value chain—namely, "raw data." Raw data is generally understood as an unprocessed collection of information that captures the state and changes of objective phenomena. Essentially, it serves as a medium for recording information. In this paper, "data" refers to any recorded information in electronic or other forms. Before processing, raw data typically lacks direct economic or practical value; it may contain errors, be incomplete, or exist in formats that are not machine-readable [3]. Once collected, organized, and processed into a structured, machine-readable format, raw data transforms into data resources capable of generating economic and social value. This transformation occurs in the "data resourceization" stage depicted in Fig. 1.

Further along the value chain, data resources can be recognized as data assets if they meet specific criteria, including a clear and undisputed origin, legal ownership or control, reliable cost or value measurement, and the potential to generate economic or social benefits. This stage is known as "narrow data assetization." Data resources confirmed as assets can then be further refined and developed into data products or services (the "data productization" stage in Fig. 1), ultimately becoming integral to large-scale social production. In contrast, the concept of "data factor" within the framework of "data factorization" is defined from an economic perspective. It refers to data and its derivative forms that market entities can access and utilize in production processes to create economic and social value [3]. As shown in Fig. 1, raw data itself does not inherently generate economic or social value.

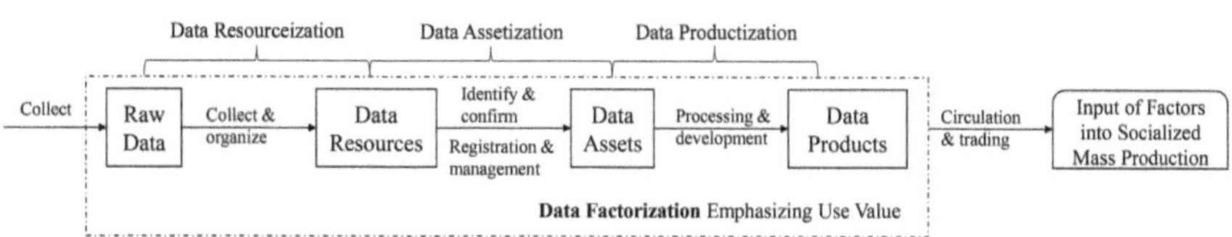

Fig. 1 The data value chain

In its unprocessed state, data is often disorganized and cannot be directly used as an input in production activities. In other words, raw data $\neq$ data factor. Data was formally recognized as a new production factor during the Fourth Plenary Session of the 19th Party Congress because it undergoes a series of processing, analytical, and transformative stages—enhancing its machine-readability, accuracy, and completeness. This process involves data resourceization, assetization, and productization along the data value chain, ultimately refining its value so that it can be effectively utilized in large-scale social production, unlocking its economic and social potential [3]. Consequently, while data resources and data products—once processed into machine-readable formats—qualify as data factors since they can be used as production inputs, data assets emphasize asset value rather than use value. Thus, data assets should not be conflated with data factors, which focus on their functional role in economic activities.

The distinction between data and data factors lies in their ability to be directly utilized as production inputs with machine-readability as a defining characteristic. To be recognized as a new type of production factor in economic theory, data factors must meet specific economic criteria. Scholars such as Tang et al. [4] have refined Berczi's "six conditions" [5] into five key requirements that differentiate data factors from traditional labor and capital. These conditions include a stable supply of data factors, effective demand, the ability to enhance output when combined with other production factors, the presence of a well-functioning market system to ensure their tradability, and a defined transaction price to regulate market supply and demand. Without these conditions, data cannot fully function as a production factor. Collectively, these requirements establish the economic foundation for the factorization of data, ensuring its role in driving economic and social value.

2.1.2 Characteristics Distinguishing Data Factor from Traditional Factors

The characteristic attributes that distinctly differentiate the data factor, as a new type of production factor, from traditional production factors can be summarized as follows: Firstly, virtuality and attachment [6]. The virtuality and attachment of the data factor make it rely on the combination of specific technologies and environments for generation, promoting its integration with digital technologies such as information and communication, thereby significantly enhancing its use value and production efficiency [7]. Secondly, the data factor features easy replicability and non-rivalry, including horizontal non-rivalry, vertical non-rivalry, and spatial non-rivalry, etc. [8], enabling it to be used simultaneously by multiple entities without mutual exclusion, which significantly reduces the costs of circulation and usage. Furthermore, the positive externality of the data factor is reflected in its positive feedback effect and network externality. Especially in terms of network externality, as the number of users and the variety of data increases, the overall value and utility are significantly enhanced [9]. In addition, the scenario-dependency of the data factor means that its application value is closely related to specific scenario environments, that is, "no application without scenarios", allowing the same data to exhibit diverse value potentials in different contexts [10]. Finally, the most remarkable characteristic that distinguishes the data factor from traditional production factors is the multiplier effect of the data factor. That is, the data factor can combine with labor, capital, land, technology, etc., to exert a multiplier-level synergistic effect. Based on the scenario-based application of data and the integrated innovation of digital technologies, it significantly improves production efficiency, promotes the efficient allocation of factor resources, and achieves the multiplier effect on output [11].

This multiplier effect—often referred to as the "data factor ×" model—facilitates cross-industry collaboration, fosters new business models, and injects fresh momentum into economic growth. It not only supports data-driven decision-making, innovation acceleration, and personalized service development but also plays a crucial role in the digital transformation of traditional industries, technological advancements, and the emergence of new economic paradigms [12, 13]. The "data factor ×" model exemplifies the vast potential of data productivity and its multi-dimensional value in large-scale economic production. These advantages have positioned data factors as a driving force behind China's modernization efforts and the high-quality development of its digital economy [12].

2.2 *The Theoretical Connotation of New Quality Productivity*

In January 2024, at the 11th collective study session of the Political Bureau of the CPC Central Committee, the General Secretary introduced the concept of new quality productivity and outlined key tasks for its development. He defined new quality productivity as advanced productivity driven by high technology, high efficiency, and high quality. Unlike traditional economic growth models, it prioritizes innovation and represents a qualitative leap in how labor, production materials, and production objects are optimally combined. This shift leads to a substantial increase in total factor productivity (TFP), with innovation as the core, quality improvement as the focus, and advanced productivity as the essence. In March of the same year, during a review session of the Jiangsu delegation at the 14th National People's Congress, he reiterated that high-quality development must remain the primary objective and that new quality productivity should be developed based on local conditions.

From the perspective of Marxist productivity theory, new quality productivity extends and evolves classical Marxist thought to fit the modern era. Classical theory defines productivity as humanity's capacity to transform nature to meet material needs, driving social progress through the effective combination of labor, production materials, and production objects [14]. In the digital economy, new quality productivity emphasizes revolutionary technological breakthroughs, innovative allocation of production factors, and deep industrial transformation. According to the General Secretary, laborers are no longer just providers of physical labor or basic skills; they are highly specialized workers capable of operating advanced technology and driving innovation. Production materials now extend beyond traditional machinery and raw materials to include cutting-edge technologies like artificial intelligence and new materials, making production tools smarter, more efficient, and environmentally friendly. Similarly, production objects now encompass not only natural resources but also data, information, and intellectual property, expanding the scope of production activities significantly.

By placing innovation at its core, new quality productivity optimizes and upgrades these three production elements, moving away from resource-intensive and environmentally damaging models. Instead, it embraces a development approach characterized by technological advancement, efficiency, and quality. This transformation enhances production efficiency and product quality while driving sustainable economic and social progress. As a result, new quality productivity serves as a key engine for long-term development, representing both an improvement in production quality and an expansion of the classical production framework.

From an economic standpoint, increasing TFP is central to the formation and growth of new quality productivity. Traditional economic growth models rely on increasing labor and capital inputs, such as in capital-intensive and labor-intensive industries. However, due to resource limitations and diminishing marginal returns, relying solely on labor and capital expansion is no longer sustainable. TFP measures output efficiency relative to all inputs, including labor, capital, and technology, emphasizing quality rather than quantity. New quality productivity aims to improve TFP through groundbreaking technology, innovative resource allocation, and industrial transformation. This shift means that economic growth no longer depends solely on traditional labor and capital but instead undergoes a fundamental restructuring through intelligent and digital technologies.

Technological innovation is the primary driver of new quality productivity, encompassing both incremental improvements and disruptive breakthroughs that redefine industries. Schumpeter's theory of "creative destruction" provides valuable insight into this process, explaining how economic progress arises from the continuous replacement of outdated systems with new ones. New quality productivity emerges through this ongoing cycle of innovation, where new technologies and business models replace obsolete ones, fostering a high-efficiency, high-quality production system. Importantly, this "destruction" is not a setback but a necessary step toward unlocking new economic growth. By reshaping industries and reallocating resources, new quality productivity accelerates economic transformation, leading to a more sustainable and higher-quality development path.

Finally, the new development philosophy offers crucial guidance for advancing new quality productivity in practice. It emphasizes innovation, coordination, sustainability, openness, and shared benefits—principles that form a comprehensive framework for modern economic growth. Innovation drives productivity improvements, coordination ensures balanced regional progress, sustainability aligns economic growth with environmental protection, openness facilitates resource allocation on a global scale, and shared development ensures that the benefits of productivity gains reach a broader segment of society, promoting equity and inclusivity.

2.3 *Logical Relationship and Preconditions for Empowerment*

2.3.1 Logical Relationship

As shown in Fig. 2, the logical relationship between data factor and new quality productivity can be summarized in three aspects.

First, the data factor is an important component of new quality productivity. In the framework of Marxist theory of productive forces, the object of labor refers to the material resources that are processed and transformed during labor to produce the goods that satisfy the living needs of workers. With the rapid development of technology, the data factor has become an indispensable resource in modern society, with its status and value increasingly prominent. The General Secretary's definition of new quality productivity is a significant extension and expansion of classical Marxist theory of productive forces; it emphasizes that new quality productivity is essentially characterized by the leap in the optimized combination of laborers, means of labor, and objects of labor. Therefore, as shown in Fig. 2, as a new quality object of labor, the data factor—after processing and transformation—is integrated into mass socialized production, not only expanding the boundaries of production activities but also profoundly changing the way value is created, making it a vital component of new quality productivity. Incorporating the data factor into the production process enhances the efficiency of utilizing information resources, thereby driving technological innovation and industrial upgrading, and ultimately boosting overall productivity.

Second, the data factor is the key driving force behind the formation and development of new quality productivity. As a new factor of production, the data factor is pivotal in empowering the formation and development of new quality productivity. As shown in Fig. 2, the data factor can empower new quality productivity by promoting revolutionary breakthroughs in technology, innovative allocation of factors of production, and profound transformation and upgrading of industries. Through the integration, analysis, and processing of massive data factors, along with scenario development and application, enterprises can precisely identify market

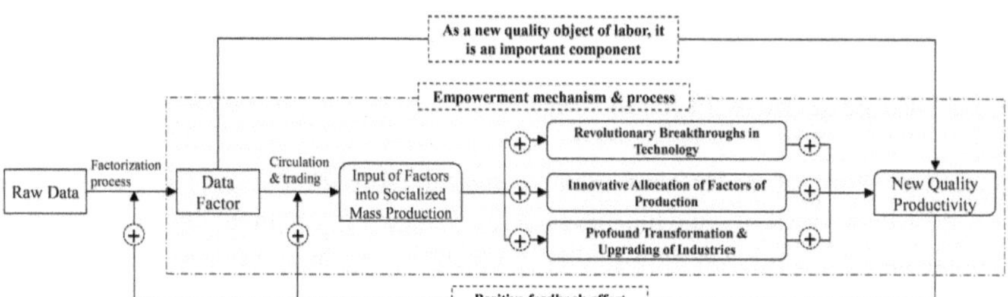

Fig. 2 The logical relationship between data factor and new quality productivity

demand, trends in technological advancement, and new forms and models of business; this not only leads to revolutionary breakthroughs in technology and spurs the emergence of new products and services but also drives fundamental changes in production methods and a qualitative leap in productivity. Moreover, the data factor can be deeply integrated with traditional factors of production; through their synergistic collaboration, resource allocation is optimized, and the efficiency of their combination is significantly enhanced. The data factor also plays an important role in the profound transformation and upgrading of industries. In the digital economy era, the data factor has become the core link connecting consumers, enterprises, and the market. By collecting, organizing, and analyzing data, enterprises can accurately grasp market demand dynamics and consumer behavior preferences, swiftly adjust product forms, structures, and functions, and continuously optimize supply chain management to enhance service quality. This data factor-based decision-making model not only improves an enterprise's market responsiveness and competitiveness but also drives the profound transformation and upgrading of the entire industry.

Third, the development of new quality productivity exerts a positive feedback effect on the demand for data factors. The formation and development of new quality productivity will significantly enhance TFP, enabling the economic system to transition from low-quality, extensive growth to high-quality, sustainable growth, which will inevitably increase the ongoing demand and reliance on data factors. Characterized by high technology, high efficiency, and high quality, new quality productivity imposes higher requirements on the optimized allocation and efficient utilization of factors of production. Therefore, to meet the developmental needs of new quality productivity, it is necessary to increase investment in and application of data factors. At the same time, the enhancement of new quality productivity will further propel the deepening of the data factorization process. The sustained demand and dependency, along with the continuous deepening of the data factorization process, form a positive feedback effect of new quality productivity development on the demand for data factors (see Fig. 2). In other words, the formation and development of new quality productivity will drive continuous innovation and refinement of the technologies for data collection, processing, analysis, and application, while the in-depth development of data factors will provide new quality factor support for the further enhancement of new quality productivity. This virtuous cycle will drive the economic system toward a state of higher technological advancement, efficiency, and quality.

2.3.2 Empowerment Preconditions

Based on a thorough understanding of the theoretical connotations of data factors and new quality productivity and the logical relationship between them, the preconditions for data factor to empower new quality productivity can be summarized into two key aspects.

Firstly, the data must complete the process of factorization. According to the definition of data factor, data is not equivalent to data factor. Unprocessed data is typically raw data—an unorganized, unrefined collection of information that does not possess direct economic or use value and is generally not machine-readable. It cannot typically serve as a production input for large-scale social production. Thus, the primary precondition for data factor to empower new quality productivity is that data must complete its transformation into data factor. This process—data factorization—entails the collection, cleaning, processing, and development of raw data into high-value, machine-readable data that can be used as an input in economic production and subsequently generate economic and social value. In his keynote speech at the China International Big Data Industry Expo during the Digital Talent Cultivation Exchange Activity in August 2024, Academician Chen Songxi pointed out that the key to data factorization lies in effective data analysis, and the core of data analysis is to assess data quality using statistical methods and related technologies, thereby extracting valuable information to accomplish the relevant empirical tasks—in other words, the so-called scenario application. Only data that has been analyzed can clearly demonstrate its quality and value, thereby possessing the potential to empower new quality productivity.

Second, data must also satisfy the conditions for factor confirmation. Once data has undergone the factorization process and transformed into a data factor ready to empower new quality productivity, it must also complete the data marketization process—in other words, the data factor must satisfy the five conditions for factor confirmation. First, there must be a stable and effective supply of data factors to ensure that the market can continuously obtain data factors to advance the practical empowerment of new quality productivity. Second, the effectiveness of demand is indispensable; only when there is actual demand can data factors be effectively utilized to realize their economic benefits. The stable existence and effectiveness of both supply and demand are crucial for building a sound data market. In addition, data factors need to combine with other factors of production to achieve a synergistic multiplier effect in overall output—namely, "Data Factor × ." In response, the National Data Administration, in coordination with relevant departments, has formulated the "Data Factor × " Three-Year Action Plan (2024–2026), which promotes industry practices through twelve key actions designed to harness multiplier effect scenarios. Furthermore, establishing a comprehensive market system is the foundation for ensuring the tradability of data factors. A robust market system will promote the smooth flow and free allocation of data factors in economic activities, enabling enterprises to acquire and utilize data factors more flexibly. Moreover, the marketization process must also ensure the legality and compliance of data factors to protect data privacy and security. Finally, clear transaction pricing, as a guide for market supply and demand, is critical to ensure the effective allocation of data factors. A reasonable pricing mechanism will incentivize the production and circulation of data factors, thereby fostering a vibrant market.

In summary, the prerequisites for empowering new quality productivity with data factors include not only the data factorization process but also the fulfillment of specific data marketization conditions. These conditions collectively form the essential foundation for data factors to effectively empower new quality productivity in the modern economy, promote the development and maturation of the data market, and further support the continuous enhancement of new quality productivity.

3 Empowerment Mechanism

3.1 Data Factor Driving Revolutionary Breakthroughs in Technology

Based on existing research, we can summarize two intrinsic mechanism pathways by which data factors drive revolutionary breakthroughs in technology: first, incorporating data factors into the innovation production function; and second, incorporating data factors into the technological progress and knowledge production function.

3.1.1 Incorporating Data Factor into the Innovation Production Function

When exploring how data factors are integrated into the innovation process to exert a driving, empowering role, this paper analyzes both horizontal and vertical innovation, revealing the mechanism through which data factor propel dominant technological breakthroughs in innovation and R&D.

(1) **Horizontal innovation**. Researchers such as Cong and his colleagues have incorporated data factors into firms' horizontal R&D activities using a Cobb–Douglas production function framework, thereby establishing one of the earliest macro-level analytical frameworks for data factor-driven horizontal innovation [8]. Subsequently, Cong et al. [15] and Professor Xie's team further examined the driving effect of data factors on horizontal innovation and their impact on economic growth. Horizontal innovation generally emphasizes cross-industry and cross-domain R&D and innovation. Within the macroeconomic framework of data growth, data factors—as new factors of production—are integrated with other traditional factors of production into firms' horizontal R&D activities, providing essential support for the research and innovation of General Purpose Technologies (GPTs). Latest GPTs, such as generative AI and large language models (LLMs), rely on massive data for training; using deep learning algorithms based on the Transformer architecture, valuable information and patterns are extracted from data factors. For example, OpenAI's latest large language models, o1-mini (Octavian-1 mini) and o1-preview (Octavian-1 preview), have both been critically enabled by data factors. Massive, high-quality multimodal data factors serve as the foundational corpus for training and fine-tuning models like ChatGPT (Chat Generative Pre-Trained Transformer), enabling them to achieve outstanding multimodal recognition, language understanding, generative capabilities (including text generation, text-to-image, and text-to-video), as well as reasoning abilities. Comprehensive tests indicate that o1-mini and o1-preview now possess reasoning and cognitive abilities comparable to graduate-level performance; continuous learning and feedback mechanisms further allow these models to be fine-tuned and optimized based on user interactions, rapidly

adapting to complex tasks. Moreover, the cross-domain integration and application of data factors enable these models to be widely deployed in industries such as education, entertainment, and technical support, thereby achieving deep horizontal technological innovation and revolutionary breakthroughs.

(2) **Vertical innovation**. Data factors also profoundly influence firms' vertical innovation activities. Based on the work of Cong et al. [8, 15] and Professor Xie's team, a vertical innovation production function incorporating data factors has been constructed to depict how data factors empower economic development by being embedded in the vertical technological innovation process. Unlike horizontal innovation, which spans various fields, vertical innovation within a firm primarily focuses on the deep improvement and breakthrough of biased technologies in a specific field. Existing literature indicates that data factors can drive biased technological innovation [16]. Decentralized data storage and sharing mechanisms have positioned data factors as a core element in building Web3 platforms, thereby promoting innovation in Decentralized Applications (DApps), particularly in FinTech innovation within Decentralized Finance (DeFi). Revolutionary breakthroughs in Web3 platform technology have significantly enhanced data transparency and transaction security, leading to increased user trust and a multiplier effect on economic efficiency [17].

3.1.2 Incorporating Data Factor into the Knowledge Production Function

Unlike the data growth framework for incorporating data factor into horizontal innovation developed by Cong et al. [8, 15] and Professor Xie's team, Jones and Tonetti earlier developed a more streamlined macroeconomic model framework of data growth by constructing a knowledge production function that incorporates data factors as the sole production factor, thereby illustrating in depth how data factors enhance knowledge levels and drive productivity improvements [18]. Abis and Veldkamp, on the other hand, constructed a multi-factor knowledge production function in the form of a Cobb–Douglas function that incorporates data factors to depict the knowledge production process driven by data factors [19]. This process is similar in essence to the DIKW (Data, Information, Knowledge, Wisdom) framework proposed by Ackoff in 1989, meaning that data can be refined into information to promote knowledge production [20]. For example, DeepMind's AlphaFold leverages vast amounts of biological data—such as protein structures and genomic information—to drive advanced biological science knowledge production and corresponding technological progress. By employing deep learning models based on Convolutional Neural Networks (CNN) and Graph Neural Networks (GNN), AlphaFold performs statistical analysis and deep learning on sample biological data factors, thereby achieving an understanding of the "protein folding problem" and producing advanced bioengineering knowledge regarding the relationship between amino acid sequences and their three-dimensional structures, which has led to a revolutionary technological breakthrough in the field of protein folding [21].

At present, the research paradigm is undergoing an unprecedented profound transformation, the core characteristic of which lies in a fundamental shift in the way knowledge is created. In the past, scientific research often relied heavily on the individual wisdom, academic intuition, and long-term accumulated experience of scientists, with new knowledge being created through their subjective initiative. However, with the rapid development of digital technologies—especially with the advent of the data economy—the research paradigm is gradually shifting from the traditional model, which depends on scientists' subjective creation of knowledge, to a more objective, quantitative, data-driven "big data + artificial intelligence" model for knowledge mining. Under this new research paradigm, the processing and utilization of data are no longer merely auxiliary or optional steps in the research process; rather, they have become an essential pathway and core element for conducting cutting-edge academic research. Data, as a new factor of production, is also emerging as a critical and constrained resource. It is foreseeable that in the future, researchers will increasingly rely on AI-based large-scale information acquisition and intelligent analysis capabilities to uncover hidden scientific laws and profound insights from massive multi-source, multimodal, high-dimensional data, thereby leading to the emergence of new scientific theories and knowledge discoveries. Interdisciplinary collaboration will become more natural, convenient, and necessary, as researchers from diverse disciplines exchange ideas directly on shared intelligent platforms, breaking down barriers to cross-disciplinary cooperation and producing more, higher-quality integrated knowledge outcomes.

3.2 Data Factor Driving Innovative Allocation of Factors of Production

The process by which data factors drive the innovative allocation of factors of production can be divided into three categories based on the manner in which data factors combine with other factors of production and their innovative configuration process: (1) data factors empowering the innovative allocation of traditional factors of production; (2) data factors empowering the new quality transformation of traditional factors of production; and (3) data factors empowering the efficient synergistic allocation of factors of production.

3.2.1 Data Factor Empowering the Innovative Allocation of Traditional Factors of Production

Essentially, data factors empowering the innovative allocation of traditional factors of production is the process of digital optimization and transformation in which virtual and dependent data factors combine with other factors of production to form a new quality configuration process of factors of production. This represents an innovative configuration transformation generated by the combination of data factors with a single factor of production.

(1) **Integration of data factors with labor**. From the perspective of enterprises, big data related to recruitment, employment, and job performance can accurately identify employees' skills and performance, thereby optimizing recruitment and training processes and enhancing talent matching. Data-enabled workflow optimization allows companies to monitor task allocation and work progress in real time, thereby boosting overall employee efficiency. In addition, by analyzing employee feedback data, enterprises can develop personalized incentive and development plans to improve employee satisfaction and retention. From the labor perspective, data factors significantly enhance the efficiency of employment and location-selection decisions. Labor can utilize data analysis tools to access comprehensive employment market information, understand industry trends, salary levels, and job demands, thus assisting job seekers in making more informed employment choices. Furthermore, data factors help labor understand employment opportunities, living costs, and regional development prospects in different areas, thereby improving the efficiency of migration and location-selection decisions. Thus, the integration of data factors with labor markedly enhances both enterprise human resource management and the efficiency of employment and location-selection decisions for labor. This mutual empowerment drives the innovative allocation of the labor factor, thereby boosting new quality productivity. For example, Alibaba employs a big data analytics platform and an Applicant Tracking System (ATS) to comprehensively assess candidates' backgrounds, skills, and behavioral data to enhance talent matching, as well as to analyze the characteristics of successful employees in order to identify outstanding candidates.
(2) **Integration of data factors with capital**. In terms of capital allocation, detailed data analysis driven by data factors significantly enhances investment efficiency and the scientific nature of decision-making for industrial capital, venture capital, and financial capital. Regarding industrial capital, companies optimize production decisions and enhance resource utilization through data analysis. In the allocation of venture capital, venture capital firms and private equity (PE) funds conduct in-depth analyses of startup market trends, financial data, and team backgrounds to effectively identify potential investment opportunities and risks—for instance, Sequoia Capital in Silicon Valley reduces risk by analyzing the market trends and financial conditions of startups. In the realm of financial capital, financial institutions use financial big data to assess and identify borrower credit risk for precise credit lending; for example, Ant Group leverages user behavior and credit data to improve the accuracy of credit decisions, thereby enhancing both the efficiency and security of fund utilization.
(3) **Integration of data factors with land**. By leveraging sensor and data analytics technologies, data factors monitor and analyze real-time information on crop growth, soil health, and climate change, thereby improving the efficiency of agricultural land allocation and enabling precision agricultural management.

For example, Huawei's smart agriculture solutions leverage data analytics and Internet of Things (IoT) technologies to drive intelligent agricultural management and enhance land efficiency. Furthermore, data factors can integrate the demands for various land uses to support scientific urban land spatial planning and resource allocation, thus enhancing the efficiency of land spatial planning.

(4) **Integration of data factors with knowledge**. According to the DIKW framework, data factors empower knowledge production by refining information. In addition, data factors facilitate the rapid dissemination and sharing of knowledge through digital platforms and social media. A data-driven knowledge learning management system can, based on employees' learning habits and current knowledge levels, personalize the recommendation of learning resources, thereby improving the targeting and efficiency of knowledge acquisition. For example, ByteDance monitors employees' learning needs and behaviors, analyzing which knowledge resources are frequently accessed, to optimize the recommendation and organization of knowledge assets.

(5) **Integration of data factors with technology**. The role of data factors in technological innovation is primarily reflected in technological iteration and upgrading. By analyzing user feedback and market data, companies can quickly adjust and optimize existing technologies to meet changing market demands, thus maintaining both foresight and flexibility. Internationally, Tesla continuously optimizes its autonomous driving system through real-time data collection and analysis, thereby enhancing vehicle safety and user experience; domestically, Xiaomi leverages user feedback data to rapidly iterate product features, continuously driving technological innovation in the smart home sector.

(6) **Integration of data factors with management**. In the field of management, data-driven decision-making enhances the scientific rigor and efficiency of management. Through data analysis, managers can make more accurate, fact-based decisions, thereby reducing uncertainty. Real-time data monitoring helps managers quickly identify issues and make necessary adjustments, boosting operational efficiency. Moreover, data analysis reveals inefficient segments in management processes, aiding companies in process optimization. For example, JD.com leverages logistics big data analytics to optimize logistics and warehousing decisions, thereby enhancing overall management efficiency and customer satisfaction.

(7) **Non-competitive integration of data factors**. The non-competitive nature of data factors—manifested in horizontal, vertical, and spatial non-competitiveness—significantly promotes the integration of data factors, creating an innovative configuration where $1 + 1 > 2$ [22]. Horizontal non-competitiveness allows data to be used simultaneously by multiple entities, fostering cross-industry sharing and innovation; vertical non-competitiveness enables data to be utilized repeatedly at different stages of the production chain, improving overall efficiency; and spatial non-competitiveness overcomes geographical limitations, facilitating cross-regional data sharing and non-competitive allocation. These characteristics make data factors a key production element in the formation and development of new quality productivity,

enhancing the allocation efficiency of traditional factors of production and supporting digital transformation and innovation across industries.

3.2.2 Data Factor Empowering the New Quality Transformation of Traditional Factors of Production

Data factors, through their integration with traditional factors of production, can not only directly enhance the allocation efficiency of these traditional factors of production but also achieve a new quality transformation of it, thereby forming new quality factors of production. It, in turn, further drives the innovative allocation of factors of production. It should be noted that the new quality transformation of knowledge and technology driven by data factors falls under the realm of revolutionary breakthroughs in technology and is not included in this section; additionally, the new quality transformation of data factors through non-competitive integration is not considered a true new quality transformation in the strict sense and is classified as part of the earlier category.

(1) **The new quality transformation of labor**. Data factors, when integrated with labor, give rise to new quality human capital. Through real-time data analysis and intelligent training, companies can accurately identify skill gaps among employees, thereby enhancing the overall quality of the workforce in a targeted manner. For example, ByteDance analyzes employees' learning needs to provide personalized training programs, which in turn enhances overall employee skill levels and job satisfaction.

(2) **The new quality transformation of capital**. Data factors empower the new quality transformation of capital through their innovative integration with capital. First, under the empowerment of data factors, smart contract technology has given rise to the concept of smart capital. For example, financial instruments traded via smart contract algorithms autonomously manage and execute fund flows by analyzing market data in real time, achieving intelligent capital management. In addition, compared to traditional capital within the conventional financial system, Web 3 tokens—empowered by data factors—exhibit characteristics such as decentralization, innovative ownership and incentive mechanisms, participatory value enhancement, and community governance, thereby driving the new quality transformation of capital.

(3) **The new quality transformation of land**. Data factors drive the new quality transformation of land primarily through data-driven environmental monitoring and ecological governance, which helps restore the soil quality and biodiversity of degraded land and improve the land ecological environment. For example, Ant Forest converts data on users' carbon reduction behaviors into virtual tree planting to generate "green energy" and, with offline funding and multi-party collaboration, facilitates tree planting and maintenance in desertified and ecologically harsh areas, thereby achieving the new quality transformation of land and promoting sustainable development as well as the awakening of public environmental awareness.

(4) **The new quality transformation of management**. Through intelligent and refined transformation enabled by data factors, traditional management models are shifting to smart, precise management, thus forming new quality management. For example, Meituan optimizes its food delivery network by using algorithms to process data from delivery orders, user preferences, and real-time traffic information, significantly optimizing delivery routes and timings, reducing delivery costs, and improving customer satisfaction.

3.2.3 Data Factor Empowering the Efficient Synergistic Allocation of Factors of Production

From an economic theoretical perspective, the efficient synergistic allocation of factors of production empowered by data factors is mainly manifested in the following three aspects:

First, reducing information asymmetry to optimize the allocation of production factor resources. Information asymmetry is one of the important causes of market failure. The widespread application of data factors can significantly reduce the problem of information asymmetry among market participants [23]. For example, by leveraging sensors and IoT technologies, companies can monitor the operating status of production lines in real time, thereby promptly adjusting resource allocation. Through market big data analysis, companies can more accurately grasp market demand, supply chain conditions, and competitors' dynamics, which enables them to make more scientifically sound decisions on the allocation of production factor resources. This transparent flow of information improves synergistic efficiency among various factors, reducing the misallocation and waste of production factor resources caused by information asymmetry.

Second, data factors promote a seamless connection among factors of production. Data factors possess high liquidity and shareability and are not limited by industry, department, or geography, enabling a seamless linkage among different factors of production. On smart digital platforms, labor, capital, technology, and other factors of production can quickly flow and be efficiently matched, thereby significantly improving the utilization efficiency of factors of production and market responsiveness.

Third, data factors promote a profound transformation in production organization. Traditional production organizations typically allocate resources and manage production on an enterprise basis; however, the application of data factors has made production organization more flexible and diverse. By building a data-driven supply chain management system, companies can achieve digitized management of the entire value chain—from raw material procurement and production processing to product sales—thereby significantly improving production efficiency and quality while reducing production costs and operational risks.

3.3 *Data Factor Driving the Profound Transformation and Upgrading of Industries*

Data factors play a crucial role in driving the profound transformation and upgrading of industries. Their impact mechanism can be categorized into three aspects: first, data factors drive the profound digital transformation of traditional industries; second, data factors promote the deep integration and upgrading of industries; and third, data factors foster innovation in new forms and models of business.

3.3.1 Driving the Profound Digital Transformation of Traditional Industries

From an economic theoretical perspective, as a new factor of production, data factors primarily contribute to industrial upgrading by promoting the deep digital transformation of traditional industries. In the traditional industry development model, issues such as information asymmetry and low resource utilization efficiency have long hindered the improvement of production efficiency and the responsiveness of enterprises to market dynamics ever since industrialization. With the introduction of data factors, traditional enterprises can utilize digital technologies such as big data analytics, the IoT, and cloud computing to collect, store, process, and analyze various types of data. This enables them to precisely grasp market demand, optimize the allocation of production factor resources, and enhance production efficiency, thereby raising TFP at both the enterprise and industry levels [24]. Specifically, companies can use data collection and analysis tools to monitor production processes, inventory status, and sales channels in real time, thus achieving refined and intelligent management of production processes. Meanwhile, intelligent decision support systems driven by data factors allow enterprises to quickly respond to dynamic market demands by adjusting product design, structure, functionality, and production plans to meet personalized and diverse market needs. This process not only propels the digital transformation of traditional industries but also fundamentally transforms production methods, enhancing the new quality productivity of traditional industries.

3.3.2 Promoting Deep Integration and Upgrading Across Industries

Data factors not only have a digital transformation impact within individual industry boundaries but also drive the deep integration and upgrading across industries through technological integration improvements and revolutionary innovation breakthroughs. According to classical industrial economics theory, synergistic effects among industries can often lead to higher production efficiency and overall benefits. Data factors facilitate rapid advancement in deep integration by promoting the sharing of resources and technological convergence across industries [25]. For example, in the field of smart manufacturing, the integration of traditional manufacturing with information technology makes production processes more intelligent and automated, driving the transformation of the manufacturing industry toward high value-added segments. For instance, Tesla's innovations in autonomous driving technology involve collecting and analyzing massive vehicle operating data and using autonomous driving algorithms to optimize driving safety and efficiency, thereby promoting the deep integration between the traditional automotive industry and the information industry. Similarly, Alibaba Cloud's smart city solutions leverage robust cloud computing and big data capabilities to integrate public data resources from transportation, environmental monitoring, and public safety, enabling precise decision-making in city management. This cross-industry and cross-sector technological integration driven by public data factors not only elevates urban governance but also fosters high-quality integration and development in related smart city industries, further empowering new quality productivity. In addition, data factors facilitate the integration of finance and industry, significantly improving the efficiency of capital allocation.

3.3.3 Fostering Innovation in New Forms and Models of Business

The empowering application of data factors has given rise to numerous new forms and models of business, injecting fresh momentum into the new quality growth of the economic system and substantially boosting new quality productivity and TFP. According to innovative economics theory, the continuous emergence of new knowledge and technology has driven the transformation of traditional industries and the development of emerging sectors. Data factors provide crucial support for revolutionary breakthroughs in technology. With the development and widespread application of technologies such as big data, IoT, and generative artificial intelligence, new forms of business empowered by data factors continue to emerge, such as smart logistics and smart finance. These new business formats not only improve production efficiency and service quality but also drive a profound upgrading of industrial structures. Moreover, data factors drive innovation in business models. In the digital economy era, data factors have become a critical enabler for precise decision-making by enterprises. Through big data analytics, companies can accurately identify and discover new market opportunities and develop entirely new business models accordingly. For example, with the maturation of generative artificial intelligence technology and the success of LLMs such as ChatGPT in the To C market, an increasing number of companies are exploring business models for AI LLMs in the To B and To SMB sectors. Tech giants like OpenAI and Google collect and analyze internal operational data and user behavior data from their clients, providing rich training samples and validation environments for AI LLMs in the To B and To SMB markets. This enables them to accurately identify enterprise needs and market trends, thereby delivering more personalized customer services and targeted solutions.

4 Current Issues in Empowerment

Based on the intrinsic mechanism and preconditions for data factors empowering new quality productivity, this paper divides the current issues in the empowerment of new quality productivity by data factors into two stages according to the position of data in the value chain and its life cycle: first, the marketization stage of data factors, that is, the stage of data factorization and circulation/trading; and second, the empowerment stage of data factors, that is, the stage in which data factors are deployed into mass socialized production to release their usage value.

4.1 The Marketization Stage of Data Factor

An essential precondition for data factors empowering new quality productivity is the completion of the factorization stage. Only when data factors are circulated to where they are needed can subsequent scenario-based applications be initiated to release their usage value, thereby forming and developing new quality productivity. Therefore, if the marketization of data factors is not smooth, that is, if there are issues in the factorization process or in the circulation of data factors—this will directly restrict the quality of subsequent empowerment. Accordingly, this paper reviews the issues that exist in the pre-, mid-, and post-processes of the factorization and circulation of data factors as follows:

First, in the data factorization stage, China currently lacks unified national standards for data cleansing, de-identification, anonymization, and metadata. In current practices, the enterprise data supplied by various companies and the public data provided by local governments are mostly of varying quality, which requires considerable resources for both the enterprises and the operators managing public data to perform data cleansing, de-identification, anonymization, and format standardization and integration. The absence of unified standards for data cleansing, de-identification, anonymization, and metadata not only increases the factorization costs of both enterprise and public data but also makes it difficult to effectively integrate data from different sources, thereby affecting data factorization and severely hindering the circulation and value release of data factors.

Second, once data has been factorized and confirmed as data factor, various issues arise during the circulation stage. According to the classification by Chen et al. [3], the current issues in the circulation stage are divided into three parts: before circulation, during circulation, and after circulation.

(1) **Before circulation**. Difficulties in establishing data rights, assessing data value, and data pricing are the main practical problems. The complexity of data ownership makes the process of confirming data rights challenging, and data enterprises, for competitive protection reasons, usually adopt a monopolistic stance toward their own data, resulting in a serious shortage of high-value data supply.

Moreover, the value of data is highly dependent on business logic, application scenarios, and analysis accuracy, which makes its valuation and pricing process full of uncertainty. This situation leads to a "lemon market" phenomenon where, due to insufficient information, buyers struggle to judge the true quality and potential value of data, further exacerbating information asymmetry and hindering the effective circulation of data. Additionally, similar to public data, issues such as insufficient sharing and a severe shortage of scientific data are also prominent; at the same time, the existence of isolated scientific data silos significantly restricts revolutionary breakthroughs in technology during the subsequent empowerment stage, thereby impeding the formation and development of new quality productivity empowered by data factors.

(2) **During circulation**. The security of data transmission and the issue of unlimited replicability are core concerns. The unique nature of data factors means they involve sensitive information such as privacy, trade secrets, and national security. Even with anonymization and encryption, potential risks of data leakage cannot be eliminated. Furthermore, the ease of data replication means that once data is removed from its original controlled environment, it may be illegally copied and misused, harming the legitimate rights of the original data owners and increasing the difficulty of tracking and evidence collection. Therefore, ensuring data transmission security and safeguarding data rights become critical issues that need to be resolved during circulation.

(3) **After circulation**. Issues regarding data usage rights and regulation become particularly prominent. Technically, data acquirers may replicate, distribute, and resell data indefinitely without permission, or even recombine data to derive illicit benefits. The process of tracking and collecting evidence in such cases is extremely complex, posing significant challenges to post-circulation regulation. Meanwhile, due to information asymmetry between buyers and sellers, disputes over the legality of data usage and the reliability of its source are difficult to resolve quickly under the current legal framework, further increasing regulatory complexity [3].

4.2 The Empowerment Stage of Data Factor

After the data factorization and circulation processes are completed, data factors then flow to the demand side, which, in conjunction with usage scenarios, develops and utilizes these factors to unlock their value and empower new quality productivity. The issues in this stage are mainly as follows.

4.2.1 Severe Lag in the Development of Data Factor Usage Scenarios

In the process of data factors empowering new quality productivity, a significant issue is the severe lag in the development of data factor usage scenarios. The economic root lies in the sluggishness of market mechanisms and information asymmetry. Specifically:

(1) **Insufficient identification of actual market scenario demands**. Although data factor resources are abundant, effectively mining and meeting genuine market scenario needs remains a critical challenge. Companies often struggle to quickly and accurately identify and develop commercially valuable data application scenarios when faced with massive amounts of data, leading to a waste of data factor resources and a lag in scenario development that severely restricts the empowerment process of new quality productivity.
(2) **Insufficient technical adaptability and innovation**. Current technologies fail to deeply develop specific scenario requirements, resulting in the shallow and superficial development and utilization of particular data factor application scenarios, and thus preventing the unlocking of the deep value of data factors that would otherwise empower new quality productivity.
(3) **Insufficient recognition of the value of data factor**. The market's understanding of the true value of data factors is limited, and there is a lack of thorough recognition and exploitation of their potential, resulting in a significant delay in the development of application scenarios.
(4) **High innovation costs**. Innovative scenario development typically requires high R&D investment and trial-and-error costs, especially when exploring revolutionary breakthroughs in technology and new forms and models of business. In the absence of a clear profit model and stable market demand, companies are often reluctant to bear these high costs and risks.
(5) **Lack of collaborative development mechanisms**. The cross-domain, cross-industry, and cross-regional application of data factors require close collaboration among various stakeholders. However, the current absence of effective collaborative development mechanisms severely contributes to the "data silo" phenomenon, most notably in scientific and public data, making it difficult to achieve scale and synergy effects.

4.2.2 High Resistance in Economic System Transformation

(1) **The cost of reallocating resources is high**. For data factors to drive the innovative allocation of factors of production and empower the formation and development of new quality productivity, it necessarily requires a reorganization and optimization of existing production resources and management processes. This process may involve the large-scale elimination of outdated equipment, significant adjustments to personnel structures, and the overhaul of production processes, all of which entail substantial sunk costs and opportunity costs. Such high costs of reallocation severely constrain the ability of data factors to drive the innovative allocation of factors of production, thereby impeding the empowerment of new quality productivity.

(2) **The rigidity of industrial structures and resistance to transformation are significant**. The entrenched industrial structures formed during the industrialization era are characterized by high fixed investments and inherent inertia, making it difficult to swiftly adapt to the revolutionary breakthroughs and innovations driven by data factor. This implies that in the process of promoting deep industrial transformation and upgrading through data factors, there is likely to be heavy resistance from vested interests and potential low consumer acceptance of new products. The inertia of the economic system is reflected in its industrial structure. The traditional manufacturing industrial structure formed during the industrialization era is typically characterized by a high degree of stability and inherent inertia, with enormous fixed investments, making it difficult to quickly adapt to the revolutionary breakthroughs in technology and innovations in new business models driven by data factors. It means that, in the process of data factors driving profound industrial transformation and upgrading, it is highly likely to face heavy resistance from vested interest groups, as well as issues such as low acceptance of new products by consumers.

4.2.3 Excessive Resource and Energy Consumption and Environmental Pollution Issues

(1) **Excessive consumption of resources and energy**. Generative AI and LLMs are among the most popular frontier fields in which data factors empower new quality productivity, serving as key technologies for unleashing the value of data factors. However, these cutting-edge digital technologies and GPTs often require exorbitant computing power, which in turn drives a massive demand for electricity and huge amounts of freshwater for cooling data centers [26]. At the same time, such high levels of energy and natural resource consumption, along with the enormous costs, can significantly hinder small and medium-sized enterprises (SMEs) from participating in the empowerment of new quality productivity by data factors.

(2) **Environmental pollution**. In addition to the excessive consumption of resources and energy, these advanced digital technologies can have adverse environmental impacts. Currently, most energy consumption still relies on traditional fossil fuels, whose excessive use leads to the emission of large amounts of pollutants into the natural environment. Furthermore, in the process of developing and efficiently utilizing data factor-driven scenarios, the manufacturing of hardware, energy consumption, and waste disposal can also adversely affect the environment. Computing centers, data centers, and servers may all generate hazardous waste during their production, and the disposal of obsolete equipment is equally concerning, particularly when it comes to new energy batteries. All these issues will exacerbate environmental pressure and are not conducive to the sustainable and green empowerment of new quality productive forces by data elements.

5 Policy Recommendations

5.1 Policy Recommendations for the Marketization of Data Factor

To address the practical issues existing in the processes of data factorization and circulation, this study recommends the following improvements:

(1) **Establishing data standardization and technical certification system**. Fully leverage the guiding role of the newly established National Data Standardization Technical Committee to build a national unified system for data standardization and technical certification. National standards covering both public and enterprise data for data cleansing, de-identification, anonymization, and metadata specifications should be researched and developed as soon as possible, thereby reducing quality disparities among data factors and improving their circulation efficiency. At the same time, to address trust, privacy, and ethical issues in data factor circulation, a comprehensive digital technology standards and certification system should be established. This includes privacy computing standards (such as secure multi-party computation, homomorphic encryption, differential privacy, and federated learning) to ensure privacy protection during data circulation; blockchain technology standards to safeguard data consistency and smart contract security; and artificial intelligence technology standards that emphasize algorithm transparency and ethics.

(2) **Improving policy supply for data factor**. It is recommended to expedite policy development in areas such as the data property rights system, data registration, valuation and pricing, revenue distribution, and trustworthy circulation and security supervision. For example, based on the concept set forth in the "Opinions on Establishing Fundamental Data Systems to Better Leverage the Role of Data Factor" regarding the data property rights system, a property rights system that separates the rights of possession, use, and management should be researched and formulated, thereby protecting the legitimate rights of rights holders and providing a legal foundation for the compliant and efficient utilization of data factors. Furthermore, exploring a "by ownership and contribution" distribution mechanism—assisted by technologies such as federated learning, multi-party game theory, and mechanism design—will ensure that data rights holders obtain economic benefits commensurate with their ownership and contributions. It is also recommended to build a regulatory sandbox to assess the impact of new technologies, innovate intelligent regulatory tools to enhance efficiency, and establish a Data Ethics Review Committee to handle ethical issues related to sensitive data and AI algorithms. Additionally, a data emergency response mechanism should be developed to address data breach incidents, and the establishment of online courts and arbitration institutions should be explored, utilizing smart contracts to automate legal procedures and ensure the legality and execution efficiency of online data litigation.

(3) **Establishing the system for the open sharing of funded scientific data**. Gradually build a consensus and mechanism whereby national and local funding agencies require funded researchers to openly share scientific data, thereby alleviating the "scientific data silos" issues. Actively explore pilot policies in research cities—such as Beijing, Shanghai, and Hefei—that implement immediate aggregation and open sharing of scientific data obtained and generated from publicly funded, non-profit research activities, and progressively expand these initiatives from non-profit research activities to all research activities, as well as from pilot cities to a wider range of non-pilot cities. This gradual deepening of open sharing practices will provide a solid foundation of scientific data to promote the informatization of research and the transformation of scientific paradigms. Both the European Union, the United States and other regions have established relevant mechanisms. For example, the Common Principles of RCUK Data Policy issued by the Research Councils UK (RCUK), the Freedom of Information Act in the United States and relevant circulars (such as Circular A-89 and Circular A-130), and the "Horizon 2020" program of the European Union have all standardized the collection, submission and public disclosure of scientific data, requiring the open sharing of scientific data funded by relevant sources.

(4) **Exploring pilot and innovation exemption mechanisms**. Considering the ambiguous policies regarding data registration, public data pricing, and revenue distribution mentioned above, local governments are encouraged, while complying with national laws and regulations, to leverage their local endowments and realities to boldly launch pilot initiatives and explore feasible pathways. At the same time, efforts should be made to establish an innovation exemption policy that provides a margin for error for exploratory practices conducted within legal and regulatory frameworks. Even if the results do not meet expectations, no accountability should be pursued, thereby effectively stimulating local government enthusiasm and creativity, accumulating pioneering experience, and helping to build a nationwide data factor policy system that promotes the compliant and efficient utilization of data factors.

5.2 Policy Recommendations to Support the Empowerment Role of Data Factor

To address the practical issues in the empowerment process of data factor, the following recommendations are proposed:

(1) **Accelerating the development of data factor scenarios**. Building on the existing national and local "Data Factor ×" competitions and AI scenario application contests, it is recommended to regularly organize data scenario development competitions in specific industries, especially public data scenario application contests in targeted sectors, by inviting enterprises, research institutions, and universities to participate. This will stimulate innovative thinking

and explore innovative applications of data factors. By establishing awards, providing funding support, and offering market access opportunities, participants will be encouraged to actively invest in data scenario development. In addition, regularly publishing demonstration cases of enterprise and public data applications across various industries will showcase the practical effects of data factors in enhancing production efficiency, optimizing business processes, and innovating service models, thereby standardizing and guiding more entities to carry out demonstration scenario applications and further innovation, ultimately promoting the prosperity and development of the data factor market.

(2) **Improving subsidy policies for upgrading data-related equipment**. Building upon the "Action Plan for Promoting Large-scale Equipment Upgrades and Consumer Goods Trade-in", with the aim of accelerating the informatization and digital transformation of scientific research, we suggest implementing standardized subsidy policies in combination with a phased implementation strategy. Specifically, we should clearly define the scope and standards of equipment and upgrade subsidies. For key equipment, such as high-performance computing devices, big data analysis platforms, and AI processing units, we need to formulate detailed subsidy catalogs and standards. These should be set reasonably according to the technological content, market value, and the contribution of the equipment to corporate digital transformation and scientific research informatization. Meanwhile, we should simplify the application process. By establishing an online subsidy application platform that provides one-stop service, we can improve the efficiency of policy implementation. Furthermore, we should adopt a phased implementation strategy. First, select some industries or regions as pilots to evaluate the effectiveness of the policy and accumulate experience. Then, based on the results of the pilots, adjust and refine the policy, and gradually expand the subsidy coverage to cover more enterprises in different industries and scientific research institutions. Ultimately, by incorporating the equipment upgrade subsidy policy into the national medium- and long-term development plan, we can establish a long-term mechanism, which will continuously promote the informatization and digital transformation of scientific research.

(3) **Promoting the construction of green digital infrastructure**. The goal of promoting the construction of green digital infrastructure is to integrate cutting-edge technologies to achieve sustainable development in the information technology sector. This includes utilizing intelligent data cleansing and integration technologies to ensure data is clean and used efficiently, as well as employing blockchain technology to strengthen data privacy protection. On the algorithmic front, it involves exploring the potential of deep learning optimization, parallel computing, and quantum computing to improve algorithm energy efficiency. At the same time, green data infrastructure should be built on the foundation of energy-saving hardware, efficient cooling systems, and high-performance computing clusters powered by green energy sources such as wind, solar, and

controlled nuclear fusion—examples include green data centers and green intelligent computing centers—coupled with intelligent scheduling systems to optimize the allocation of computing resources. In addition, enhancing the application of edge computing, 5G, and next-generation communication technologies will reduce data transmission delays and energy consumption. Moreover, establishing a circular economy will ensure that hardware devices are effectively recycled and reused at the end of their lifecycle, thereby reducing energy consumption, carbon footprint, and pollutant emissions associated with the development and application of data factors. Through these comprehensive measures, an efficient, low-carbon, and environmentally friendly digital ecosystem can be formed, laying a solid foundation for the empowerment of green productivity by data factors.

References

1. Shi D, et al. (2023) Accelerating the formation of new quality productivity. People's Daily, 2023-11-24(9) (in Chinese).
2. Feng Y, Lin H (2024) Data enablement of new quality productivity: Theoretical logic and practical path. Economist (5):15–24 (in Chinese).
3. Chen G, Yan B, Tang K (2024) Data elementization and assetization: theoretical analysis and practical exploration. International Economic Review (5):153–176 (in Chinese).
4. Tang K (2023) Data assetization. Beijing: People's Publishing House, Beijing (in Chinese).
5. Berczi A (1981) Information as a factor of production. Business Economics, 16(1):14–20.
6. Cai J, Liu Y, Gao H, et al. (2022) The approach of data factor participating in value creation: a general equilibrium analysis based on the general theory of value. Journal of Management World 38(7):108–121 (in Chinese).
7. Xu X, Zhang Z, Hu Y (2022) Research on deriving measures of data assets in national accounts. Journal of Management World 38(2):16–30, 2 (in Chinese).
8. Cong LW, Xie D, Zhang L (2021) Knowledge accumulation, privacy, and growth in a data economy. Management Science 67(10):6480–6492.
9. Xu X, Li K, Tian X (2021) Research progress on data as a factor of production. Economic Perspectives (4):142–158 (in Chinese).
10. Veldkamp L (2023) Valuing data as an asset. Review of Finance 27(5):1545–1562.
11. Ouyang R, Liu Y (2024) Theoretical mechanism, constraints, and policy suggestions for the multiplication effect of data elements. Research on Financial and Economic Issues (3):3–18 (in Chinese).
12. Jin X, Tan X, Li H (2024) Enabling the development of real economy through the multiplier effect of data elements: function mechanism and path selection. Information Studies: Theory & Application 47(6):31–38 (in Chinese).
13. Wu J, Tao C (2024) Activating data elements empowers thousands of businesses: policy interpretation of "The 'Data Elements ×' Three-year Action Plan (2024–2026)". Information Studies: Theory & Application 47(3):16–19 (in Chinese).
14. Li Z, Liao X (2023) The theoretical, historical, and realistic "triple" logics of developing "new quality productivity". China Review of Political Economy 14(6):146–159 (in Chinese).
15. Cong LW, Wei W, Xie D, et al. (2022) Endogenous growth under multiple uses of data. Journal of Economic Dynamics and Control 141:104395.
16. Liu W, Jia W (2023) Biased technological progress, factor allocation, and economic growth. Modernization of Management 43(1):7–18 (in Chinese).

17. Wu S, Su Q (2011) Process of data integration in web3.0. Library and Information Service 55(24):112–115 (in Chinese).
18. Jones CI, Tonetti C (2020) Nonrivalry and the Economics of Data. American Economic Review 110(9):2819–2858.
19. Abis S, Veldkamp L (2024) The changing economics of knowledge production. The Review of Financial Studies 37(1):89–118.
20. Ackoff RL (1989) From data to wisdom. Journal of applied systems analysis 16(1):3–9.
21. Wang T, Li J (2022) Application of deep learning in protein structure prediction and its inspirations. Acta Polymerica Sinica 53(6):581–591 (in Chinese).
22. Cai S, Weng X (2024) A new theoretical framework for the economics of data. Research on Financial and Economic Issues (5):33–48 (in Chinese).
23. Akerlof GA (1970) The market for "lemons": quality uncertainty and the market mechanism. The Quarterly Journal of Economics 84(3):488–500.
24. Shi D, Sun G (2024) Data elements and new quality productive forces: a perspective from total factor productivity of enterprises. Economic Theory and Business Management 44(4):12–30 (in Chinese).
25. Huang X, Gao Y (2023) Technology convergence of digital and real economy industries and enterprise total factor productivity: research based on Chinese enterprise patent information. China Industrial Economics (11):118–136 (in Chinese).
26. Li P, Yang J, Islam MA, et al. (2023) Making AI less "thirsty": Uncovering and addressing the secret water footprint of AI models. arXiv:2304.03271.

Ke Tang Professor at the Institute of Economics, School of Social Sciences, Tsinghua University, and the Dean of Zhishan College. His main research areas include commodity markets (including the data trading market), financial technology, and the digital economy. He has published numerous papers in journals such as the *Journal of Finance, Review of Financial Studies, Management Science*, PNAS, *Social Sciences in China, and Economic Research Journal.* He has been selected as one of the Highly Cited Chinese Researchers by Elsevier from 2020 to 2024.

Gang Chen Postdoctoral/Assistant Researcher at the Institute of Economics, School of Social Sciences, Tsinghua University. Main research areas include public data authorization operations and pricing, quantitative spatial economic theory and models, digital economy, and green development.

The New Research Informatization Infrastructure Platform Drives the Paradigm Shift in Scientific Research

Fangyu Liao, Yang Wang, Rongqiang Cao, Bo Zhang, Zhenyu Li, Huajin Wang, Xin Chen, Yangang Wang, and Xin Wei

Abstract The research informatization infrastructure platform serves as an essential foundation for contemporary scientific research and a critical indicator of national scientific and technological innovation capacity. Nations globally emphasize the construction and investment in research informatization infrastructure platforms, resulting in numerous significant accomplishments. The swift advancement of information technology and industry has prompted a transformation in the research informatization infrastructure platform. The novel research informatization infrastructure platform aims to enhance the support for scientific research by prioritizing technological advancements in areas such as cross-domain global intelligent scheduling, massive data storage and efficient data collection, intelligent computility, domain-specific large artificial intelligence (AI) models, and AI-oriented high-quality scientific data, among others. The construction and application of the platform will realize transparent coherence between the physical world and the digital world, accelerate the realization of critical scientific and technological breakthroughs, and push forward the shift of intelligent scientific research paradigm.

Keywords New research informatization infrastructure platform · Research paradigm · Intelligent scientific research · Artificial intelligence

1 Introduction

On 31 January 2024, General Secretary Xi Jinping stressed at the eleventh collective study of the Political Bureau of the Communist Party of China Central Committee (CPC Central Committee), that "it is important to accelerate the development of new qualitative productivity and solidly promote its high-quality development." In

F. Liao (✉) · Y. Wang · R. Cao · B. Zhang · H. Wang · X. Chen · Y. Wang · X. Wei
Computer Network Information Center, Chinese Academy of Sciences, Beijing, China
e-mail: fyliao@cnic.cn

Z. Li
Institute of Computing Technology, Chinese Academy of Sciences, Beijing, China

China's e-Science Blue Book 2025, https://doi.org/10.1007/978-981-96-8252-2_4

general, the new qualitative productivity is an advanced form of productivity characterized by innovation-led development, breaking away from traditional economic growth models and productivity development paths. It features with high technology, high efficiency and high quality, aligning with the new development concept. In recent years, Artificial Intelligence (AI) technology has been rapidly advanced and widely applied in scientific research, sparking a wave of transformation in the intelligent scientific research (AI for Science, AI4S) paradigm [1–3]. Under this new research paradigm, the new research informatization infrastructure platform will enable the emergence of machine intelligence and the intelligent integration of human, machine and things, driving the transformation of research paradigms and promoting the development of new qualitative productivity.

This paper begins with the concept and value of the new research informatization infrastructure platform, analyzing how it can drive the transformation of the intelligent scientific research paradigm. In response to the needs of the intelligent scientific research paradigm, we present key technologies that need to be broken through. Finally, we look ahead to the future development direction of the new research informatization infrastructure platform.

2 The Connotation and Scientific Value of the New Research Informatization Infrastructure Platform

The research informatization infrastructure platform not only includes hardware facilities such as networks, supercomputers, and storage that support scientific research, but also encompasses system middleware, basic software, application software closely integrated with disciplinary development, and scientific data resources deployed on these hardware facilities [4–6].

Countries around the world attach great importance to the construction of research informatization infrastructure platforms, viewing it as key measures to maintain global technological leadership and enhance national competitiveness. The European Union has proposed and built the European Open Science Cloud (EOSC) [7], which integrates existing European informatization infrastructures, including Pan-European Data Infrastructure, European Grid Infrastructure, Partnership for Advanced Computing in Europe (PRACE), and Pan-European Research and Education Network, forming a unified research informatization infrastructure platform. This platform enables long-term lightweight management of scientific data assets and provides users with services for scientific data storage, management, analysis, and reuse. The United States has established high-speed connections between all Department of Energy (DOE) laboratories and DOE-supported universities through the Energy Sciences Network (ESnet), interconnecting with over 100 other networks to achieve high-speed data transmission and complete ultra-high-speed transmission experiments. "CSTCloud" as a national-level research informatization infrastructure platform, focuses on the open aggregation and cloud services of scientific and technological resources. It has established an open overall architecture and technical specification system, and initially built new national-level research informatization infrastructure platforms integrating networks, data and computing. It has aggregated 315 PFlops of computing resources, 150 PB of storage resources and PB-level data resources. It has deployed over 60 integrated service platforms and more than 1,000 scientific research software, becoming a uniquely tailored cloud for researchers.

In recent years, with the widespread applications of AI technology in scientific research, major scientific breakthroughs have increasingly relied on advanced informatization technologies and methods [8]. At the end of 2021, Google's DeepMind team used the AlphaFold2 algorithm [9] to successfully predict the structures of over 2 million proteins from about 1 million species in just 18 months. In May 2024, AlphaFold3 emerged, enabling humans to predict the structures and interactions of nearly all important biological molecules with unprecedented atomic precision. The 2024 Nobel Prize in Physics was awarded to scientists for their fundamental discoveries and inventions in machine learning using artificial neural networks, and the 2024 Nobel Prize in Chemistry was awarded to scientists for their achievements in protein structure prediction. The awarding of the Nobel Prizes in Physics and Chemistry signifies that the development of AI technology is transitioning from a critical breakthrough phase to a new stage with broader societal impacts.

As AI continues to be integrated and applied to scientific research, future major scientific and technological infrastructures will be closely integrated with AI, forming the new research informatization infrastructure platform. Compared to traditional platforms, the new research informatization infrastructure platform will integrate high-speed networks, massive storage, intelligent computing power and AI models, forming an infrastructure capability that supports the scientific data system. It will achieve ubiquitous, cross-domain, high-speed connectivity and global intelligent scheduling of research elements, supporting the entire lifecycle activities of scientific data transmission, storage, analysis and computing, as well as the elemental circulation, driving the transformation of the intelligent scientific research paradigm and promoting technological innovation in the AI era.

3 The New Research Informatization Infrastructure Platform Drives the Paradigm Shift in Scientific Research

3.1 The Historical Evolution of Research Paradigms Promoted by Research Informatization Infrastructure Platforms

The research paradigm refers to the way the scientific community conducts scientific research during a specific historical period, aligning with the inherent laws of technological innovation [10]. In the history of human scientific research, it is widely acknowledged that there have been four distinct research paradigms. Early scientific research was primarily characterized by the recording and description of natural phenomena, such as the observation of the natural world by Galileo, known as "Experimental Science," representing the first research paradigm. Later, due to limitations in conditions, experimental science struggled to achieve a more precise and complex understanding of natural phenomena. As a result, scientists began to eliminate secondary distractions, retaining only key factors and simplifying experimental models, eventually summarizing through calculations. This marked the transition to the second research paradigm, the "Theoretical Science" stage, which persisted until the late nineteenth century.

As the difficulty and economic investment required to verify classical and modern physical theories increased (often requiring large-scale facilities), scientific research entered a phase of challenging problems. In 1954, the von Neumann architecture computer emerged, enabling researchers to use computational power based on large-scale parallel computer architectures. By designing algorithms and programs, they could simulate and calculate complex phenomena, providing clear explanations and applications for intricate problems. As computer simulations increasingly replaced experiments in analyzing and solving scientific problems across various fields, computers began to play a significant role in scientific research, ushering in the third research paradigm, known as the "Computational Science" (see Fig. 1).

With the continuous increase in data volume and variety, how to handle and utilize complex big data has become a scientific research challenge. In response, Turing Award winner Jim Gray delivered a speech titled "Revolution in Scientific Methods" at the 2007 National Research Council-Computer Science and Telecommunications Board (NRC-CSTB) conference. He proposed that, in response to the explosive data growth, computers should not be limited to simulations but should also be capable of analyzing and summarizing data to derive theories [2]. Consequently, scientists began to integrate various informatization technologies, such as high-speed

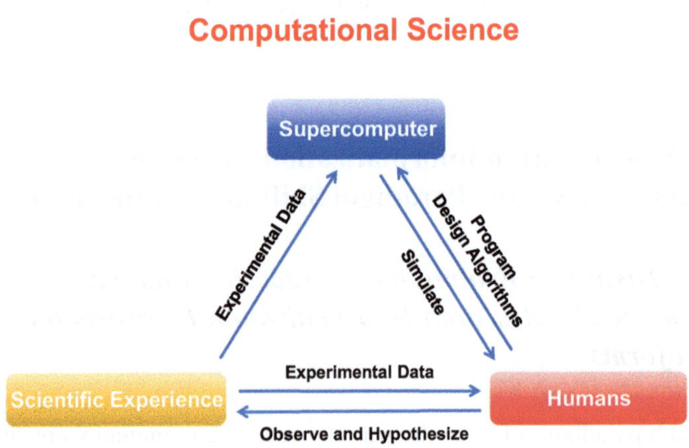

Fig. 1 Computational science

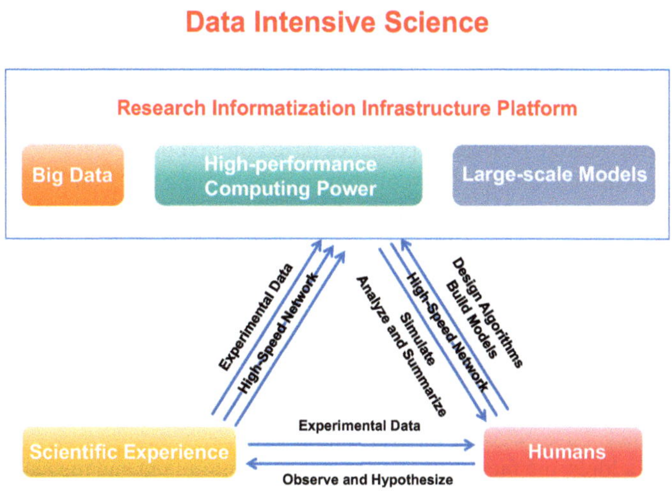

Fig. 2 Data intensive science

networks, advanced algorithms and big data, to collaborate to solve research challenges. Research informatization infrastructure platforms, centered on data and integrating high-speed networks, powerful computational resources and model libraries, gradually began to play a crucial role in scientific research. Tasks that once required tedious experimental processes by scientists like Newton and Einstein could now be accomplished using advanced computational tools and analytical models provided by research informatization infrastructure platforms, significantly reducing human resource and improving research efficiency. Since then, the fourth research paradigm "Data-Intensive Scientific Discovery" emerged as a distinct research paradigm from the third research paradigm (see Fig. 2).

3.2 The New Research Informatization Infrastructure Platform Leads the Intelligent Scientific Research Paradigm Shift

With the construction and deployment of large-scale scientific facilities and the continuous accumulation of experimental data, the volume of data generated in scientific research has grown explosively. In the process of handling and applying complex big data, the fourth research paradigm encountered numerous unresolved issues, including data uncertainty and complexity, the explosive growth of data dimensionality and the boundaries of data scales. etc. [11, 12] Scientists began to seek more effective solutions to address the uncertainty and complexity of big data. As technological innovation progressed, AI became fully integrated into scientific, technological and engineering research, sparking a wave of transformation in research paradigms. Cheng Xueqi and Li Guojie tentatively called it the "intelligent scientific research paradigm" [2, 12]. In a 2024 article published in the Bulletin of the Chinese Academy of Sciences, Academician Li Guojie pointed that the intelligent scientific research paradigm was a fusion of the previous four research paradigms, particularly the integration of model-driven approaches based on first principles and data-driven approaches, with interdisciplinary collaboration becoming the mainstream research method. Facing the highly complex combinatorial explosion problems, the intelligent scientific research paradigm would focus on complex systems as the primary study object and rely more on large platforms characterized by big models, closely integrating scientific research with engineering practice.

As scientific research continues to deepen, the paths and modes of scientific research have undergone fundamental changes. Major scientific discoveries and technological transformations increasingly depend on research informatization infrastructures, which have become the key support for breaking through the scientific frontiers and addressing the economic and social development and the national major scientific and technological problems [11]. Moreover, with the rapid development of information technologies such as the Internet of Things (IoT), big data and AI, "data and intelligence" are profoundly influencing and driving the transformation of research paradigms. Intelligent systems now play a role in scientific research for exploration and realization, with scientists no longer solely relying on computers and algorithms to enhance data processing efficiency. Instead, they are integrating human brains with computers to form a "human-in-the-loop" research model (see Fig. 3). As the infrastructure leading the transformation of research paradigms, the research informatization infrastructure platform will also undergo model changes, evolving into the new research informatization infrastructure platform to effectively address challenging combinatorial explosion problems and accelerate the transformation of research paradigms.

4 The Transformation of Intelligent Scientific Research Paradigm Requires the Support of the New Research Informatization Infrastructure Platform

In February 2023, when presiding over the third collective study of the Political Bureau of the CPC Central Committee, General Secretary Xi Jinping emphasized that "we should collaborate to build a system of national laboratories with Chinese characteristics, lay out and construct research centers for basic science, and deploy ahead the new research informatization infrastructure platform, so as to form a powerful backbone network for basic research." With the continuous deepening of scientific research and the explosive growth of research data, the capabilities of existing research informatization platforms are struggling to meet the demands brought by the development of new research paradigms. To address the needs of the new research paradigms, China urgently needs to upgrade the technical framework

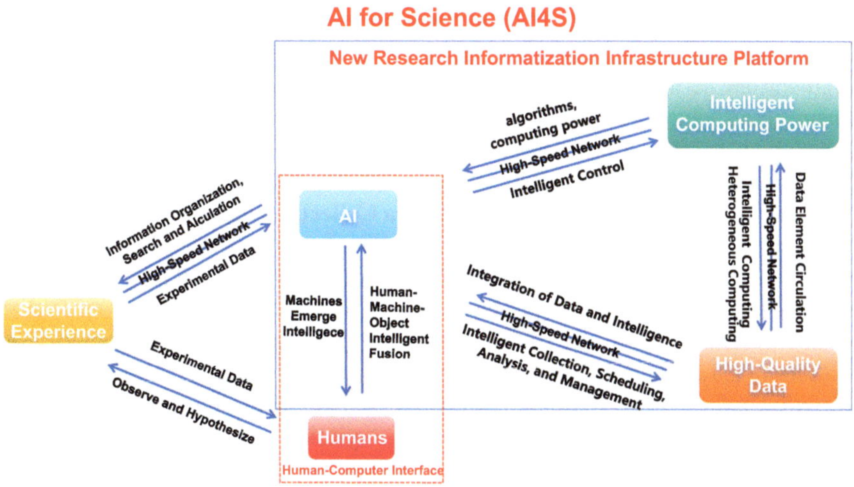

Fig. 3 AI for science (AI4S)

of existing platforms and construct a new research informatization infrastructure platform that aligns with the requirements of the times. This platform will enable the intelligent scheduling of global resources and the efficient supply of data, network, and computing resources. The key technologies for the new research informatization infrastructure platform, tailored to the new research paradigm, include the following aspects.

Cross-Domain Global Intelligent Scheduling. In the third and fourth research paradigms, scientific activities were conducted at the locations where scientific data was generated and stored, resulting in minimal demand for cross-domain data transmission. Therefore, scientific data was transmitted offline, with stable data generation patterns and transmission requirements, and transmission time measured in days. Intelligent scientific research, centered around AI models, requires massive amounts of data to train general or domain-specific models, making cross-domain data transmission as one of its key characteristics. To meet the demands of large-scale scientific data storage, cross-domain transmission and efficient data retrieval, the new research informatization infrastructure platform needs to model the dependencies among various components in the computing-network integrated infrastructure, including data storage, underlying computing, information communication, model training and knowledge invocation. With this basis, we then develop key technologies that enable and support global optimal data transmission paths, cost-effective resource scheduling and computing-network integration. These technologies include multi-cloud resource aggregation and sharing scheduling, data storage resource scheduling and sharing. By fully decoupling and offloading computing tasks and intelligently mapping them to transmission paths and software/hardware platforms, scientific data can be efficiently processed while being rapidly transmitted across networks, bridging the performance gap between network transmission and data computation. Through intelligent software/hardware scheduling and coordination, the platform can overcome the bottleneck of low throughput in traditional high-entropy networks (where multiple service flows share resources in a best-effort manner) and achieve low-entropy transmission for computing-network collaborative scheduling, enhancing the determinism of network transmission and exponentially improving Energy Efficiency Ratio (EER).

Massive Data Storage and Efficient Data Collection. In the era of large models, the number of parameters has grown from billions to hundreds of billions or even trillions. The datasets have expanded from text corpora to including images and videos as training samples, with data volumes increasing from terabytes (TBs) to petabytes (PBs). New large models, configured with hundreds of billions or even trillions of parameters, allowing a single training node to process 20,000 images per second, require 80,000 IOPS per node. For example, the training data volume for GPT-5 alone reaches 4PB. Traditional storage systems cannot meet such demands. Under the intelligent scientific research paradigm, the storage systems of intelligent computing centers need to achieve capacities of tens to hundreds of petabytes, with IOPS reaching the tens of millions level, latency reaching sub-millisecond level, and total read and write bandwidth reaching tens of gigabytes per second or even hundreds of gigabytes per second. To achieve high-throughput read/write performance, the computing and storage areas of intelligent computing centers are interconnected via ultra-high-bandwidth high-speed switching devices, and technologies such as RDMA and NVMe-oF are used to directly transfer data to the storage area, reducing data copying and exchange operations, enabling high-performance network data access and exchange for storage devices. Additionally, the storage system provides simultaneous data access services through multiple distributed all-flash storage nodes equipped with NVMe flash media to meet the concurrent access requirements of large-scale computations.

Intelligent Computing Power. During the third and fourth research paradigms, computing power was primarily characterized by high parallelism and high throughput in CPU-based high-performance computing and cloud computing. With the advent and rapid development of the intelligent scientific research paradigm, computing power represented by GPUs and other accelerator cards will play an increasingly important role in AI computing technologies. In computing infrastructures that integrate CPUs, GPUs and other components, the proportion of GPU computing power is expected to significantly increase. In other words, intelligent computing under the intelligent scientific research paradigm will be characterized by GPU computing and the organic integration of computing software and hardware. This necessitates that the new research informatization infrastructure platform meets the intelligent development needs throughout the entire scientific research process, including scientific data acquisition, large-scale parameter learning and model reasoning. In terms of original algorithms, methods and theoretical research, breakthroughs in intelligent computing systems are required, including heterogeneous computing scheduling technologies at different granularities (within chips, across multiple cards and multiple nodes), to improve the efficiency of data preprocessing, training and inference for general large models and domain-specific models in scientific research. The platform should also significantly expand the scale of basic arithmetic libraries and the arithmetic capacity for large model training, enhance the hardware system's adaptability to AI computing, so as to support the key processes such as efficient AI model development, debugging, training and inference. Additionally, the platform should develop quantum computing basic algorithm software to support the integration of quantum computing and traditional computing.

Domain-Specific Large Models. Such models usually refer to AI models with a large number of parameters designed to address domain-specific scientific research problems. Notable examples include the AlphaFold 2 [9] model for protein structure prediction and the NowCastNet [13] model for short-term precipitation forecasting. In terms of model training, compared to general large models, domain-specific large models in scientific research exhibit significant localization and end-to-end characteristics [14]: they start from high-quality training data tailored to specific problems (rather than general corpora) and undergo end-to-end training based on the transformer architecture (rather than multi-stage training), allowing the large models to directly fit the correlations inherent in the training data for specific research objects; In terms of model inference and application, compared to general large models, domain-specific large models in scientific research heavily rely on highly specialized knowledge bases to enable intelligent analysis of scientific data based on retrieval-augmented generation (RAG), automate the orchestration of complex workflows for scientific data analysis, and execute entire workflows seamlessly online through the automated matching and scheduling of network, computing, and data resources. This reduces the necessity for domain scientists to possess advanced programming skills and minimizes manual intervention. Therefore, in addition to providing general training corpora, the new research informatization infrastructure platform should particularly strengthen its ability to supply high-quality training datasets and professional knowledge bases tailored to specific scientific research problems, and enhance research on end-to-end large model training frameworks. Key research areas include the acquisition and organization of large-scale scientific data, the unified modeling of massive heterogeneous scientific data and knowledge, the design of domain-specific large model architectures for specific research problems, end-to-end domain-specific large model training and fine-tuning frameworks, and knowledge distillation and reasoning. Furthermore, establishing domain-specific large model libraries and question-answering scientific data analysis service platforms are essential steps in building the new research informatization infrastructure platform.

High-Quality Scientific Data for AI. High-quality scientific data reflects natural laws, and AI-Ready datasets are essential for enabling AI systems to understand, process and discover new scientific principles. Unlike internet text, speech and image, primarily used by current general AI models, scientific data offers greater diversity in modality, higher value density, and a more comprehensive real-world representation. Despite its potential, existing scientific datasets face several challenges, including resource dispersion, low knowledge representation, inconsistent standards and insufficient sharing mechanisms. To address these challenges, it's essential to integrate key technologies such as scientific data registration, rights confirmation, authentication, evaluation and trusted circulation. These technologies will facilitate the connection of high-quality scientific data resources and the construction of a scientific data elementization foundational service platform based on identifiers and blockchain. This platform will enable the trusted circulation, associative organization, and knowledge-based integration of scientific data, thereby forming a high-quality AI-Ready scientific data supply capability that is knowledge-embedded, model-integrated, and intelligently scheduled and circulated. Additionally, the platform will establish high-value, high-reliability, and high-impact scientific databases, providing high-quality data supply for the intelligent scientific research paradigm.

The new research informatization infrastructure platform, which integrates high-speed networks, massive storage, intelligent computing power, and AI models, will serve as the foundational infrastructure to support the scientific data system. It will enable the ubiquitous, cross-domain, high-speed connection and global intelligent scheduling of scientific research elements, thereby leading, driving and supporting the transformation of the intelligent scientific research paradigm.

5 Suggestions for Advanced Deployment of New Research Informatization Infrastructure Platform

To accelerate the implementation of the directive from General Secretary Xi Jinping on "advanced deployment of a new research informatization infrastructure platform" and to meet the strategic goals of China becoming a leading innovative nation by 2035 and a world scientific and technological power by 2050, China should moderately advance the construction of a new research informatization infrastructure platform. This will achieve a high level of self-reliance and strength in the new research informatization infrastructure platform, more effectively support technological innovation and national economic and social development, and contribute Chinese technological strength to the community of shared future for mankind.

(1) **Fully Utilize National Existing Informatization Resources, Strengthen Co-construction and Sharing of Research Informatization Software and Hardware Facilities**

It is recommended that the nation fully utilize the capabilities of existing informatization infrastructure such as supercomputing centers, intelligent computing centers, the East Data West Computing project, and scientific data centers. High-speed networks should be used to achieve interconnection and interoperability of informatization resources, providing non-profit information infrastructure resources service for scientific research activities. At the same time, by integrating and optimizing existing informatization resources, and promoting the research and application of key technologies such as AI, quantum computing and PB-level data storage and transmission, the construction of research informatization software and hardware facilities should be strengthened. Through strengthening the support and guidance of relevant policies and regulations, the co-construction and sharing of research informatization software and hardware facilities can be achieved.

(2) **Strengthen Independent Innovation in Software and Algorithms, Improve the Application Efficiency of Hardware Resource**

China should fully leverage its hardware resource advantages, focusing on AI and future computer systems, to strengthen the basic research of scientific computing software. Relying on national supercomputing centers and the new generation of AI open innovation platforms, and in collaboration with relevant national strategic technological force, the scientific computing software adapting to the domestic heterogeneous systems should be conducted to achieve rapid development and independent innovation of software on domestic heterogeneous systems; Close coordination with the domestic E-level and 10E-level computer development plans should be maintained, focusing on national major tasks of tackling key scientific and technological problems, major scientific and technological infrastructures, and key areas related to national development and security. Innovative AI software and algorithms based on domestic hardware architectures should be proposed to support China's high-level self-reliance and self-strengthening in science and technology.

(3) **Build an Integrated Research Informatization Platform for the New Research Paradigm, Lowering the Threshold for Researchers**

To meet the development needs of the new research paradigm, China should break through the existing information technology system architecture that separates transmission, storage and computation. By integrating key technologies such as intelligent computing power, PB-level data storage and high-throughput read/write capabilities, cross-domain hardware-software integrated scheduling, large-scale models for vertical domains and high-quality data resources tailored for AI, it is essential to proactively deploy a new research informatization infrastructure platform that combines hardware and software. This platform will enable efficient adaptation and integrated intelligent scheduling of resources such as hardware, software and data. For scientific research in various disciplines, a specialized AI4S platform should be constructed through distributed deployment and logically unified access of hardware and software systems. This will support researchers in using the platform in a one-stop, intelligent manner, thereby lowering the barrier to the application of AI technologies.

(4) **Prospectively Explore New Technologies and Methods, Build a Forward-leading AI4S Testbed**

It is recommended that the nation seize the historical opportunity presented by the rapid development of new technologies such as AI and quantum computing. By accelerating the development of cutting-edge experimental platforms for scientific research, we can create a real-world testing environment for the verification of advanced information technology innovations and the demonstration of the new research informatization infrastructure platform. This includes providing a AI4S testbed that spans multi-modals such as chips, equipment, algorithms and software, while integrating full-chain capabilities such as AI, high-speed networks, massive storage and intelligent computing. Such efforts will provide robust support and assurance for the construction of the new research informatization infrastructure platform in China.

(5) **Innovate Diversified Operational Models for the Research Informatization Infrastructure Platform, Promote Healthy and Sustainable Development**

AI has brought vast industrial development opportunities to various sectors. It is recommended that, during this critical period of national transformation and development, China should prioritize public welfare fields such as scientific research. By actively innovating operational models and encouraging multi-party investments from government agencies, enterprises, institutions and research organizations, a diversified operational approach can be adopted to jointly construct and manage the research informatization infrastructure platform. This collaborative effort will allow all stakeholders to benefit from the advantages brought by the new research paradigm to scientific activities, achieving complementary advantages between government and social capital and resources, and promoting the healthy development of the platform.

References

1. Zhang J R, Sun M G, Han T (2023). Research on scientific research data under the trend of intelligent scientific research[J]. Science Focus, 18(04): 49–61.
2. Li G J (2024). AI4R: The fifth scientific research paradigm[J]. Bulletin of Chinese Academy of Sciences, 39(1): 1–9.
3. Chinese Academy of Sciences (2018). The road to building a powerful country in science and technology: China and the world[M]. Beijing: Science Press: 424–455.
4. Wang Y, Zhou Y C, Wang Y G, et al (2022). Promoting infrastructure construction in advance to support sci-tech self-reliance and self-strengthening at higher level[J]. Bulletin of Chinese Academy of Sciences, 37(5): 652–660.
5. Ye Y J (2015). Strengthen construction of science & technology infrastructure to promote management of science & technology resource[J]. China Science & Technology Resources Review, 47(2): 1–6.
6. Liao F Y, Hong X H, Wang Y, et al (2019). The data and computing platform is an important infrastructure which drives modern scientific research development[J]. Frontiers of Data and Computing, 1(5): 2–10.
7. Wen L M, Li Y, Guo L (2021). The practice progress and future exploration of open science at home and abroad[J]. Library and Information Service, 65(24): 109–122.
8. Li S S (2019). Data and computing are powerful drivers of science and technology innovation[J]. Frontiers of Data and Computing, 1(5): 1.
9. Jumper J, Evans R, Pritzel A, et al (2021). Highly accurate protein structure prediction with AlphaFold[J]. Nature, 596: 583–589.
10. Chen T (2020). Promote the upgrading of scientific research paradigms and strengthen the national strategic scientific and technological forces[J]. China Awards for Science and Technology, (8): 67–68.
11. Li Y L, Wei Q (2023). The transformation of scientific research paradigm driven by the digital platform of "future laboratory"[J].Think Tank of Science & Technology, (4): 49–57.

12. Cheng X Q, Mei H, Zhao W, et al (2020). Data science and computing intelligence: concept, paradigm, and opportunities[J]. Bulletin of Chinese Academy of Sciences, 35(12): 1470–1481.
13. Zhang Y C, Long M S, Chen K Y, et al (2023). Skilful nowcasting of extreme precipitation with NowcastNet[J]. Nature, 619 (7970): 526–532.
14. Abramson J, Adler J, Dunger J, et al (2024). Accurate structure prediction of biomolecular interactions with AlphaFold 3[J]. Nature, 630(8016): 493–500.

Fangyu Liao the Chief Scientist and Research Professor at the Computer Network Information Center of the Chinese Academy of Sciences. His primary researches focus on the information technology strategy, the infrastructure architecture for scientific research informatization, and the data security governance. Additionally, he serves as the Deputy Director of the Cyberspace and Information Consultation Expert Committee of the Chinese Academy of Sciences, the Deputy Director of the Computer Security Professional Committee of the China Computer Federation, and the Deputy Editor-in-Chief of the journal China Scientific Data. He has successively led multiple national and Chinese Academy of Sciences informatization special projects and published relevant monographs and research papers.

Yang Wang the Director of the Development Strategy and Evaluation Center for Informatization at the Computer Network Information Center of the Chinese Academy of Sciences. His primary researches focus on the information technology strategy and the infrastructure architecture for scientific research informatization. He has led and participated in more than 20 research projects, including the Chinese Academy of Sciences informatization special projects and advisory projects, the initiatives for the National Information Expert Consultation Committee, and the projects funded by the China Internet Development Foundation. Also he has published more than 20 academic papers in SCI/EI-indexed and Chinese core-listed journals.

Rongqiang Cao an Associate Researcher at the Computer Network Information Center of the Chinese Academy of Sciences. In July 2010, he graduated from the Graduate School of the Chinese Academy of Sciences with a PhD in Computer Software and Theory, and has been working in the Computer Network Information Center of the Chinese Academy of Sciences ever since. His major is high performance computing and intelligent computing and currently he is mainly engaged in the research and construction of artificial intelligence platform systems, including heterogeneous resource scheduling, computing and data fusion, and distributed computing. He has published more than 20 papers and completed more than 10 projects.

Bo Zhang the Project Researcher and Deputy Director of the Department of Science and Technology Cloud Operation and Technology Development, Computer Network Information Center, Chinese Academy of Sciences. He has long been engaged in the planning, design, construction and operation management of cloud computing, massive data storage and supercomputing infrastructure environment of the Chinese Academy of Sciences.

Zhenyu Li a Full Professor at the Institute of Computing Technology of the Chinese Academy of Sciences. His main researches focus on the network architecture and the networked systems. He is also an Adjunct Professor at the University of the Chinese Academy of Sciences, and has severed as an organizational or program committee member in international conferences such as SIGCOMM, CoNEXT, and INFOCOM.

Huajin Wang an Assistant Professor at the Computer Network Information Center of the Chinese Academy of Sciences. His research areas are big data technology and database technology. He has published many scientific data-related papers and participated in the development of open source big data software such as BigFlow, PandaDB and PackOne. He is also an executive member of the CCF Opensource Development Committee and a member of the W3C RDF-DEV Communication Group.

Xin Chen the Deputy Director of the General Data Center of the Chinese Academy of Sciences and is responsible for coordinating and providing data infrastructure service for the scientific data centers of the Chinese Academy of Sciences in this role. She received her PhD degree in computer science in 2011. Since then she has been working in Computer Network Information Center of the Chinese Academy of Sciences,mainly engaged in scientific data management, analysis and processing related research. Additionally,She's also a Co-Chair of RDA Fair Digital Object Fabric IG.

Yangang Wang a Research Professor at the Computer Network Information Center of the Chinese Academy of Sciences. His primary researches focus on the construction of artificial intelligence computing and data service platforms, as well as the development of artificial intelligence application software and parallel application software for scientific discovery. He has published over 80 academic papers in international conferences and journals such as PPoPP and SC, holded more than 30 authorized patents, and authored three books. He has led several research projects, including those under the National Key R amp D Program of China.

Xin Wei an Engineer at the Computer Network Information Center of the Chinese Academy of Sciences. Her researches focus on the information technology strategy,the infrastructure architecture for scientific research informatization,and open science. She has participated in more than 10 research projects, including the Chinese Academy of Sciences informatization special projects, the major project on tobacco big data, and the the Chinese Academy of Sciences advisory projects.

AI for Scientific Research: Development Trends and Governance Suggestions

Jiaofeng Pan, Shengyin Wang, and Jing Wu

Abstract The application of artificial intelligence (AI) in global scientific research is deepening and has become an important driving force for research innovation. This article outlines the strategic policy layouts of various governments in the field of "AI for Science" (AI4S) from the perspectives of frontier scientific research, research infrastructure construction, and talent cultivation. It analyzes the socialization trends of innovation brought about by AI4S and summarizes the entire process of AI's role in identifying research problems, conducting experimental analysis, fostering collaboration, disseminating research findings, and managing research evaluation. Finally, the article addresses the challenges and trends in the development of AI4S, proposing relevant questions and countermeasures from the angles of human-centered approaches, ethical considerations, and regulatory improvements, with the aim of promoting sustainable innovation and secure development in AI4S.

Keywords Artificial intelligence · AI for science · Innovative socialization · Strategies and policies · Governance

1 Introduction

The current wave of Artificial Intelligence (AI) is sweeping across the globe, profoundly reshaping various sectors of the economy and society and ushering in a disruptive transformation. In the field of scientific research, AI is making inroads into multiple areas of research practice, prompting a paradigm shift in technological innovation while accelerating interdisciplinary and cross-domain integration and breakthroughs. With the advent of models and algorithms such as AlphaFold (which accurately predicts protein structures), GNoME (which facilitates the discovery of new materials), and AlphaDev (which improves the efficiency of sorting algorithms), these successful applications demonstrate that AI has enormous potential to advance

J. Pan · S. Wang · J. Wu (✉)
Institutes of Science and Development, Chinese Academy of Sciences, Beijing, China
e-mail: wujing@casisd.cn

China's e-Science Blue Book 2025, https://doi.org/10.1007/978-981-96-8252-2_5

the speed, scope, and precision of scientific exploration, yielding insights that traditional methods find difficult to uncover, and assisting scientists in probing frontier research domains [1–3]. Consequently, "AI for Science" (AI4S) has emerged. AI4S aims to leverage machine learning, deep learning, and other AI technologies to accelerate scientific discovery, solve complex scientific problems, and thereby drive technological innovation.

2 Global Strategic Layout and Initiatives for AI-Driven Scientific Research

In recent years, the world's leading science and technology (S&T) powers have been intensively deploying strategies for "AI-driven scientific research" and accelerating the formulation and implementation of related policies. Their goal is to maintain and expand their competitive edge in the rapidly evolving AI era of international S&T innovation. From a national strategic standpoint, countries are strengthening AI4S efforts by addressing frontier research agendas, infrastructure development, talent cultivation and reserve, as well as establishing governance and regulatory frameworks. The overarching objective is to eliminate barriers to AI adoption in scientific research and fully unlock its potential for speeding up scientific discovery, driving innovation, and enhancing the quality of research outputs, thereby boosting national competitiveness [4–8].

2.1 Frontier Scientific Research Layout

"AI for Science" has emerged as the new frontier of AI development. With intensifying global technological competition, countries now recognize that a comprehensive approach to AI technology research and development (R&D) and its vertical applications is key to realizing the potential of AI4S. Consequently, many nations are increasing investments in AI4S, prioritizing advanced AI research and actively supporting domain-specific research and applications. These efforts aim to accelerate innovation and enhance the progress of scientific research at the national level.

(1) **Key Technologies**

In its technical reports, the United States Department of Energy (DOE) summarizes several directions in AI technology that effectively advance scientific research. Specifically, these directions include: next-generation surrogate models that transcend traditional simulation paradigms; AI foundation models for scientific knowledge discovery, integration, and synthesis; advanced property inference and inverse design; AI-based design, prediction, and control of complex engineered systems; AI and robotics for autonomous discovery; AI for programming and software engineering.

(2) **Vertical-Domain Applications**

Countries worldwide are intensifying deployments in both fundamental scientific fields—such as mathematics, physics, chemistry, astronomy, and biological sciences—and applied sectors, including meteorology, materials science, agriculture, energy, drug discovery, and the social sciences. The aim is to build up an AI4S frontier research and development (R&D) system that deepens scientific inquiry and enhances innovation capacity in all sectors. In China, in March 2023, the Ministry of Science and Technology and the National Natural Science Foundation jointly launched a special initiative on "AI-Driven Scientific Research". This program targets critical questions in core disciplines—such as mathematics, physics, chemistry, and astronomy—while focusing on scientific needs in drug discovery, gene research, biological breeding, and new material development. Through this initiative, China is building a robust AI4S frontier R&D framework that advances scientific research across multiple domains.

2.2 *Research Infrastructure Layout*

AI4S is prompting governments and research institutions worldwide to ramp up investments in AI4S infrastructure. The primary focus is on supporting the construction, expansion, and upgrading of research facilities, thereby effectively reducing research costs and barriers while enhancing efficiency. Overall, AI-related research can be broadly categorized into the following three key areas:

(1) **Computing Power Infrastructure**

To overcome limitations in data processing capacity and access to computational resources, countries across the globe are building supercomputers—examples include Frontier and Summit in the United States, Lumi in Finland, Jupiter in Germany, and Japan's upcoming "zetta-scale" (on the order of 10^{21}) Fugaku. These machines serve a critical role in AI model training and scientific simulation. Meanwhile, China has set up multiple National Supercomputing Centers throughout the country in locations such as Wuxi, Tianjin, Jinan, Shenzhen, Changsha, and Guangzhou. These centers form a robust foundation for large-scale data analysis, simulation, and scientific computing. Notably, in 2023, China's Ministry of Science and Technology launched a special program on "AI-driven Scientific Research", calling for the development of dedicated AI4S platforms in key research areas and accelerating the construction of next-generation national open innovation platforms for AI public computing power. As of December 2024, China has established 25 of these platforms nationwide, providing computing resources, software, and other support needed to advance AI4S research.

(2) **Data Infrastructure**

High-quality data is a prerequisite for applying AI in scientific research. Accordingly, countries are actively setting up data centers and promoting the openness of scientific data to increase the supply of quality datasets. For instance, the European Union continues to develop the European Open Science Cloud (EOSC), which integrates dispersed scientific data across Europe. In China, 20 National Scientific Data Centers and 31 National Biodiversity Germplasm and Experimental Material Resource Repositories have been established at the national level. Moreover, universities such as Peking University, Tsinghua University, and Fudan University are building their own campus-level scientific data centers. These facilities provide essential support for storing, sharing, and analyzing scientific data.

(3) **Integrated Research Service Platforms**

In recent years, countries have accelerated the development of integrated research service platforms. These platforms consolidate AI4S resources and offer unified cloud services, facilitating shared access to data, algorithms, and computing power across institutions. Representative initiatives include the National Artificial Intelligence Research Resource (NAIRR) in the United States and the Artificial Intelligence for Europe (AI4EU) project in the European Union. Both seek to establish AI ecosystems that combine high-quality datasets, computing resources, models, algorithms, and software to advance AI4S innovation. In China, the Chinese Academy of Sciences continues to develop "China Science and Technology Cloud" and "Global Open Science Cloud" [9]. Through a high-speed, dedicated research network, these projects connect the nation's key innovation resources—major national research facilities, national laboratories, national supercomputing centers, national scientific data centers, field stations, research institutes, and universities—thereby enabling resource aggregation, circulation, and collaboration, and providing a unified cloud service for researchers nationwide.

2.3 *Talent Development and Reserve Layout*

To address the urgent need for high-level AI talent—particularly those with interdisciplinary backgrounds—nations have introduced a series of policy measures for AI-for-Science talent cultivation. These initiatives span multiple domains, including basic education, higher education, and overseas talent introduction.

(1) **Basic Education**

In order to gain an early advantage and nurture research and innovation talent with solid STEM (Science, Technology, Engineering, and Mathematics) fundamentals, many countries have integrated AI education into the basic education (K-12) stage. Specific measures include: increasing investment in K-12 education, reinforcing STEM curricula, introducing compulsory courses in statistics and computer science principles at the junior and senior high school levels, encouraging students from diverse academic backgrounds to pursue advanced AI studies, hiring AI educators and enhancing in-service teacher training.

In December 2024, China's Ministry of Education issued new directives aimed at bolstering AI education in primary and secondary schools. These directives set forth objectives such as building a systematic AI curriculum framework, developing universal AI teaching resources, creating ubiquitous AI teaching environments, and substantially expanding the supply of AI teachers. These initiatives aim to foster students' scientific curiosity and digital competencies, thereby nurturing the talent pipeline needed for future innovation.

(2) **Higher Education**

In the higher education arena, countries worldwide have continued to reinforce systematic policy support, focusing on training high-level talent. Such policies include redesigning university courses, building AI-specific academic disciplines, strengthening AI skill instruction, promoting cross-disciplinary AI research, establishing STEM scholarships, broadening the diversity of AI4S talent pools, offering summer internships in national laboratories, fostering government-industry-academia-research cooperation for talent development, and constructing a multi-tiered AI talent development system [10]. As early as 2018, China's Ministry of Education launched the "Action Plan for Artificial Intelligence Innovation in Higher Education Institutions," proposing to set up AI as a first-level discipline, advance "new engineering" programs to form an "AI+X" training model, promote both textbook and online open-course development, and improve industry-academia-research cooperation for talent development. Building on these efforts, both the National Natural Science Foundation of China and the Academic Degrees Committee of the State Council officially introduced "Artificial Intelligence" and "Intelligent Science and Technology" as first-level disciplines. Numerous universities have since established AI colleges and expanded graduate enrollment, yielding significant progress.

(3) **Overseas Talent Introduction**

To attract outstanding AI4S talent globally, nations have adopted various measures, such as increasing talent budgets and project funding, improving scientific research infrastructure, eliminating barriers in academic careers, offering more appealing professional development pathways, relaxing immigration policies, establishing world-class AI research centers, and refining corporate AI innovation assessments. China also actively promotes international exchange and cooperation, for instance by creating the "Silk Road" Scholarship under the Chinese Government Scholarship framework to support international students in AI-related fields. The government likewise encourages and supports Chinese students pursuing AI studies in leading countries abroad and, through the UNESCO China Entrepreneurship Education Alliance, advances international exchange and collaboration in AI-driven innovation and entrepreneurship.

(4) **Building Exchange and Collaboration Platforms**

China's Ministry of Science and Technology has highlighted, through its AI Special Program, the need to involve more scientists and researchers from fields such as mathematics and physics in related studies, aiming to cultivate and assemble cross-disciplinary R&D teams. Domestic AI4S forums are emerging, such as the AI for Science Forum—jointly hosted by the Department of Computer Science at Peking University and Beijing AI for Science Institute (AISI)—has convened experts and researchers from academia, industry, and various other sectors, collectively examining the application and future prospects of AI in scientific research.

2.4 Governance Standards and Framework

Establishing robust AI governance standards and regulatory mechanisms—ensuring high levels of security, transparency, trustworthiness, accountability, and explainability for AI models—is pivotal for the sustainable development of AI4S. In response to the ethical and security risks posed by AI algorithms and AIGC, countries worldwide are expediting the formulation of pertinent laws, regulations, technical standards, and guidance documents. For instance, the European Union introduced the world's first AI regulatory framework, the EU AI Act, while in 2024, China likewise issued a series of policies such as the Basic Security Requirements for Generative AI Services and the AI Safety Governance Framework. These AI governance measures not only provide a solid foundation for the standardized application of AI but also generate significant spillover effects in the realm of scientific research sector.

In the domain of scientific research, the European Union issued the Guidelines on the Responsible Use of Generative AI in Scientific Research in March 2024, with the goal of ensuring that AI-driven scientific research adheres to principles of safety, transparency, and ethics. Correspondingly, the Supervision Department of China's Ministry of Science and Technology augmented the Guidelines on Responsible Research Conduct (2023) with provisions on the usage of generative AI. Moreover, domestic universities and research institutes have successively introduced their own guidelines on employing generative AI. For example, in September 2024, the Chinese Academy of Sciences released the Integrity Reminder on the Regulated Use of AI Technology in Scientific Research, setting forth principles for researchers—such as verifying the authenticity of AIGC, thoroughly disclosing instances of AI usage, employing only AI service tools that have been filed and registered with national authorities, and avoiding infringements on intellectual property rights and data security requirements.

3 Innovation Socialization as a New Trend in AI4S Research

Sociological studies indicate that socialization refers to the process through which individuals enter a new social environment and acquire the knowledge, skills, and behaviors necessary for functioning within and becoming part of that environment [11]. In the present context, as breakthroughs in artificial intelligence, big data, and related technologies accelerate, the digital realm is emerging as a new environment and frontier for human activities. In the scientific research arena, digital research infrastructure, digital collaboration networks, digital modeling software, and digital documentation are together constructing a comprehensive digital research environment. This environment propels the expansion—and even migration—of traditional scientific research, once bound to physical spaces and physical research instruments, into digital spaces. Consequently, it exerts a profound influence on how researchers acquire knowledge and adapt their research behaviors within this digital setting, thereby presenting a trend of "innovation socialization" in AI4S research. In other words, innovation socialization is the process by which researchers, drawing on digital technologies, and utilize the digital research environment to acquire knowledge and conduct research activities. This process extends beyond simply mastering technological tools, involving multidimensional transformations across every stage of scientific innovation—spanning problem identification, experimental analysis, research collaboration, dissemination of findings, as well as research management and evaluation. Such transformations enable researchers to establish novel forms of interaction within digital spaces, broaden the scope of traditional research practices, and drive both knowledge innovation and technological advancement.

3.1 AI-Driven Problem Identification in Scientific Research

The application of AI algorithms and tools is bringing transformative changes to the process of discovering scientific research problems. With robust capabilities for analyzing multimodal data, AI technology can support the summarization, integration, and analysis of vast amounts of scientific literature, thereby significantly improving researchers' reading efficiency and interdisciplinary thinking. As a result, researchers can gain inspiration for research topics in a broader range of fields. Furthermore, employing machine learning and deep learning for in-depth mining of spatiotemporal big data can effectively promote the identification and resolution of social demands as well as social issues.

3.1.1 Conversational Interaction Improves the Speed and Precision of Knowledge Acquisition

Scientific literature is the primary source for research inspiration and scientific knowledge, yet effectively utilizing these resources can be a daunting task. Traditionally, researchers have had to spend significant amounts of time searching, reading, and summarizing vast quantities of information. However, with the advent of large language models (LLMs) such as ChatGPT, the way researchers utilize scientific literature is now undergoing a major transformation. For example, the "Spark Research Assistant"—a large-scale scientific literature model jointly developed by the National Science Library of the Chinese Academy of Sciences and iFlytek—was officially launched in October 2023. This platform offers functionalities such as literature reading, summary generation, Q&A about papers, and multi-paper comparisons, providing strong support for analyzing scientific literature.

Capitalizing on their outstanding natural language understanding capabilities, these LLMs can thoroughly examine complex questions posed by researchers, quickly synthesize information and perspectives from various sources, and generate potential answers—much like interactive discussions and Q&A between a teacher and a student. This conversational approach surpasses traditional keyword-based knowledge searches by allowing researchers to interact with AI at any time, continuously refining responses, significantly shorten the time needed to obtain key information, and meet personalized and precise research needs.

3.1.2 Big-Data Analysis of Literature Promotes Systematic, Interdisciplinary Thinking in Scientific Topic Selection

By thoroughly mining and analyzing massive amounts of scientific literature, LLMs can efficiently carry out tasks such as literature aggregation, review generation, topic inspiration, rapid trend monitoring, and knowledge graph construction. These functions can inspire researchers to think more comprehensively and multidimensional when selecting topics. For example, knowledge graphs make it possible to organize literature in a structured, systematic way, helping researchers to quickly grasp interconnections among large bodies of scientific knowledge and effectively identify research hotspots, emerging trends, and knowledge gaps. This, in turn, can lead to the discovery of potential innovative topics. One study has demonstrated how LLMs assist researchers in conducting literature reviews and, through automated literature analysis and synthesis, help identify research gaps and possibly overlooked key points in related fields [12].

3.1.3 Spatiotemporal Big-Data Analysis Accelerates the Identification of Potential Societal Issues

The widespread use of mobile devices, social media, sensors, and the Internet of Things has generated abundant spatiotemporal big data, which is an important reference for reflecting socioeconomic activities and urban operations. By leveraging AI technologies such as machine learning and deep learning to model and analyze these high-dimensional datasets, researchers can rapidly detect potential social and scientific problems. For instance, in the field of public health, AI-driven analyses of medical data, health monitoring information, and social media content can provide real-time forecasts of disease transmission trends and high-risk areas, allowing for optimal allocation of medical resources. In environmental monitoring, integrated analyses of air quality data, water quality data, and weather information make it possible to predict future pollution and environmental risks more accurately, supporting the formulation of preventive measures. In social governance, analyzing community safety data alongside social media sentiment helps gauge residents' sense of security and satisfaction with public services, guiding improvements in public safety measures and enhancing social service quality. These AI-empowered applications provide decision-makers with scientific bases for more timely and effective policymaking, thereby improving urban operations and advancing residents' quality of life.

3.2 AI-Driven Experimental Analysis in Scientific Research

Recent findings demonstrate that employing AI technology can solve challenging problems in basic research with unprecedented speed and precision. This progress is attributable to the transformation of traditional scientific experimentation driven by AI-driven computer simulation and analysis, which has accelerated the exploratory processes of scientific experiments. At the same time, AI, as a general-purpose technology, is becoming a bridge across different disciplines and institutions. This makes scientific discovery increasingly reliant on cross-domain collaboration and open sharing, thereby promoting extensive coordination and integration within the global research community in the realm of scientific experimentation.

3.2.1 Driving the Shift in Experimental Science

In recent years, numerous cases of AI-driven scientific experimentation have continuously emerged, demonstrating the feasibility of transitioning experimental science from conventional trial-and-error methods to a "data + computation" predictive model. These successes also prove that AI-driven scientific computing enables researchers to achieve breakthroughs with unprecedented efficiency, thereby expanding our understanding of nature and its underlying mechanisms. Below are a few typical examples.

In the field of biology, it used to take weeks or even months to obtain high-resolution protein structures using traditional experimental methods. However, DeepMind's AI system AlphaFold has revolutionized the process, using computational simulations to accurately predict the 3D structures of proteins. Within just five years, AlphaFold has undergone two major iterations Currently, AlphaFold 3 can precisely predict the structures and interactions of all biomolecules, including proteins, DNA, and other molecular components, which could drastically impact drug research and development.

In the field of materials science, DeepMind's GNoME has identified up to 2.2 million theoretically stable, yet experimentally unrealized crystal structures, of which 380,000 are likely to be successfully synthesized in the future. This discovery boosts the number of known stable materials by nearly tenfold, significantly accelerating the pace of new material discovery.

In the field of computational physics, China's DP Technology team launched DeePMD in 2017, successfully employing the Summit supercomputer to simulate the trajectories of hundreds of millions of atoms—while maintaining first-principles accuracy. This marks the first time that high-precision, high-performance molecular dynamics computations have been achieved on such a massive scale. The feat reduced what was originally anticipated to be a 60-year simulation to just one day, breaking a 35-year barrier that confined molecular dynamics simulations to smaller systems of merely a few thousand atoms. This breakthrough has laid a crucial foundation for further advances in chemistry, biology, materials science, and other related fields.

All of these breakthroughs were realized through cross-disciplinary collaboration. Furthermore, they have been, to varying degrees, made openly accessible to the global scientific community, encouraging worldwide researchers to further advance these efforts and, in turn, more broadly serve humanity and enhance human well-being.

3.2.2 Emergence of AI Scientists

In the field of chemistry, "AI scientists" capable of conducting independent research are gradually reshaping the fundamental nature of scientific experimentation. For example, in 2023, the U.S. Lawrence Berkeley National Laboratory successfully developed the Automated Laboratory (A-Lab) system, which can process materials samples at a rate 50–100 times higher than a human experimenter. The A-Lab operates continuously 24/7 without human intervention, carrying out literature reviews, designing plans, using robotic arms to synthesize compounds, adjusting formulas, and repeating experiments until it either meets its target or exhausts all potential formulations. Impressively, this system managed to create 41 new materials on its own in just 17 days.

In China, in 2022, Professor Jiang Jun's team at the University of Science and Technology of China unveiled a data-driven, end-to-end "machine chemist," integrating two mobile robotic operators with 15 intelligent chemistry workstations. This system successfully undertook a series of complex tasks previously only achievable by humans—reading extensive chemical literature, acquiring prior chemical knowledge, proposing scientific hypotheses, designing experimental procedures, conducting experiments, building theoretical-feedback models, analyzing data, and generating new hypotheses—thus blending theory and experiment seamlessly. By reviewing a vast amount of literature to identify five non-noble metals, and incorporating 20,000 sets of theoretical data along with 207 sets of experimental data, the system constructed an intelligent theoretical-experimental model. Through Bayesian optimization, it pinpointed the best high-entropy catalyst composition from among 550,000 potential metal mixtures, compressing a traditional brute-force process that would have taken an estimated 1,400 years into just five weeks [3].

These cases exemplify a future in which routine experiments can be delegated to robots, saving considerable human and material resources. This, in turn, allows researchers to devote themselves to more creative and human-oriented tasks—such as innovative design and collaborative teamwork—thereby advancing the socialization and openness of scientific research.

3.3 AI-Driven Collaboration in Scientific Research

As an advanced technology that transcends disciplinary and domain boundaries, AI's intrinsic capacity for knowledge convergence will catalyze paradigm shifts in research collaboration, fostering deeper integration and synergy among scholars from diverse fields. Moreover, this collaborative model extends beyond traditional research institutions and universities, creating opportunities for broader societal engagement and encouraging the public and grassroots organizations to participate in scientific exploration.

3.3.1 From Hierarchical, Siloed Innovation to Cross-Disciplinary Convergence

AI technology excels at tackling highly complex, interdisciplinary scientific challenges, prompting AI-driven research to place greater emphasis on deep collaboration across multiple fields. Taking the AlphaFold research team as an example, their paper published in *Nature* features 34 authors, 19 of whom share first authorship. These contributors are prominent scholars spanning machine learning, speech and computer vision, natural language processing, molecular dynamics, life sciences, high-energy physics, and quantum chemistry. AlphaFold's success illustrates that future major scientific breakthroughs will increasingly depend on multidisciplinary collaboration, emphasizing cross-field synergy and shared success. By contrast, solitary efforts or reliance on a single senior expert (the so-called "1+N" innovation model) are becoming ever less capable of addressing the complex, intersecting nature of today's scientific problems.

3.3.2 From a "Small Workshop" to "Platform-Based" Model

Countries worldwide are actively advancing AI infrastructure and steadily increasing investment to facilitate the integration, sharing, and efficient utilization of AI resources, thereby lowering the threshold for AI4S. These infrastructures encompass AI algorithms, high-performance computing capabilities, AI-enabled automated experimental platforms, high-quality databases, and integrated intelligent research collaboration systems. Representative examples include the EU's EOSC and AI4EU, the United States' NAIRR, and the Chinese Academy of Sciences' "China Science Cloud" and "Global Open Science Cloud." By offering convenient access to and sharing of data, documentation, code, and research findings, these platforms play a critical role in fostering openness and cooperation in scientific research, promoting the reuse of results, eliminating redundant work, and ultimately accelerating scientific progress.

3.3.3 From "Elitism" to "Mass Participation"

With the advancement and widespread adoption of platform-based research tools—such as cloud computing platforms, online collaboration systems, and open-source machine learning frameworks—the barrier between scientists and the public is steadily diminishing. Traditionally, scientific research has largely been confined to a small group of highly specialized experts, offering limited opportunities for broader engagement. Today, however, the public can actively participate in data collection, experimental design, and outcome analysis through citizen science projects, crowdsourcing platforms, open-source initiatives, and competitions, leading to a new "mass innovation" model. This approach not only enables more individuals to contribute to scientific endeavors but also expands the talent pool, leveraging the collective wisdom of society and injecting renewed vitality into the scientific enterprise. For instance, Zooniverse—the world's largest citizen science platform—has attracted millions of participants and leverages human–machine collaboration to help scientists develop AI models for automated object identification. Major projects on this platform include "Galaxy Zoo," which classifies millions of stellar images in the Milky Way, and "Snapshot Serengeti," which promotes wildlife conservation research through animal behavior monitoring. By distributing research tasks among a broad public, these citizen science projects effectively reduce dependence on professional researchers, driving the decentralization of scientific research.

3.4 AI-Driven Dissemination of Scientific Research Findings

The widespread adoption of AI technologies has accelerated scientific knowledge contributions, fostering a shift from sharing only final results to sharing the entire research process, and propelling the resolution of real-world societal challenges.

3.4.1 Advancing Efficient Presentation and Dissemination of Research Outcomes

Large language models (LLMs) possess remarkable capabilities in paper drafting, language refinement, and cross-lingual translation. They can rapidly generate text while ensuring coherence and fluency, thereby serving as a convenient tool for efficiently presenting research findings. According to a report by the Stanford Institute for Human-Centered Artificial Intelligence (HAI), in the computer science domain, LLMs contributed to the drafting of 17.5% of papers and 16.9% of peer-review content, respectively [13, 14]. Moreover, LLMs offer substantial support to early-career researchers and non-native English speakers, enhancing their ability to articulate complex ideas, accelerating the production and dissemination of scientific papers, and ultimately expanding both the efficiency and scope of global scientific exchange.

3.4.2 Transforming Outcome Sharing Into Process Sharing

Traditionally, research outcomes have been shared through papers and patents, focusing on the dissemination of final results. However, with the rise of open access and open-source communities, sharing intermediate processes such as research data, code, and models has become increasingly prevalent, facilitating broader dissemination and deeper application of scientific findings. For instance, on open-source code hosting platforms like GitHub, researchers can share code and models for AI projects, enabling version control and collaborative development. This approach allows project contributors to receive rapid feedback and improvement suggestions, facilitating the swift iteration of AI technologies while ensuring transparency and reproducibility of the research process. Additionally, GitHub's extensive technical learning resources lower the entry barrier, empowering more researchers to rapidly translate AI technologies into practical applications.

3.4.3 Accelerating Societal Problem-Solving through Outcome Sharing

AI technologies broaden the dissemination of research outcomes, foster interdisciplinary collaboration, expedite the translation of findings, and enhance societal impact, thereby better serving society and addressing social challenges. For instance, during the COVID-19 pandemic, researchers worldwide accelerated vaccine and drug development by sharing data, algorithms, and models. In weather forecasting, a research team from Huawei Cloud developed the high-resolution AI meteorological forecasting system "Pangu Meteorological Model," which offers second-level global weather forecasts at a resolution of $0.25^\circ \times 0.25^\circ$ for four surface variables and five atmospheric variables [15]. This achievement is now operational at the European Centre for Medium-Range Weather Forecasts (ECMWF), providing new solutions for humanity in the face of climate change and natural disasters.

3.5 AI-Driven Management and Evaluation in Scientific Research

3.5.1 Trends Toward Flattened Research Organizational Structures

AI provides robust technological support for optimizing research management processes. Through data analysis, it streamlines redundant management layers, reduces hierarchical structures, and grants frontline researchers greater autonomy and decision-making power. This flattened organizational structure enhances management efficiency and stimulates innovation within research teams. For instance, the Academia Resource Planning (ARP) project at the Chinese Academy of Sciences offers an integrated management platform for over 180 research institutions, facilitating the allocation of human resources, funding, and research infrastructure to optimize resource configuration and research management workflow. Additionally, leveraging AI's smart processing capabilities has enhanced the aggregation and analytical exploration of management data, offering services like intelligent document recognition and risk management. This has played a critical role in improving the information management level of the Chinese Academy of Sciences, optimizing resource allocation, and supporting management decision-making.

3.5.2 Trends in Precision Research Management

AI technology facilitates the fine-grained management of the entire project lifecycle, enabling real-time tracking of key elements, including project progress, resource allocation, and risk management. Additionally, the in-depth analysis of talent data enables the construction of detailed talent profiles, providing valuable support for talent development and selection. This data-driven precision management enhances the transparency and accountability of research activities, facilitating their scrutiny and evaluation by all sectors of society.

3.5.3 Performance Evaluation Driven by Social Impact

In the context of AI for Science (AI4S), performance evaluation driven by social impact becomes increasingly feasible. By utilizing a multidimensional set of indicators such as citations (to measure academic value), downloads (to reflect application potential), and social media interactions (indicating societal impact and policy contributions)—a more comprehensive assessment of research can be achieved. Unlike traditional academic evaluations, this model broadens the focus beyond mere recognition within the academic community to emphasize the societal value of research, encouraging researchers to pay greater attention to social needs and challenges and to actively address societal issues, thereby promoting the real-world application of research results.

4 Governance Challenges and Suggestions for AI Applications in Research

The innovative socialization process under AI4S encompasses not only the transformation of scientific knowledge acquisition, generation, dissemination, and application value but also the redefinition of values, behavioral norms, and ethical standards within the digital research environment. As AI technology is increasingly applied in research, new ethical dilemmas related to technology and research ethics have emerged, posing significant governance challenges for AI4S. In light of this trend towards innovative socialization, AI4S should advocate for a human-centered and open-sharing value system. This approach will encourage researchers to uphold social responsibility and ethical principles while pursuing innovation and efficiency, ensuring that technological advancements serve the common good and promote sustainable societal development.

4.1 Human-Centered Approach to Meeting Diverse Innovation Needs

As AI applications and service platforms significantly transform research, it is crucial for AI for Science (AI4S) to facilitate a transition from "specialized" to "user-friendly" platforms that prioritize the needs of researchers and other users. Nevertheless, several challenges remain:

(1) **Inadequate Understanding of Researchers' Needs**: Various systems lack a proper understanding of researchers' needs, resulting in a disconnect between the services offered and actual requirements. This includes insufficient relevance of service content to genuine research activities and a mismatch between service formats and researchers' usage habits and workflows, which limits practical application.
(2) **Underutilization of Research Data's Value**: The deeper value of research data remains largely untapped. On one hand, data is viewed as a core asset of research teams and organizations, but limited incentives for data sharing have resulted in persistent data silos between departments and organizations. On the other hand, while research service platforms have managed to aggregate data effectively, the services offered based on this data remain insufficient, and its potential value is not fully leveraged.
(3) **Digital Literacy Gap Among Researchers**: There is a polarization in researchers' proficiency with emerging digital tools. Some researchers effectively utilize these tools for data analysis, literature management, and experimental design, enhancing their research efficiency. Conversely, others, hindered by lack of training or resistance to new technologies, find it challenging to access

and analyze data. This disparity exacerbates inequalities in research capabilities, outputs, and quality, impacting funding, project applications, and career opportunities.

To address these challenges, the following suggestions are proposed:

(1) **Enhance User Experience on Research Service Platforms**: Improve the user experience by allowing researchers to personalize the system based on their research characteristics and work habits, such as customizable dashboards tailored to various fields. Additionally, provide intelligent support, including intelligent recommendations and data visualization tools, to help researchers swiftly identify and access valuable data and information.
(2) **Optimize Service Supply for Typical Domain Scenarios**: Based on the existing big data platform, conduct in-depth research on typical application scenarios across various fields, develop diverse domain models and software, and establish digital and intelligent research environments to better meet researchers' needs for in-depth utilization of domain-specific big data.
(3) **Enhance Researchers' Digital Literacy**: Conduct targeted digital skills training aligned with the specific characteristics and needs of different research institutions to fully leverage the potential of AI in research and promote the efficient conduct of research work.

4.2 Emphasizing Ethics to Promote Responsible AI Application

While the widespread application of AI technology has brought numerous conveniences, it has also revealed several issues and risks, leading to a serious crisis of scientific credibility in the research field.

(1) **The Challenge of AI's "Black Box" Nature to Research Transparency**: The "black box" nature of AI is a key factor affecting research transparency. Many AI models, especially deep learning models, have complex and unexplainable internal mechanisms, making it difficult for researchers to understand their decision-making processes. Additionally, AI model training typically relies on large amounts of data. If the sources, quality, and processing of this data are unclear, it can lead to biased or misleading conclusions, making it difficult for other researchers to validate and replicate relevant experiments. Thus, the lack of interpretability and transparency directly undermines the reliability of research results, which raises concerns about the credibility of scientific inquiry.
(2) **The Challenge of AI "Hallucinations" and Resulting Dependency**: Currently, AI often generates content that appears real but is actually inaccurate or fabricated, which can be highly misleading. As model optimization and content generation quality continue to improve, researchers may develop excessive trust in AI tools. This overreliance can result in a neglect of a deep understanding and application of foundational theories, as well as a diminished

emphasis on traditional scientific methods and human judgment. Such "technological dependency" may weaken researchers' professional skills and critical thinking abilities, stifle their motivation to explore new methods and theories, and potentially leave them ill-equipped to address challenges.

(3) **The Challenge of AI "Deepfakes" to Research Authenticity**: There are widespread concerns in the scientific community about the potential dangers of AI propagating misinformation and fraudulent research, as scientific inquiry relies on truthful and reliable data and conclusions. AI technologies can easily generate false research findings or fabricate data, which can mislead researchers about their research directions, lead to a waste of research resources, and ultimately impact scientific progress while undermining public trust in science.

(4) **The Challenge of Algorithmic and Data Biases in AI to Fairness**: Improper use of AI technologies can lead to algorithmic biases, which stem from biases in algorithm design and insufficient data representativeness. Such biases can have serious consequences; for example, in medical research, if the training data for an algorithm does not adequately represent certain minority groups, it may result in inaccurate diagnoses and treatment delays for these populations, threatening their health and well-being.

To address these challenges, the following suggestions are proposed:

(1) **Enhance Guidelines for AI Technology Use**: To uphold scientific integrity, it is essential to emphasize the responsibilities of academic and research institutions in strictly adhering to ethical standards, technical protocols, and legal regulations. Additionally, researchers should be guided to use AI responsibly, with a focus on issues such as interpretability and transparency.

(2) **Cultivate Critical Thinking and Skills Among Researchers**: Educational and research institutions should strengthen training for students and researchers to enhance their critical understanding and application of AI tools. Researchers are encouraged to continue thinking deeply about foundational theories and to thoughtfully integrate traditional scientific methods with advanced AI technologies.

(3) **Establish a Credit Evaluation System for Researchers**: Implementing a dynamic system for evaluating researchers' credibility and academic conduct will serve as a crucial basis for project applications and promotion assessments. This will incentivize researchers to continually improve academic integrity and the quality of their research, fostering a healthy research ecosystem.

(4) **Develop Ethical Guidelines and Regulatory Mechanisms for AI in Research**: It is vital to guide researchers and developers in adhering to ethical standards in AI research and application. Higher education institutions and research organizations should establish ethics committees or review processes to ensure compliance with ethical standards in AI applications, ensuring that research decisions align with scientific norms, ethical standards, and societal values. At the same time, care must be taken to prevent excessive regulation from stifling research innovation.

4.3 Enhancing Regulation to Ensure Safe and Compliant Use of AI

As AI technology continues to advance, its application in scientific research faces increasingly complex security challenges, encompassing areas such as data security and cybersecurity, and other related aspects.

(1) **Challenges in Data Privacy Protection**: Scientific research often involves data that contains personal and sensitive information, such as participant data from clinical trials, genomic data, and survey data from social science studies. The unauthorized disclosure of such data not only constitutes an infringement on individual rights that may precipitate legal proceedings and grave ethical consequences, but also erodes public confidence in research institutions and related organizations. This, in turn, can diminish participant engagement and negatively impact future data collection endeavors.
(2) **Increased Risk of Cyberattacks**: With technological advancements and the exponential growth of data, the risks of cyberattacks and data breaches have intensified disproportionately. Research institutions, as high-value targets, face persistent threats from attack vectors including malware deployment and phishing campaigns, leading to the unauthorized system infiltration and potential exfiltration of confidential research datasets.
(3) **Regulatory Lag caused by Rapid Technological Development**: The swift advancement of AI technology often leaves existing technical standards, laws and regulations, and ethical oversight frameworks struggling to keep pace with technological progress. This lag creates regulatory gaps that may lead to potential risks. For instance, in scientific research, when AI systems produce errors or biases which cause severe consequences, defining the boundaries of responsibility between human researchers and AI systems becomes an extremely complex issue. Moreover, significant disparities in AI regulatory policies across nations and regions impose additional compliance challenges on research institutions involved in international collaborations.

To address these challenges, the following suggestions are proposed:

(1) **Enhance Regulations for Research Data Security**: Establish rigorous data management rules covering data collection, storage, use, and destruction. Ensure personal privacy and sensitive information protection throughout the entire data lifecycle while preventing data misuse. These regulations should incorporate best practices for data processing to enhance researchers' capability in safeguarding scientific data and privacy security.
(2) **Develop a Robust Cybersecurity Framework**: Design and implement systematic cybersecurity policies and standards to identify and assess potential cyber risks. Strengthen data encryption, access control, and user authentication mechanisms while improving emergency response mechanisms and incident handling processes. Additionally, enhance researchers' cybersecurity awareness and

skills through regular training programs to ensure their ability to recognize and address cyber threats, thereby effectively maintaining data security.

(3) **Establish a comprehensive regulatory system tailored for AI research and applications**: Building on existing laws and regulations, conduct in-depth analyses to clearly define the specific legal needs related to AI research and application, continuously refining relevant legislation. Specify data usage guidelines, algorithm transparency requirements, and principles of accountability to clearly delineating the responsibilities of human researchers and AI systems during research process. Implement detailed accountability allocation mechanisms to ensure traceability and liability attribution in case of errors. Furthermore, promote consistency in AI regulatory policies across nations to ensure compliance in international collaborations.

References

1. E Weinan. AI helps to establish a new paradigm for scientific research[J]. Bulletin of the Chinese Academy of Sciences, 2024, 39(1):10–16.
2. Li Guojie. AI4R: The fifth scientific research paradigm[J]. Bulletin of the Chinese Academy of Sciences, 2024, 39(1):1–9.
3. Yang Jinlong, Jiang Jun. Embracing the New Research Paradigm—The scientific research revolution brought about by artificial intelligence[J]. Science and Society, 2023, 13(3):11–22.
4. Qin Hao. Strategic Goals, Measures and Experience Analysis of American Government's Artificial Intelligence[J]. Journal of China Academy of Electronics and Information Technology, 2021, 16(12):8.
5. Qi Shuo, Li Shixin, Yang Yimeng. Strategic Layout and Future Prospects of Artificial Intelligence in Global Perspective[J]. World Sci-Tech R&D, 2024(4):442–455.
6. Gao Wen. Analysis of AI Cutting-Edge Technologies and High-Quality Development[J]. Current Affairs Report, 2023, 6:96–113.
7. Bai Lu. From U.S. Government Policies: How to Unleash the Potential of AI-Driven Scientific Research Innovation [R]. Beijing: Development Research Center of the State Council, 2024.
8. Wang Zhaoran. New State and Deployment of AI for Science in U.S.: Take United States Department of Energy as an Example[J]. Global Science, Technology and Economy Outlook, 2024, 39(6):8–14.
9. Li Jianhui, Zhang Lili, Wen Liangming. Global Open Science Cloud Initiative for Scientific Collaboration[M]. Chinese Academy of Sciences. China's e-Scientific Blue Book (2022). Beijing: Electronic Industry Press, 2022.
10. Cui Dan, Li Guoping. New Global Trends in AI Talent Cultivation and Education Policies [N]. Guangming Daily, 2024-3-21.
11. Brim O G. Socialization through the life cycle[G] Socialization After Childhood: Two Essays. New York: Wiley, 1966: 1–49.
12. Liang W, Izzo Z, Zhang Y, et al. Monitoring AI-Modified Content at Scale: A Case Study on the Impact of ChatGPT on AI Conference Peer Reviews[J]. arXiv:2403.07183, 2024.
13. Liang W, Zhang Y, Wu Z, et al. Mapping the Increasing Use of LLMs in Scientific Papers[J]. arXiv:2404.01268, 2024.
14. Zhu Q, Zhang F, Huang Y, et al. An all-round AI-Chemist with a scientific mind[J]. National Science Review, 2022, 9(10).
15. Bi K, Xie L, Zhang H, et al. Accurate medium-range global weather forecasting with 3D neural networks[J]. Nature, 2023, 619: 533–538.

Jiaofeng Pan Director and Professor of the Institutes of Science and Development, Chinese Academy of Sciences; Dean of the School of Public Policy and Management, University of Chinese Academy of Sciences; Chairman of the Chinese Association of Development Strategy Studies; Editorial Board Member of the Bulletin of the Chinese Academy of Sciences. Mainly engaged in research on science and technology strategy planning, innovation policy, and theoretical/methodological studies of think tanks.

Shengyin Wang Postdoctoral Researcher at the Institutes of Science and Development, Chinese Academy of Sciences. Mainly engaged in research on AI for think tank and geospatial-temporal data mining.

Jing Wu Professor at the Institutes of Science and Development, Chinese Academy of Sciences; Executive Director of the Advisory Research Support Center, Academic Divisions of the Chinese Academy of Sciences; Associate Director of the Department of Think Tank Science and Engineering, School of Public Policy and Management, University of Chinese Academy of Sciences. Mainly engaged in macro policy simulation and analysis, as well as research on digital economy and digital transformation policies.

Infrastructure Developments

AI Computility Grid—An Artificial Intelligence Empowered Scientific Research Platform

Ninghui Sun and Sa Wang

Abstract With the 2024 Nobel Prizes in Physics and Chemistry both awarded to achievements in artificial intelligence (AI) and related fields, AI for Science (AI4S) has undeniably become the new generation of scientific research paradigms. Leveraging the AI4S paradigm and AI-empowered capability platform, scientists can significantly improve research efficiency and expand the frontiers of scientific understanding. This article begins by reviewing the evolution of scientific research paradigms and the development of computing infrastructure tailored for scientific research. It summarizes the evolution from supercomputing centers to intelligent computing centers, and ultimately to the emergence of AI computility grid. This paper explores the current technical challenges faced by AI computility grids and introduces two typical cases of how the computility grid, *Info-Superbahn*, empowers scientific research.

Keywords AI4S · AI computility grids · Scientific research paradigm · AI4R

1 Introduction

The advent of artificial intelligence technologies, such as deep learning algorithms and generative large language models, has sparked a new wave of AI enthusiasm across the world. In 2024, the Nobel Prize in Physics was awarded to artificial neural networks, while the Chemistry Prize was bestowed upon AI-assisted protein structure prediction and protein design. These accolades underscore that AI-empowered scientific research has become the new scientific research paradigm [1, 2]. The associated intelligent computing platforms have emerged as significant new capabilities in scientific research, following in the footsteps of supercomputers and large-scale

N. Sun (✉) · S. Wang
Institute of Computing Technology, Chinese Academy of Sciences, Haidian, Beijing, China
e-mail: snh@ict.ac.cn

S. Wang
e-mail: wangsa@ict.ac.cn

China's e-Science Blue Book 2025, https://doi.org/10.1007/978-981-96-8252-2_6

scientific facilities. These platforms are set to enhance research efficiency by orders of magnitude and significantly expand the frontiers of scientific knowledge.

Early scientific discoveries primarily relied on experimentation and the intellect of the human mind, with scientists making bold conjectures and meticulously verifying them. Information technology, whether in the form of computing techniques or data science, played only a supplementary and validating role. Artificial intelligence holds significant advantages in areas such as memory, handling high-dimensional complexity, comprehensive perspective, depth of reasoning, and conjecture capabilities compared to humans [3]. The pivotal challenge of this wave of AI-empowered revolution in scientific research is whether AI can take the lead in scientific discoveries and technological inventions, such as uncovering laws of physics, predicting protein structures, designing high-performance chips, and efficiently synthesizing new drugs. Artificial intelligence large language models possess comprehensive data and a god's-eye view, enabling them to look further ahead than humans through deep learning capabilities. If AI models continue to evolve and achieve the leap from inference to reasoning, they will have the potential to possess the imagination and scientific conjecture abilities akin to those of a top-tier human scientist like Einstein. This advancement is bound to significantly enhance the efficiency of human scientific discoveries and push beyond the boundaries of human cognition.

Meanwhile, the IT infrastructure dedicated to scientific research has been undergoing continuous evolution. Initially, it was characterized by costly, exclusive supercomputing centers. Later, grid computing technology made them public, platformized and user-friendly. Nowadays, propelled by advancements in artificial intelligence, intelligent computing centers have emerged, further evolving into a novel capability platform known as the computility grid. The AI computility grid is capable of providing infinitely scalable computility supply, unified data encapsulation and sharing, as well as automation acceleration throughout the entire scientific research process, thereby further expediting the entire cycle of exploration, experimentation, discovery, and innovation for scientists.

However, numerous technical challenges remain, primarily including:

(1) There is lack of a unified abstraction and encapsulation for scientific data, to adapt to various AI models across different disciplines, and enhance training and inference efficiency.
(2) Large language models for research need 64-bit high precision computation, which can be seamlessly integrated into high-performance computing workflows and meet the precision requirements of scientific research.
(3) AI computing resources in computility grids are geographically distributed, independently managed, and heterogeneous distributed, which are quite challenges to manage. We need a unified abstraction, e.g. task closure, for different users or large language models to encapsulate their computing demands and tasks. Then the computility grid can schedule the task closure to the suitable node.
(4) The automation of the entire scientific intelligence process tailored for AI4S, deeply integrating the full cycle of large language model data collection,

data cleansing, pre-training, fine-tuning, evaluation, deployment, and inference with the scientific research process, to achieve full automation of scientific intelligence workflows.

As an AI capability platform that empowers scientific research, the AI computility grid closely integrates intelligent computing with scientific computing and scientific data, holding significant implications for scientific research. It plays a crucial role in every stage of the research process and is poised to propel scientific inquiry into an entirely new stage.

2 The Evolution of Scientific Research Paradigms

Modern scientific research originated from the 16th–17th-century Scientific Revolution, with Galileo and Newton as its founding figures. The widely known story of Newton and the apple illustrates how Newton observed not only falling apples but also the universal tendency of unsupported objects to fall toward the Earth. Newton formulated the law of universal gravitation by generalizing these observations, describing the mutual attraction between objects based on their mass and distance. This law successfully explained both celestial motion and terrestrial gravity, marking **the first paradigm of scientific research: experimental inquiry based on observation and induction**.

The second paradigm of scientific research is theoretical research based on scientific hypotheses and logical deduction, the fundamental methodology in research fields like mathematics. For example, Euclid began his geometric studies by proposing axioms and hypotheses such as, *a straight line can be drawn between any two points*, and *all right angles are equal*. He then constructed the Euclidean geometry system, which became the cornerstone of classical geometry and profoundly influenced mathematical development, through this rigorous deductive logic.

The first two paradigms dominated scientific inquiry for centuries until the mid-twentieth century, when the advent of computers introduced **the third paradigm: computational simulation of complex phenomena**. In aircraft design, to evaluate aircraft performance under various flight condition, engineers may use computational fluid dynamics software to simulate aerodynamic performance of aircrafts in addition to wind tunnel experiments. By setting parameters like aircraft shape, speed, and air density, computers simulate distribution of airflow near the aircraft, predicting aircraft performance such as lift, drag and stability. This allows further optimization of aircraft design for safety and efficiency. Similarly, material scientists employ finite element analysis to model stress distribution and deformation of materials under varying conditions, providing data to help design and optimization of materials.

The twenty-first century ushered in a data revolution driven by rapid development in sensor technology, mass construction of large-scale scientific facilities, and further popularization of the internet, giving rise to **the fourth paradigm: data-intensive research**, in fields like high-energy physics, astrophysics and biology.

At CERN's Large Hadron Collider, petabytes of data are generated per second from collision of proton beams with very high level of energy. These data are collected by sensors and analyzed by scientists. Through complicated data process algorithms and high-performance computing technology, possible signatures of Higgs boson were identified from mass data, resulting in the final discovery of Higgs boson, making a breakthrough in high-energy physics.

Astronomers leverage radio and space telescopes to collect galactic amounts of astronomical data, like spectrum, brightness distribution and motion trail of celestial bodies. Analysis of these data may lead to discovery of new types of galaxies and deeper understanding of cosmic evolution. Research on dark matter's existence is a typical example of analyzing astronomical data. Evidence of dark matter's existence can be inferred by analyzing rotational data of galaxy, since the rotational speed of galaxy doesn't match theoretical data calculated by considering only visible matters.

The rapid advancement of gene sequencing technology has enabled biologists to acquire vast amounts of genetic sequence data. For instance, in the Human Genome Project, scientists obtained sequence information encompassing approximately 3 billion base pairs. Analyzing these massive genetic datasets facilitates the study of the genetic foundations of diseases. By comparing the gene sequences of healthy individuals and those afflicted with specific diseases, researchers can identify genetic mutations associated with those conditions. In cancer research, for example, analyzing gene sequencing data from large cohorts of cancer patients has revealed numerous mutations linked to cancer initiation and progression, providing critical insights for cancer diagnosis, treatment, and the development of targeted therapies.

This wave of AI technological revolution, centered on data-driven approaches and artificial neural networks, builds upon high-performance scientific computing and data-intensive research. By integrating large-scale scientific models, it advances novel HPC+AI computational methodologies to simulate cross-scale complex systems [4]. Supported by next-generation infrastructure in computility and scientific data, the dawn of **intelligentize research methods** has emerged, promising widespread breakthroughs in scientific inquiry and positioning itself as the likely **fifth paradigm of research** [5]. Currently, intelligentize approaches have already achieved remarkable results in accelerating drug research, optimizing the design of new materials and predicting performance, and other applied domains.

Drug development is a lengthy, complex, and costly process. The traditional drug development process encompasses multiple stages, including drug discovery, preclinical trials, and clinical trials, each of which can be improved through intelligent technologies. For instance, during the drug discovery phase, screening potential therapeutic drug molecules from vast compound libraries constitutes a time-consuming and labor-intensive process. With the emergence of HPC+AI scientific computational methodologies, high-performance computing can accelerate the simulation of interactions between drug molecules and targets. Technologies such as molecular docking simulations enable rapid evaluation of a drug molecule's activity and selectivity. In the clinical trial phase, AI leverages extensive biomedical data, including disease genomics, drug molecule-protein interactions, and clinical datasets, to perform machine learning and predict therapeutic efficacy against specific diseases.

Meanwhile, AI can help optimize clinical trial design to enhance both efficiency and accuracy.

Traditional material design and performance prediction often require extensive experimentation and trial-and-error processes. Designing a new material with specific properties, such as high strength or high conductivity, can be highly time-consuming and resource-intensive. HPC+AI high-performance computational methodology can simulate the physical and chemical properties of materials under varying conditions, using molecular dynamics simulation software to study molecular behavior and interactions. By applying artificial intelligence to machine learning on data encompassing material composition, structure, and performance, large-scale material science models can be developed to predict the properties of novel materials. AI further aids in optimizing material design proposals, significantly reducing the number of experiments required.

3 The Evolution of Computing Infrastructure for Scientific Research

The early form of computing infrastructure for scientific research was primarily supercomputing centers. High-performance computing (HPC) technology originated from scientific research needs, with new technologies first applied to scientific research. Supercomputing centers played a vital role in fields like scientific and engineering computing, numerical simulation, weather forecasting, and geological exploration. Supercomputing centers were extremely expensive and operated as private computing resources in the early days, making them largely inaccessible to ordinary researchers. The emergence of grid computing technology transformed supercomputing centers into public, user-friendly platforms, driving their evolution toward fundamental computing infrastructure.

Currently, China's computing infrastructure for internet applications, government, and enterprise informatization relies mainly on public cloud services, while public computing infrastructure for scientific research remains centered on supercomputing centers. Emerging forms include intelligent computing centers and computility grids. Historically, China's scientific research computing platforms have been dominated by supercomputing centers. In recent years, with the explosive growth of AI technologies, intelligent computing centers have also become critical infrastructure for empowering scientific research. As the demand for computility from AI applications like large language models grows exponentially, single-type computing resources or isolated data centers struggle to meet these needs. The challenge lies in integrating geographically dispersed, heterogeneous, and independently managed computing resources into a unified "large computer" through networking, providing abstracted and cohesive computing services. This is both the goal of computility networks and the future trend of computing infrastructure.

3.1 Supercomputing Centers

The United States pioneered the development of public-service-oriented National Center for Supercomputer Applications (NCSA), which provided computing resources to tens of thousands of scientists, engineers, and graduate students across the U.S. Later, grid computing technology [6] enhanced its scalability and usability, leading to the establishment of a comprehensive national supercomputing public service system in the U.S. Notable projects include the National Technology Grid (NTG), Distributed Terascale Facility (DTF), NASA's Information Power Grid (IPG), the Department of Energy's ASCI Grid, and the Department of Defense's Global Information Grid (GIG).

In China, the development of supercomputing centers dates to the late 1990s. In 1998, under the State High-Tech Research and Development Program (863 Program) initiative on "Intelligent Computer Systems," China established the National High-Performance Computing Environment. During this period, five high-performance computing centers were established, forming the prototype of China's HPC infrastructure.

As of 2024, the Ministry of Science and Technology has approved the establishment of 14 national supercomputing centers (including those under construction), as shown in Table 1.

To improve the utilization of national supercomputing resources, China independently developed the National Supercomputing Infrastructure Support Software System, known as the China National Grid (CNGrid). This system enables the aggregation, management, and sharing of distributed, heterogeneous, dynamic, and autonomous resources, forming the largest national-level supercomputing infrastructure. It provides over 10 billion core-hour of computility annually, completes 20 million tasks per year, and maintains an overall availability rate greater than 99%.

Supported by the 863 Program, key national R&D initiatives, and local government funding, China's supercomputing infrastructure has achieved significant progress over two decades. It now serves diverse scientific fields, including space astronomy, nuclear energy and new energy, materials science, meteorology and climate change, aerospace, drug discovery and bioinformatics, and industrial simulation.

3.2 Intelligent Computing Centers

With the rapid advancement of AI and breakthroughs in generative AI and large language models, demand for intelligent computility has surged exponentially. According to OpenAI, since 2012, the computility required for AI applications has doubled every 3–4 months. This massive demand has driven the rapid growth of intelligent computing centers. China has built several notable facilities, including

Table 1 List of national supercomputing centers approved by the ministry of science and technology (including under construction) as of 2024

No	Name	Year established	Current operational computing resources
1	Computer Network Information Center, Chinese Academy of Sciences	1995 (Approved in 2023)	Dongfang supercomputer
2	Shanghai Supercomputer Center	2000	Magic Cube III: 3.3 Pflops Magic Cube II: 399 Tflops
3	National Supercomputer Center in Tianjin	2009	Tianhe-1A: 4700 Tflops Tianhe-3 exascale prototype system: 146 Pflops
4	National Supercomputing Center in Changsha	2010	Tianhe supercomputer: 1372 Tflops
5	National Supercomputing Center in Shenzhen	2009	Dawning 6000: 1271 Tflops
6	National Supercomputing Center in Jinan	2011	Sunway BlueLight: 1100 TFlops Sunway exascale prototype system: 3.13 Pflops
7	National Supercomputer Center in Guangzhou	2013	Tianhe-2: 100.679 Pflops
8	National Supercomputing Center in Wuxi	2016	Sunway TaihuLight: 125.436 Pflops
9	National Supercomputing Center in Zhengzhou	2020	Songshan supercomputer: 100 PFlops
10	National Supercomputer Center in Kunshan	2020	Dawning 7000
11	National Supercomputing Center in Chengdu	2020	Dawning 7000: 170 Pflops
12	"Wuzhen Light" Supercomputing Center	2021	Silicon Cube: 181.9 PFlops
13	National Supercomputing Center in Xi'an	2021	Dawning 7000
14	National Supercomputing Center in Taiyuan	2022	Dawning 7000: 300 Pflops

Pengcheng Cloud Brain, SenseTime's AI Computing Center, and Cambricon's intelligent computing platform.

Pengcheng Cloud Brain, developed by Peng Cheng Laboratory in Shenzhen, utilizes Huawei's domestically produced software and hardware systems. Pengcheng Cloud Brain II delivers 1 exaflops of half-precision computility and hosts groundbreaking AI models such as:

(1) Pengcheng·PanGu: The world's first fully open-source 200-billion-parameter Chinese pre-trained language model.

(2) Pengcheng-Baidu·Wenxin: The world's first knowledge-enhanced 100-billion-parameter language model.
(3) Pengcheng·Dasheng: A 1-billion-parameter vision model with internationally advanced performance.

SenseTime's New Generation AI Computing and Empowerment Platform is the first ultra-large AI computing center operational in East China. Phase I of the center covers 130,000 m^2 and houses 5,000 server cabinets.

Cambricon, in collaboration with local governments, has established multiple intelligent computing centers, supporting innovations in fintech, smart manufacturing, retail, healthcare, and transportation, including:

(1) Hengqin Advanced Intelligent Computing Center: The first AI computing platform in the Guangdong-Hong Kong-Macao Greater Bay Area.
(2) Fengdong AI Computing Innovation Center: The first AI infrastructure in Northwest China.
(3) Nanjing AI Computing Center: The most powerful operational intelligent computing platform in the Yangtze Delta.

These facilities support innovations across various sectors including fintech, smart manufacturing, intelligent retail, smart healthcare, and intelligent transportation.

4 Computility Grids

The concept of the computility grid can be traced back to the 1960s when American scholar John McCarthy proposed the idea of Utility Computing, suggesting that *computing may someday be organized as a public utility just as the telephone system is a public utility*. In the field of supercomputing, the Grid Computing proposed by Ian Foster and Carl Kesselman also gained widespread popularity. For many years, China's high-performance computing technology initiatives have supported research in this area. The startup company Beijing PARATERA Tech Corp., Ltd. has utilized such technologies to serve a large number of supercomputing users, playing a significant role in popularizing supercomputing in China. Cloud computing is also a case of this idea, revolutionizing the enterprise data center market and providing a scalable, highly reliable computing foundation for internet applications. In recent years, with the official launch of China's "Eastern Data and Western Computing" initiative, the computility grid has once again become a hot topic in both industrial and academic circles.

4.1 The Concept and Characteristics of the Computility Grid

As shown in Fig. 1, compared to cloud computing, the computility grid further integrates geographically distributed, heterogeneous, and independently managed

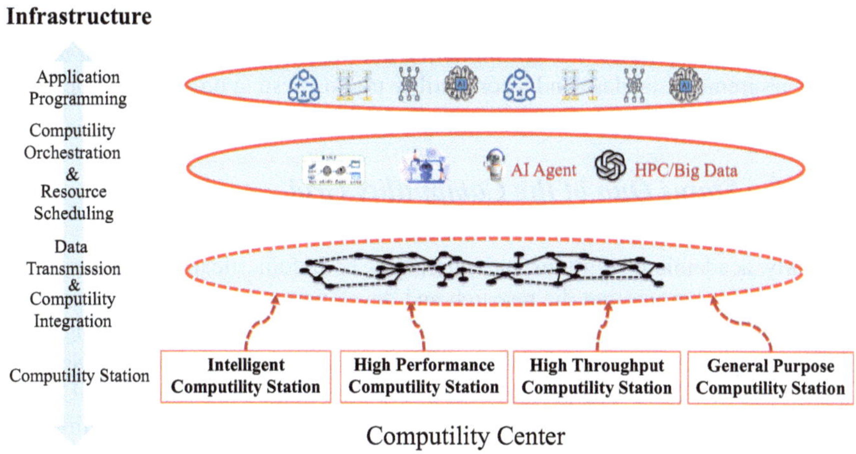

Fig. 1 The layered and decoupled architecture of the computility grid

resources. Through technologies such as grid integration, scheduling, application templates, and model factories, it logically forms *One Large Computer*. In terms of computing types, it can be more diverse and better adapted, while significantly reducing computing costs through scale expansion. It also provides better usability with low-code orchestration methods. The computility grid includes the following main characteristics:

(1) **Computility Integration**. Due to inconsistent interfaces in multi-cloud architectures, cloud platforms' computility and data resources have formed many *computility silos*, limiting broader resource consolidation and more efficient data usage. Computility networking adopts a *federated* business model to connect these *computility silos*, enabling inter-cloud interconnectivity and interoperability, and builds a unified cloud on top of a cross-domain distributed resource pool.

(2) **Computility Scheduling**. The widespread and diverse computility supply from national-level hubs, regional computility centers, city-level computility centers, and edge-level computility has brought unprecedented challenges to applications. Computility scheduling is key to achieving supply-demand matching, efficient delivery, and high-quality service. It needs to support fine-grained scheduling within a single server, coordinated scheduling between general-purpose computing and accelerators, large-scale (thousands of servers) scheduling within a single data center, cross-domain multi-data-center scheduling, and the integrated scheduling of computing, storage, network, and security resources.

(3) **Unified Measurement of Computility Usage**. Traditional computility service measurement is server-centric and based on usage duration, which cannot establish a unified measurement standard for the geographically distributed, heterogeneous, and independently managed computility resources of a computility grid.

The computility grid should achieve unified measurement of multi-level heterogeneous computility through modeling and normalization, enabling a unified measurement standard and a computility trading system on the user side.

4.2 The Status Quo of the Computility Grid

Currently, academic institutions and enterprises both domestically and internationally are actively promoting the research and development of computility grids. For instance, the Info-Superbahn computility grid [7] by the Institute of Computing Technology, Chinese Academy of Sciences (ICT, CAS), the China Computility Network (C^2NET) [8] by Peng Cheng Laboratory, and Sky Computing [9] by the University of California, Berkeley, are representative examples of current computility grid initiatives.

Proposed by ICT, CAS, the Info-Superbahn computational grid is a large-scale integrated platform initiated in 2019 to address demands in artificial intelligence (AI), edge-cloud distributed collaborative computing, and related scenarios. Designed as a multifunctional infrastructure for computational resource scheduling, AI model and dataset aggregation, and large-scale model development and application, it combines scientific research innovation with user-centric services. The platform integrates foundational resources including computility, advanced algorithms, massive datasets, high-precision models, development tools, efficient training environments, and seamless inference frameworks. By the end of 2024, it has interconnected 1,236 PFLOPS of AI computing capacity across four intelligent computing stations in Nanjing, Kunshan, and Zhengzhou, leveraging hardware from NVIDIA, Huawei, Sugon, and Cambricon. Currently, the Info-Superbahn supports 20+ large-scale AI models across seven research domains: natural language processing, multimodal learning, cybersecurity, 3D modeling, biomedicine, chip design, and AI education. Notable applications include the Biological Sequence Foundation Model by the Institute of Zoology of CAS, the Medical Large Language Model by Beijing University of Posts and Telecommunications, the Immune Protein-Antibody Structure Prediction Model by ICT, CAS and the *Shennong* Large Language Model by China Agricultural University. This platform exemplifies China's strategic advancement in AI infrastructure, fostering interdisciplinary collaboration and accelerating breakthroughs in both foundational research and industrial applications.

Under the support of China's National Development and Reform Commission, Peng Cheng Laboratory launched the *China Computility Grid* initiative in 2019. The project has achieved significant milestones, including the development of a core software stack and distributed scheduling platform compatible with multiple heterogeneous AI chips. It has now aggregated an impressive AI computility totaling over 2.3 exaFLOPS at half-precision floating-point accuracy. In collaboration with various institutions, Peng Cheng Laboratory has proposed the *Artificial Intelligence Computility Grid* series of standardization frameworks as well. This initiative aims to stratify different types of heterogeneous intelligent computing resources into standardized tiers, facilitating systematic encapsulation, unified data definition, and coordinated resource allocation across the network.

Sky Computing is a new concept proposed by the University of California, Berkeley. As a traditional leader in computer systems, Berkeley has consistently driven technological advancements from early supercomputing to cloud computing, big data, and the AI era. Its established labs like AMPLab and RISElab have incubated influential open-source projects such as Spark, Mesos, and Ray, which gained significant recognition in both academia and industry. Notably, the Spark system once sparked a wave of technological advancements in the big data field. In 2022, Berkeley established the Sky Computing Lab. Researchers believe that cloud computing has entered the multi-cloud era, requiring novel technical architectures to address challenges like cross-cloud collaboration and resource management. The goal is to enable users to deploy tasks without concern for specific cloud platforms, making computility truly function like utility resources such as electricity and water. Currently, the Sky Computing Lab has launched core projects including SkyPilot, Skyplane, and Basil, exploring computility grid technologies across inter-cloud coordination, resource control, and privacy protection.

4.3 How Info-Superbahn Empowers Scientific Research

Taking two typical application cases of the Info-Superbahn computility grid [10], protein sequence design and the *Shennong* Large Language Model, as examples, this demonstrates how the computility grids supports AI in empowering scientific research. By leveraging the computility grids' technologies, other researchers can easily replay such large-scale, cost-effective AI-empowered research.

Proteins are fundamental components that execute biological functions within living organisms, playing essential roles in biological processes such as catalysis, immunity, and signal transduction. Protein sequence design is generally regarded as the inverse problem of protein structure prediction. Specifically, it involves designing sequences that can fold into the target protein structure and possess the desired protein function based on a given three-dimensional protein structure. This process is a critical step in de novo protein design; once the backbone structure is generated, designing the optimal sequence becomes crucial. Protein sequence design has significant applications in fields such as drug design and enzyme engineering. Given that the number of possible protein sequences and structures exceeds the number of particles in the universe, achieving accurate and robust protein sequence design remains a major challenge. The Institute of Computing Technology, Chinese Academy of Sciences, has developed Carbon Design, a protein sequence design software capable of accurately and robustly designing protein sequences. It can be widely applied to various protein design scenarios and can predict the functional impact of protein mutations. The Info-Superbahn provides a one-stop, end-to-end support system for the CarbonDesign tool, including one-click generation from protein structure files to protein sequences and 3D structural visualization. It also enables one-click fine-tuning of models based on user-provided private data and facilitates seamless integration of multiple design software tools through a drag-and-drop interface. Meanwhile, the AI Computility Grid enables elastic scaling of GPU resources, automatically adjusting the number of service replicas according to the workload of protein sequence design tasks, thereby enhancing overall concurrency performance.

The *Shennong* Large Language Model developed by China Agricultural University can provide a wide range of functions, including agricultural knowledge Q&A, semantic understanding, text summarization, and decision reasoning. It has also made significant breakthroughs in multimodal interactions such as image, sound, video, and document processing, as well as intelligent reasoning. The model can support various agricultural application scenarios, including breeding, planting, farming, agricultural remote sensing, and meteorology. The Info-Superbahn provides full-process support for the *Shennong* Large Language Model, including model weight adjustment from the training to the inference stage, inference code management, and rapid cloud deployment of inference services.

5 Challenges and Opportunities for the Computility Grid

The research and development of the Computility Grid is still in its early stages and has not yet matured. As the scale of parameters in applications such as large AI models continue to grow, the demand for computility is increasing correspondingly. AI applications have become the primary killer application for the Computility Grid. The future development focus of the Computility Grid will mainly center on how to integrate distributed intelligent computing centers into a unified network, transforming AI computility into a fundamental infrastructure, essentially creating an AI-empowered Computility Grid.

The computility grid can provide scientific intelligent research with infinitely scalable computility supply, unified data encapsulation and sharing, as well as automation acceleration throughout the entire research process, further speeding up the entire cycle of exploration, experimentation, discovery, and innovation for scientists. However, it currently faces numerous technical challenges, some of which are listed below as scientific issues.

5.1 A Unified Abstraction and Encapsulation for Scientific Data

The sources of data for scientific research are not primarily from the Internet, but rather large-scale scientific facilities, experimental equipment, and field stations. These data often vary greatly in format, structure, and representation. Through unified abstraction and encapsulation, data can be transformed into a standardized format, enabling data from different sources to be understood and processed in a uniform manner.

Scientific data that has undergone unified abstraction and encapsulation can better meet the input requirements of various AI models, making it AI-ready. When data has a unified format and encapsulation, the model training process becomes simpler and more efficient, as researchers do not need to spend significant time and effort preprocessing data in different formats. Additionally, during data transmission, if the data is not uniformly processed, the receiving end would need to expend considerable time and computational resources on format conversion. By converting data into a standardized format, the receiving end can avoid complex format conversion operations, further enhancing the performance of cross-domain training.

5.2 64-Bit High-Precision Scientific Large Language Model

Scientific research typically demands high-precision computations, with data processing accuracy requirements reaching 64 bits, and in some cases, even 80 bits. In scientific computing, many algorithms rely on iterative methods for calculation. If the computational precision is insufficient, errors will be generated during each iteration. These errors accumulate as the number of iterations increases, and the model cannot converge eventually. For instance, in numerical weather prediction, precise simulation of atmospheric physical processes is essential. Numerical weather prediction models involve a vast array of complex computations, such as solving atmospheric dynamic equations and simulating the processes of water vapor condensation and evaporation. If the computational precision of the model is low, the predicted weather data will increasingly deviate from actual conditions over time, potentially rendering the forecast results entirely unusable. High-precision calculations ensure that the model's output is closer to the true value, thereby guaranteeing the quality of scientific research.

In the realm of large language models, text-to-image models, and other internet-based applications, the requirement for computational precision in models is not as stringent, with 16 bits, or even 8 bits and 4 bits, being sufficient. However, the AI computility grid necessitates 64-bit high-precision scientific large language models, which are integral throughout the entire process of pre-training, fine-tuning, and inference.

5.3 *A Unified Abstraction for Geographically Distributed, Independently Managed, and Heterogeneous Distributed AI Computility*

With the continuous growth in the scale of artificial intelligence models, the demand for AI computility in scientific computing has also exploded. The GPT-3 model used by the first-generation ChatGPT has a parameter scale of 170 billion, requiring 570 GB of data and tens of thousands of V100 GPUs to support its training and inference. The complex inference and training algorithms, along with the massive data processing, impose higher demands on computational resources, posing significant challenges to computing infrastructure. Traditional scientific research relied solely on one supercomputing center for high-performance computing. However, as supercomputing evolves into intelligent computing, the integration of HPC+AI necessitates the aggregation of geographically distributed, independently managed, and heterogeneous distributed intelligent computing resources into a single image. This is essential to meet the exponentially growing intelligent computing demands of scientific AI models and the diverse computing needs of individual tasks.

In the computility grid, what is transmitted is not only data information but also *task closures* which describe intelligent computing requirements and data operations. This is necessary to support the expression of diverse intelligent computing resource demands for various AI tasks. Therefore, it is essential to design a unified abstraction capable of describing the diverse intelligent computing needs of different AI applications, serving as the basic unit for task scheduling when utilizing intelligent computing resources.

The computility grid needs to logically abstract away the differences for upper-layer applications by integrating geographically distributed, independently managed, and heterogeneous intelligent computing resources. This involves enabling global naming, unified abstraction, centralized scheduling, and single metrics for intelligent computing resources. These characteristics are fundamental requirements for the infrastructuralization of computility.

5.4 *The Full-Process Automation of Scientific Intelligence for AI4S*

As shown in Fig. 2, the computility grid empowers AI4S not merely by executing large scientific models, but more crucially through deep integration of the full workflow—encompassing data collection, data cleansing, pre-training, fine-tuning, evaluation, deployment, and inference of large language models—with scientific research processes. This achieves full-process automation of scientific intelligence. In the phase of scientific problem identification and formulation, it leverages robust data processing and analytical capabilities to mine problem clues from massive datasets, transcending traditional cognitive boundaries. For expanding hypothesis exploration

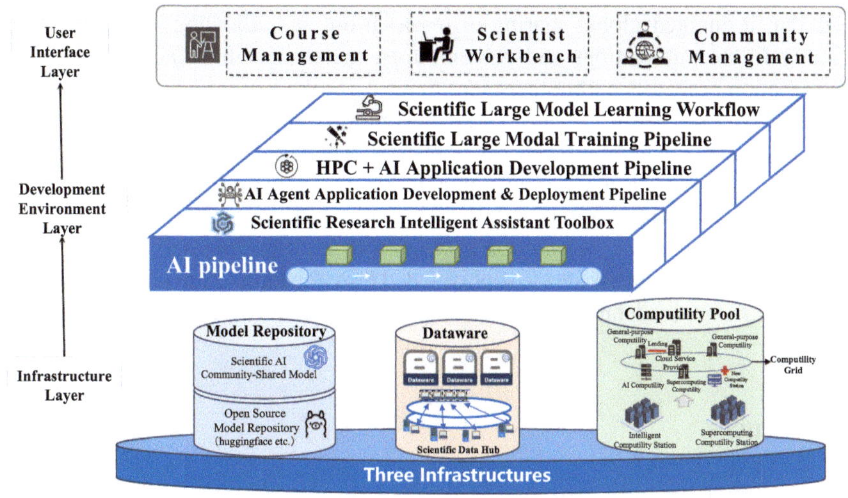

Fig. 2 The full-process support of scientific intelligence for AI4S

space, it generates diverse possibilities through model-driven approaches, broadening the dimensionality of scientific hypotheses. During novel scientific perspective generation, it synthesizes multi-source data and information to produce innovative insights. The automation of testing and validation phases enhances efficiency while minimizing human errors. When evaluating new perspectives, it provides objective assessments grounded in scientific data and AI model outputs. These interconnected phases form a closed-loop scientific research system, driving scientific endeavors toward greater efficiency and innovation.

6 Conclusions and Future Prospects

Amid intensifying global technological competition, artificial intelligence is empowering scientific research at an unprecedented pace and depth, marking a pivotal moment of profound historical significance. Nations worldwide are actively exploring strategic deployments to establish leadership in this emerging domain. The entity that first holds the core technologies and critical capability platforms of AI-empowered scientific research will dominate future scientific inquiry and steer the trajectory of scientific advancement.

The essence of informatization lies in enhancing operational efficiency—a principle equally applicable to scientific processes. Historically, information technology has driven two paradigm shifts: First, it gave rise to computational disciplines (Comp-X) such as computational physics, computational chemistry, and computational biology, providing novel methodologies for expanding the frontiers of scientific understanding. Second, through web technologies and search engines like Google

Scholar, it enabled global sharing of research outputs, allowing scientists to efficiently locate publications and collaborators, thereby establishing comprehensive information networks within the scientific community.

The AI computility grid, as a next-generation capability platform, fundamentally integrates scientific intelligence with established research paradigms, including scientific theory, experimentation, computation, and data management. This convergence propels scientific research into an era of interdisciplinary convergence characterized by *big brain*, *big facilities*, *big machines*, *big data*, and *big models*. By lowering the technical barriers to sophisticated research tools, it unveils new pathways for solving unprecedented scientific complexities that defy traditional approaches.

References

1. Guojie Li.. Intelligentized Research (AI4R): The Fifth Paradigm of Scientific Research. Bulletin of the Chinese Academy of Sciences, 2024, 39(1): 1–9.
2. Weinan E. AI Facilitates the Creation of a new Scientific Research Paradigms. Bulletin of the Chinese Academy of Sciences, 2024, 39(1): 10–16.
3. Wang, H., Fu, T., Du, Y. et al. Scientific discovery in the age of artificial intelligence. Nature 2023, 620: 47–60.
4. Guangming Tan, et al. Computing Systems for Simulation Intelligence. Bulletin of the Chinese Academy of Sciences, 2024, 39(1): 17–26.
5. Xueqi Cheng, Hong Mei, Wei Zhao, et al. Data Science and Computational Intelligence: Connotation, Paradigm, and Opportunities. Bulletin of the Chinese Academy of Sciences, 2020, 35(12): 1470–1481.
6. Xuehai Hong, Zhuoqun Xu, Wenkui Ding. A Review of Grid Computing Technology and Its Applications. Computer Science, 2003, 30(8): 1–5.
7. Zhiwei Xu, Guojie Li, Ninghui Sun. A New Type of Information Infrastructure: High-Throughput Low-Entropy Computility Grid (Info-Superbahn). Bulletin of the Chinese Academy of Sciences, 2022, 37(1): 46–52.
8. Wen Gao. Opportunities and Challenges of China's Computing Power Network. Communications of the China Computer Federation, 2023, 19(1): 31–36.
9. Chasins Sarah, Cheung Alvin, Crooks Natacha, et al. The sky above the clouds. arXiv preprint arXiv:2205.07147. 2022.
10. Xiaohong Wang, Sa Wang, Hongwei Tang, Xiaohui Peng. Building the Technological Foundation for the National Strategy of "New Infrastructure": Practices and Reflections on the Construction of the "Info-Superbahn" Comprehensive Test Field. Bulletin of the Chinese Academy of Sciences, 2021, 36(9): 1066–1073.

Ninghui Sun is an academician of the Chinese Academy of Engineering, specializing in computer architecture. As a senior researcher and doctoral supervisor, he has dedicated his career to the study of high-performance computing. He has held prestigious positions including Director of the Institute of Computing Technology, Chinese Academy of Sciences and currently serves as Chairman of the Academic Committee at ICT, CAS and President of the China Computer Federation. His major contributions include participation in the development of the Dawning-1 parallel computer and the Dawning-1000 massively parallel computer. He has led the development of the Dawning 2000~6000 series of high-performance computers, new high-throughput computers, and specialized intelligent computers. He has received the National Natural Science Foundation of China (NSFC) Distinguished Young Scholars Fund, the China Youth Science and Technology Award, the title of "China's Top Ten Outstanding Youths," the 2005 Chinese Academy of Sciences Outstanding Scientific and Technological Achievement Award, four Second Prizes of the National Science and Technology Progress Award, two Best Paper Awards from top conferences, and one National Patent Gold Award.

Sa Wang is an Associate Professor at the Institute of Computing Technology, Chinese Academy of Sciences (ICT, CAS). He was a member of the Youth Innovation Promotion Association of CAS. His research interests include cloud computing, operating systems, and performance analysis. He has received honors such as IEEE Micro's Top Picks, the Alibaba Best Collaboration Award, and the Meituan Excellence in Innovation Award.

Autonomous Scientific Knowledge Discovery Platform

Bowen Zhou, Lei Bai, Ning Ding, and Biqing Qi

Abstract This article briefly explores the evolution of scientific paradigms, tracing the transition from traditional inductive Scientific Discovery. It first provides a detailed introduction to recent advancements in the field of AI for Science, highlighting significant breakthroughs in physical science, life science, and earth science. Then, the article introduces the concept of Autonomous Scientific Knowledge Discovery Platform, detailing its system architecture across three layers: the foundational platform, the multi-agent knowledge discovery core, and the human-machine collaborative research engine. Finally, it discusses the challenges in building such platforms and emphasizes the importance of creating open, large-scale, and interdisciplinary platforms for advancing technological innovation.

Keywords Scientific paradigms · Artificial intelligence · AI for science · Autonomous scientific knowledge discovery

Human beings' exploration of the origin and evolution of all things in the universe is an inexhaustible driving force behind the progress of civilization. From Newton's formulation of the law of universal gravitation to Einstein's discovery of the theory of relativity, from Darwin's theory of evolution to Crick and Watson's discovery of the DNA double helix structure, from Faraday's discovery of the theory of electromagnetic induction to the proposal of Maxwell's equations, and from Lorenz's butterfly effect to Wallace-Broecker's discovery of global warming, the continuous exploration and understanding of the natural world has accumulated more and more scientific knowledge for mankind, and has promoted mankind to influence the world through creation and action, pushing human society into an era of rapid development. However, with the continuous deepening of the understanding of the world and the evolution of the scale of measurement tools, human beings'

B. Zhou (✉) · L. Bai · B. Qi
Shanghai Artificial Intelligence Laboratory, Shanghai, China
e-mail: zbw@pjlab.org.cn

B. Zhou · N. Ding
Tsinghua University, Beijing, China

China's e-Science Blue Book 2025, https://doi.org/10.1007/978-981-96-8252-2_7

exploration and understanding of the world are gradually developing in the direction of "expanding to the extreme macro, deepening to the extreme micro, moving towards extreme conditions, and focusing on extreme interdisciplinary integration", which brings huge challenges of data volume explosion and combination relationship explosion.

The pace of global scientific breakthroughs has slowed down, urgently necessitating the formation of new research paradigms. Driven by the wave of digitalization, information intelligence technology is transforming the world at an unprecedented speed. However, a 2023 *Nature* cover article reported that while the volume of scientific publications and patents has surged dramatically, the scientific breakthroughs are slowing down [1]. One of the fundamental reasons for this slowing pace of breakthroughs is the "information cocoon" phenomenon that has emerged through scientific development. With the continuous accumulation and professional segmentation of scientific knowledge, researchers tend to focus narrowly on their own sub-fields, thus forming relatively closed "information cocoons" with information overload and high barriers between cocoons. In these cocoons, scientists struggle to keep pace with the latest progress in their own fields, and it is even more difficult to pay attention to the relevance and potential cross-innovation opportunities in other fields. Therefore, the exchange and integration of scientific knowledge is greatly restricted, and the speed of breakthrough innovation is significantly reduced. In addition, there is also the problem of information overload in the cocoon. Various disciplines are rapidly generating a large number of research papers and data, resulting in information overload, which makes it difficult for scientists to absorb all relevant literature and research development, and to a certain extent limits their ability to acquire new knowledge and innovation.

Artificial intelligence for science is accelerating the process of scientific innovation and breakthroughs. In this context, artificial intelligence (AI) has become a powerful tool for processing scientific data, accelerating scientific computing, and even assisting scientists in forming hypothesis and building automated workflow. The integration of AI with fundamental science is increasingly becoming a shared goal for scientists in both fields. The 2024 Nobel Prize in Physics was awarded to John Hopfield and Geoffrey Hinton, pioneers in machine learning and neural networks, in recognition of their fundamental contributions to the field of machine learning. The 2024 Nobel Prize in Chemistry was awarded to David Baker, Demis Hassabis, and John M. Jumper for their groundbreaking work in leveraging AI to advance protein structure prediction. It can be seen that AI is widely combined with the physical world and exerting profound impacts on discoveries in fundamental science.

This article first provides an overview of the inception and development trajectory of the "AI for Science (AI4S)" concept, then introduces the concept and system composition of the scientific knowledge autonomous discovery platform, and finally looks forward to and summarizes future development.

1 Proposal and Development of AI4S Concept

Early scientific research activities can be traced back to the sixth century BC in ancient Greece, where thinkers and scientists such as Aristotle and Euclid made important pioneering contributions. The foundation of modern scientific research emerged during the sixteenth–seventeenth century Scientific Revolution, with figures like Galileo and Newton recognized as the founders of modern scientific inquiry. For centuries leading up to the mid-twentieth century, scientific methodologies primarily included experimental studies based on observation and induction, as well as theoretical research based on scientific hypotheses and logical deduction. With the widespread adoption of electronic computers in the mid-to-late twentieth century, computational calculations, simulations, and modeling gradually became essential tools in scientific research [2]. Over the past two decades, the advancement of scientific research and the popularization of scientific instruments have led to a surge in large-scale data, propelling data-driven scientific research to the forefront and catalyzing its rise. As a digital representation of real-world entities, phenomena, and behaviors, data encapsulates the underlying laws governing the physical world. Analytical techniques that uncover hidden correlations within data have expanded new perspectives for scientific exploration [3]. However, data-driven scientific research itself cannot derive the fundamental principles governing these correlations, nor can it transcend interdisciplinary information barriers, thereby limiting its capacity to reveal essential truths. Furthermore, as accessible and usable big data continues to grow, challenges such as uncertainty, complexity, the curse of dimensionality, and scale limitations demand more advanced tools. Scientists urgently need a new scientific research paradigm that is closer to the essence of data and intelligence and more effectively understands complexity and uncertainty.

In this context, AI—a technology adept at learning high-dimensional representations and uncovering hidden patterns from massive datasets—has gradually entered scientists' vision. Many scholars have begun advocating for an AI-driven research paradigm known as "AI for Science" (AI4S). Professor Weinan E, an academician of the Chinese Academy of Sciences, pointed out in his report "*AI for Science: An Ongoing Technological Revolution*" that AI4S aims to address the challenges of data scarcity and the lack of analytical tools in data-driven methodologies through deep learning. It will propel scientific research from a "cottage industry" model to a "platform-driven" paradigm, solving common challenges across diverse research fields. Academician Guojie Li from the Institute of Computing Technology of the Chinese Academy of Sciences believes that the characteristics of the fifth paradigm are mainly reflected in the following six aspects: "(1) AI fully permeates the fields of science, technology and engineering research, realizing the automated processing of knowledge and the intelligence of the entire scientific research process; (2) Human-machine integration, where machine intelligence becomes central to research, incorporating implicit knowledge and machine speculation; (3) Focus on complex system research, effectively addressing the combinatorial explosion problem associated with high computational complexity; (4) Emphasis on non-deterministic challenges, with probabilistic and statistical reasoning playing enhanced roles in scientific research; (5) Interdisciplinary cooperation becomes the dominant mode of scientific research cooperation, realizing the integration of first-principles models and data-driven; (6) Scientific research relies more on platforms with large models as the core, and scientific research is closely linked to engineering practice" [2].

Over the course of AI4S development in recent years, a notable advancement has been the emergence of vertical domain-specific models tailored for scientific and engineering data (as shown in Table 1). These models are designed and optimized to address specialized datasets in targeted fields, enabling tasks such as prediction, reconstruction, and classification. By significantly enhancing performance or efficiency of specific stages in existing research workflows, these specialized models have achieved remarkable progress (see Table 1 for some of the results). In the field of material science, domain-specific models like DeePMD [4], GPIP [5], and GNoME [6] have greatly improved the efficiency from molecular dynamics simulation to crystal structure design. Leveraging AI technologies and high-throughput computing, these deep learning-based approaches overcome limitations in scale and computational simulation, accelerating material screening processes, enabling targeted material design, and optimizing the full-chain development of materials and devices. These progress provide a robust material foundation for industries such as renewable energy and information technology. In the field of Earth sciences, specialized models such as Pangu [7], GraphCast [8], and Fengwu [9] use AI technologies to model large amounts of meteorological reanalysis data, identify complex atmospheric patterns, improve the accuracy and efficiency of weather forecasts, enhance the ability to predict extreme weather events such as typhoons and rainstorms, and further assist in climate change research [10]. The life sciences have witnessed groundbreaking achievements with AlphaFold [11], an AI system that tackles protein structure prediction with atomic-scale precision, reducing computational time from years to minutes. Other models such as RoseTTAFold [12], SCUBA [13], and Evo [14] have further advanced the field, enabling novel disease target discovery and accelerating drug design. In electromagnetic simulation, AI models can quickly simulate and optimize complex electromagnetic field distributions, improving simulation and design efficiency. For instance, DeepMind's 2022 work published in Nature demonstrated the magnetic control of tokamak plasma using deep reinforcement learning [15]. In addition, AI technology has also achieved a series of representative results in electromagnetic compatibility analysis, astronomical detection, fluid simulation, industrial design, etc.

While the aforementioned advancements have generated a significant academic impact, we must also recognize that in current AI4S research, AI technology and scientific research remain an orthogonal relationship, that is, AI technology plays an instrumental role in scientific research. This orthogonality parallels how computing technology accelerated domain-specific calculations and simulations, or how internet technology revolutionized knowledge dissemination and retrieval. In other words, AI has yet to achieve deep integration with scientific knowledge discovery, representing an revolutionary leap in research tools rather than a systematic transformation of scientific paradigms. Nevertheless, as AI technology continues to advance, its future

Table 1 Examples of domain-specific models in AI for science

Domain	Model name	Function	Institution
Material science	DeePMD	Machine learning methods for molecular dynamics simulations in materials and chemical systems. Artificial neural networks are used to learn the potential energy surface of the system, describing the relationship between atomic positions and the system's potential energy	DP Technology
	GNoME	The materials exploration graph network GNoME has discovered up to 2.2 million theoretically stable crystalline structures, the vast majority of which have not yet been experimentally realized	Google
	MatterGen	A generative large model for inorganic materials design. MatterGen is a diffusion model specifically engineered to generate novel and stable materials	Microsoft Research
	MatterSim	Capable of accurately and efficiently simulating materials and predicting their performance across a wide range of elements, temperatures, and pressures. It predicts atomic-level energy and stress in materials while reducing data requirements by 90–97%	Microsoft Research
	Darwin	A specialized language model for physics, chemistry, and materials science applications, leveraging open-source scientific FAIR datasets and scientific literature datasets. By learning from extensive molecular database structures, it provides viable compounds and their properties, accelerating the discovery and development of novel materials	University of New South Wales, etc.
	GPIP	A geometric learning framework for MLIPs, GPIP, which utilizes unlabeled configurations to enhance the performance of MLIPs	Shanghai Artificial Intelligence Laboratory, etc.
Earth science	FourCastNet	The first AI model to achieve high-resolution (0.25°) global medium-term weather forecasts	NVIDIA
	PanGu	The first high-resolution (0.25°) AI medium-range weather forecast model with accuracy comparable to IFS (European Centre for Medium-Range Weather Forecasts Integrated Forecast System)	Huawei

(continued)

Table 1 (continued)

Domain	Model name	Function	Institution
	GraphCast	AI medium-term weather forecast model predicts the next 10 days in 60 s, and 90% of the forecast content exceeds existing human weather forecasts	Google
	FengWu	The first global medium-range weather forecast model built on multimodal and multi-task deep learning methods with an effective forecast period of more than 10 days and a modeling resolution exceeding 10 km	Shanghai Artificial Intelligence Laboratory, etc.
	Aurora	The first large-scale climate model to predict and mitigate extreme weather events	Microsoft Research
Life science	AlphaFold 1–3 series	Protein structure prediction model, the latest AlphaFold3, can predict the structure and interaction of all life molecules with unprecedented accuracy	Google
	Evo	A genomic foundation model capable of multimodal and multiscale learning, which performs prediction and generation tasks spanning molecular to genome-wide scales. Trained on whole-genome data, it generalizes the three fundamental modalities of the central dogma of molecular biology and excels at multi-element generation tasks	Stanford University
	RoseTTAFold	A protein structure prediction model based on deep learning that can process protein sequence and structure information and generate high-precision protein three-dimensional structure predictions	University of Washington
	scGPT	Foundational models designed for single-cell transcriptomics, chromatin accessibility, and protein abundance	University of Toronto
	HyenaDNA	A foundational model for raw DNA sequences	Stanford University

impact on scientific innovation is poised to grow exponentially, potentially emerging as a catalyst for revolutionary breakthroughs in scientific exploration.

2 Concept of Autonomous Scientific Knowledge Discovery Platform

While domain-specific AI models for scientific research are flourishing, breakthroughs in general artificial intelligence continue to emerge. OpenAI's recently released o1 system has significantly enhanced the reasoning capabilities of foundational language models, achieving state-of-the-art performance in text comprehension tasks across physics, chemistry, and life sciences, exerting substantial impact on both industry and academia. As researchers deepen their exploration of foundational language models and domain-specific scientific language models, their potential for scientific discovery is being progressively uncovered [16–19], attracting global attention. Recent work by a joint team of Tsinghua University and the Shanghai Artificial Intelligence Laboratory has found that foundation language models have the potential to become an important tool for knowledge interaction and discovery, and through systematic evaluation, they provided the first verification of the ability of foundation language models as scientific hypothesis generators in the biomedical field [20]. Through a large number of experiments and questionnaire evaluations, researchers at Stanford University found that foundation language models can propose better scientific research ideas than human experts [21]. The SciAgents [22] multi-agent framework proposed by researchers at Massachusetts Institute of Technology attempts to generate and optimize cross-disciplinary scientific hypotheses based on knowledge graphs. Coscientist, an AI system developed by Carnegie Mellon University, can autonomously complete the entire process from information retrieval to experimental execution and data analysis [23]. The system successfully designed and synthesized common drug molecules such as aspirin, acetaminophen and ibuprofen; it designed an experimental scheme for palladium-catalyzed cross-coupling reactions in less than four minutes and successfully reproduced this important and complex chemical reaction. Researchers at the University of Oxford proposed the concept of "AI Scientist" [24], the core of which is a system for idea generation, experimental iteration, and paper writing driven by a foundation language model. In the field of algorithm discovery, researchers from Stanford University proposed MLAgentBench [25] to promote autonomous algorithm design and improvement for specific goals.

These exploratory works have initially demonstrated the potential of deeply integrating of foundation language models and domain tools, bringing new possibilities for scientific knowledge discovery. However, such works remain in prototype stages or are confined to narrow subfields. Against this backdrop, building open, large-scale, interdisciplinary autonomous scientific knowledge discovery platforms has emerged as a transformative approach to merging AI with scientific research. Unlike AI applications that accelerate specific computational tasks, scientific knowledge discovery and formalization constitute a multi-stage iterative reasoning process. This process encompasses multiple stages, starting with observing and perceiving data (phenomena), then combining existing knowledge for deep reasoning to make a professional-level understanding of scientific knowledge and propose scientific hypotheses, then from scientific hypotheses to designing computational simulations or wet-lab experimental plans, then from experimental results analysis to optimizing scientific hypotheses and experimental processes, and finally to practical application. Throughout this cycle, scientists employ diverse methodologies and technologies to analyze observed phenomena, propose explanatory hypotheses/theories, and iteratively refine them through validation, ultimately forming widely accepted scientific knowledge. A fully autonomous discovery platform, therefore, must automate and intelligently execute this entire workflow, enabling systemic integration of human-like reasoning, experimental iteration, and knowledge synthesis.

From the perspective of scientific research evolution, interdisciplinary integration has become a critical driver of scientific and technological innovation, while the isolated development of single disciplines increasingly appears outdated. In 1900, there were about 500 disciplines, and by 2000, this number had grown to approximately 5,000, a tenfold increase [2]. Meanwhile, knowledge barriers between disciplines have intensified, making it exceedingly difficult for researchers to master expertise across multiple fields and slowing progress in cross-disciplinary collaboration. Foundation language models naturally have the ability to learn existing literature and materials across various disciplines, so they are expected to discover difficult problems in the scientific field and propose innovative solutions through interdisciplinary knowledge, thus breaking through the knowledge cocoon. In addition, the current scientific research process has increasingly higher requirements for researchers' professionalism, theoretical foundation, and practice skills, so the most cutting-edge research often requires a long period of talent training. The research process itself also faces additional challenges, such as high uncertainty and protracted experimental timelines, hindering the achievement of scalability. For instance, in Earth system modeling, advancing global weather forecasting accuracy by a single day demands a decade-long collaboration involving hundreds of top atmospheric physicists and engineers. By integrating scientific literature, datasets, foundation language models, domain-specific models, embodied intelligence, and research tools into autonomous scientific knowledge discovery platforms, the efficiency of scientific research and knowledge discovery can be accelerated in many ways, such as reducing human errors, achieving more accurate and reproducible experimental management, and enabling more efficient parallel scientific research to achieve economies of scale. This paradigm shift—decoupling humans from labor-intensive exploration—promises to advance research intelligence toward unprecedented automation and efficiency. It may fundamentally disrupt traditional scientific methodologies, catalyzing transformative breakthroughs and laying the groundwork for accelerated progress across scientific frontiers.

From the perspective of AI development, scientific knowledge discovery is a typical and complex scenario and task where human intelligence is highly concentrated, serving as a core benchmark for evaluating Artificial General Intelligence (AGI). AGI, as the ultimate goal of AI research, is measured not only by its efficiency and precision in executing specific tasks (such as image recognition, language translation, or prediction) but more importantly by its ability to emulate human-like complex logical reasoning, continuous learning [26], innovative thinking, and flexible adaptation across broad knowledge domains and dynamic environments. Using scientific knowledge discovery as the core scenario for verifying the capabilities of AGI is not only of great significance in terms of providing indicators and guidance for the development of AI technology, but also has immense scientific value and future-oriented industrial potential. Furthermore, the spirit of exploring the unknown and daring to innovate reflected in the process of scientific knowledge discovery is also the core value that AGI should pursue. A genuinely capable AGI system should autonomously perform literature review, hypothesis generation, experimental design, optimization, verification, and result analysis, ultimately producing original theories or discoveries within specialized fields—a process that inherently mirrors the essence of scientific research. Consequently, developing autonomous scientific discovery platforms could not only accelerate progress in natural sciences but also serve as a milestone scenario to validate AGI's competencies in domain knowledge comprehension, specialized data modeling, and complex problem-solving.

3 System Architecture of the Autonomous Scientific Knowledge Discovery Platform

Although exploratory efforts have been made in developing autonomous scientific knowledge discovery platforms, achieving paradigm-shifting breakthroughs remains imperative to further accelerate scientific innovation. Current challenges persist in constructing systematic, integrated, interdisciplinary, and open-access autonomous scientific discovery platforms. Combining the scientific knowledge exploration and discovery process and the technological progress of AI, this article proposes a conceptual system architecture for such platforms, comprising three components: the autonomous scientific knowledge discovery platform base, the multi-agent knowledge discovery brain, and the human–machine collaborative research engine (as shown in Fig. 1).

3.1 Autonomous Scientific Knowledge Discovery Platform Base

The platform's foundational base supplies data, computing power, and tools, including an interdisciplinary literature library, cross-domain databases, domain-specific models, specialized toolkits, and computational infrastructure. In addition, for disciplines (such as chemistry and materials) that rely on wet-lab experiments to verify ideas, the base can also be equipped with a robotic laboratory platform where intelligent robots can perform laboratory operations.

The interdisciplinary literature library encompasses various types of scientific documents stored in textual formats across disciplines, including papers, patents, reports, and web data. For example, the arXiv platform has accumulated over 520,000

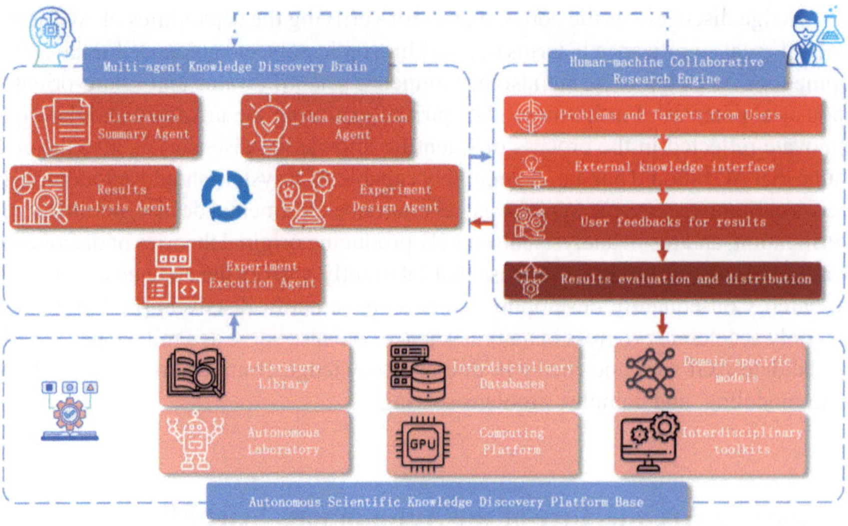

Fig. 1 Architecture of the autonomous scientific knowledge discovery platform

published and unpublished preprints, which encapsulate the breadth of human-explored knowledge. A central challenge in constructing such repositories lies in automatically detecting contradictions within the literature and identifying erroneous knowledge, thereby preventing the autonomous scientific discovery process from being compromised by misinformation.

The interdisciplinary database serves to store various types of unstructured scientific data generated from scientific research and industrial R&D activities, which describe objective phenomena and system states, encompassing raw observational data, experimental results, experimental records, survey data, and simulation data. Compared to textual data and conventional industry data, professional scientific data are significantly larger and more complex. First, their sources and types are more diverse, originating from various instruments, sensors, simulations, and other means, resulting in substantial differences in formats and semantics. Second, scientific data demand exceptionally high quality, requiring objectivity, accuracy, completeness, and high resolution, as these attributes directly determine the credibility of computational results. Third, the volume of scientific research data is huge, such as experimental observation data and simulation data. They are usually massive in scale and grow rapidly, far exceeding general commercial data. China's Five-hundred-meter Aperture Spherical radio Telescope (FAST) generates up to 500 TB of raw data daily [27], while the European Centre for Medium-Range Weather Forecasts (ECMWF) has already stored over 500 PB of raw observational data, with daily additions of approximately 300 TB. Finally, scientific data exhibit richer dimensionality compared to other data types. Examples include meteorological, geographical, and biological data, which involve multiple dimensions such as time, space, and species, leading to more complex structures. Consequently, the construction of interdisciplinary databases necessitates the development of unified storage formats for cross-domain scientific data, data cleaning methodologies, and scientific data quality enhancement techniques (e.g., data imputation) to improve the quality and usability of scientific databases.

Domain-specific models and toolkits are key components for achieving automated scientific research. These tools and model libraries provide the necessary infrastructure and algorithmic frameworks for autonomous scientific discovery platforms to execute diverse research tasks. Domain-specific vertical models encompass pre-trained models tailored to specialized research fields (e.g., physics, biology, chemistry, atmospheric sciences—see Table 1), which accelerate scientific workflows and deliver domain-specific insights, thereby supporting the platform's capacity for innovation within targeted disciplines. Furthermore, interdisciplinary toolkits integrate data mining utilities, statistical analysis packages, machine learning frameworks, and visualization tools capable of processing and analyzing complex datasets from multiple domains. By integrating these advanced models and tools, autonomous scientific discovery platforms gain the capability to independently design experimental, analyze data, and generate knowledge without human intervention.

3.2 *Multi-agent Knowledge Discovery Brain*

The multi-agent knowledge discovery brain consists of five key parts: the literature summarization agent, scientific research idea generation agent, experimental plan design agent, scientific experiment agent, and result analysis and idea optimization agent.

Accurate scientific literature summarization and related information extraction capabilities are the basis of scientific knowledge discovery and should also be one of the key capabilities of scientific knowledge discovery platforms. This capability involves the automatic reading, comprehension, and summarization of a large amount of scientific literature. By leveraging natural language processing (NLP) and machine learning models, such platforms can efficiently filter, categorize, and summarize thousands of research papers, distilling key findings and track emerging trends. Current foundation language models, such as AutoSurvey, which addresses the complexity of scientific papers through a two-stage generation process, demonstrate preliminary success in this domain. In addition, the platform needs to identify the connections between documents and build a structured knowledge network to connect knowledge points scattered in different disciplines and reveal the potential connections and interactions between them. By integrating literature summarization and knowledge structuring, the platform can explore new scientific research problems within a broader knowledge context, fostering interdisciplinary innovation. Structured knowledge can also provide an efficient query and retrieval tool for scientific research idea design and iteration, enabling it to quickly find the required information and data in the scientific research process. Future research may focus on more effective methods for discovering cross-disciplinary connections and how to use these connections to inspire new research directions and innovative ideas.

The proposal of innovative scientific ideas or hypotheses is the key to scientific research and knowledge discovery. By analyzing existing literature libraries, the platform can propose more interdisciplinary scientific research ideas. For example, AI can combine knowledge of biology and materials science to design new biomaterials [22]. This interdisciplinary capability will promote new scientific breakthroughs and help solve complex global problems such as climate change and energy crises. Currently, with the improvement of sensor technology and data collection capabilities, the amount of data in various disciplines is growing exponentially. AI can already automatically process and analyze massive amounts of data. Combined with the interdisciplinary scientific knowledge contained in large models, the autonomous scientific knowledge discovery platform has the potential to discover phenomena that are difficult for humans to perceive, and extract valuable patterns and hidden laws from them, thereby proposing better scientific hypotheses.

After scientific research ideas or hypotheses are proposed, the multi-agent knowledge discovery brain uses the ability of interdisciplinary knowledge maps and literature summarization to autonomously propose innovative experimental designs. By deeply analyzing the potential connections between different disciplines, the platform can design forward-looking experiments to verify new scientific hypotheses. In contrast, human scientists typically depend on manually synthesized literature reviews to attempt innovative proposals when designing experiments. However, the inherent limitations of human information processing capacity may constrain experimental methodologies to researchers' personal expertise and experience. Consequently, the traditional experimental design process is not only time-consuming and labor-intensive, but also susceptible to cognitive blind spots. In comparison, autonomous scientific experimental design not only improves the efficiency and innovation of experimental design, but also holds the potential to provide new ideas and methods for solving complex scientific problems.

The scientific experiment agent is responsible for conducting simulations or generating control instructions for the robot experimental platform according to the experimental plan. In theoretical disciplines such as computer science, physics, and mathematics, the scientific experiment agent generates corresponding simulation verification codes to systematically verify the designed theoretical plan. This automated process not only improves the accuracy of experiments, but also accelerates the efficiency of theoretical verification. For experimental disciplines such as chemistry, biology, and materials science, the scientific experiment agent can guide robotic platforms to perform wet-lab experiments with precision by executing detailed experimental protocols, accurately carrying out preparation and characterization processes. The work of Professor Jun Jiang's team at the University of Science and Technology of China shows that through multi-robot collaboration, the execution time of chemical experiments can be reduced by more than 40% [28]. Through this efficient automated control, experimental repeatability and reliability can be greatly improved. In addition, the fully automated experimental process significantly improves the progress of scientific research, reduces the errors that may occur in human operation, and at the same time maximizes the safety of experimental personnel while minimizing the risk of direct contact with hazardous materials. This multi-agent knowledge discovery brain provides an innovative method for scientific research and will greatly promote experimental research in various disciplines.

After obtaining the experimental results, the multi-agent knowledge discovery brain can quickly and accurately analyze the experimental data through highly automated data processing and pattern recognition technologies. It then filters research hypotheses based on experimental results, automatically analyzing data and evaluating results to eliminate low-success-rate proposals while providing optimization recommendations for high-potential hypotheses. Concurrently, it enhances experimental designs through long-range feedback mechanisms, iteratively refining scientific assumptions to ensure both precision and innovation in research directions. In this process, since the entire scientific discovery process involves multiple stages, and the success rate of scientific hypotheses is inherently low, a key challenge remains: how to perform long-term optimization based on sparse positive feedback on the basis of existing memory mechanisms, reinforcement learning, and model integration.

3.3 Human-Machine Collaborative Research Engine

Ensuring the openness and security of the scientific knowledge autonomous discovery platform is a necessary condition for accelerating scientific progress and validating their performance. By building a human-machine collaborative research engine, researchers from all disciplines and industries around the world can propose their own research questions and goals, and leveraging the platform's fully automated experimental workflow to systematically verify these hypotheses. At the same time, for the lack of literature and data on the platform, users can also choose their own literature and data that are most relevant to the research questions and goals to supplement the platform's knowledge base, forming a crowdsourced literature and data sharing mechanism. In the process of scientific experiment design and optimization, researchers can provide personal experimental feedback through the interactive interface, forming a human-in-the-loop laboratory design and iteration method, which better guides the platform to continue to explore in depth along the research direction required by researchers, or terminate the exploration in time when negative experimental results continue to be generated, so as to avoid waste of resources. Throughout this process, personalized research support can also be provided according to personal research interests and styles. By analyzing the researcher's work habits, research topics and historical data, customized experimental plans and data analysis tools can be designed, making scientific research more efficient and creative. Finally, the platform, under researcher's supervision, can also enable the autonomous writing, evaluation and sharing of scientific papers through the researcher interaction interface, thereby promoting academic exchange and cooperation, and driving the continuous advancement of scientific knowledge.

4 Prospect and Conclusion

The paradigm of scientific research and knowledge discovery has been constantly evolving with the deepening of human understanding of the world and the development of technology. With the advancement of AI technology, the integration of scientific research and AI technology is becoming increasingly profound, significantly affecting and reshaping the paradigm of scientific research and industrial R&D. At the same time, universities, scientific research institutions and enterprises are actively exploring interdisciplinary research and cooperation, promoting the deep integration of AI and scientific research, and gradually forming a diversified industrial ecosystem covering multiple disciplines, multiple fields and multiple application scenarios. In the process of deep integration of AI and scientific research, the construction of an autonomous scientific knowledge discovery platform based on foundation language models, scientific domain-specific models and embodied intelligence is expected to become one of the most important directions in the future. By automating workflows such as literature summarization, hypothesis generation, autonomous experimentation, and iterative hypothesis refinement, the autonomous scientific knowledge discovery platform is expected to go beyond mere enhancements in computational and experimental capabilities. Instead, it may catalyze the emergence of autonomous knowledge discovery, creating a self-reinforcing cycle: breakthroughs in AI-driven discovery feed back into improved computational and experimental tools, which in turn accelerate further discoveries. Such a paradigm could ignite an explosive growth in human knowledge, propelling revolutionary advances across fields like biopharmaceuticals, materials science, energy transformation, and climate response.

At the same time, we must also recognize that building a large-scale interdisciplinary autonomous scientific knowledge discovery platform still presents significant challenges, mainly in the following aspects:

(1) Alignment of scientific professional data with language data. The proposal of cutting-edge scientific ideas and hypotheses often requires in-depth analysis of scientific data in conjunction with existing knowledge. Although many domain-specific scientific models and vertical language models have been developed, the technical pathways for effectively aligning domain-specific scientific models with vertical language models remain unclear. For example, aligning current foundation language model with life science data (protein, DNA, RNA, etc.) and earth science data (atmospheric and ocean observation data, astronomical observation data, etc.) remains a major challenge, especially in fine-grained alignment. It is necessary to resolve the differences between data formats, dimensions, and semantic representations to ensure both accuracy and reliability of the data. Addressing these issues requires collaborative efforts among experts from interdisciplinary fields to develop technologies and methods that can effectively integrate different types of data.

(2) Interdisciplinary scientific idea generation. Although current research indicates that foundational language models have the potential to propose more creative scientific hypotheses, systematic analysis remains largely confined to the field

of computer science, with findings suggesting that the practical feasibility and diversity of ideas generated by these models still require improvement. Furthermore, producing innovative yet actionable interdisciplinary scientific concepts requires advanced and complex reasoning capabilities. Developing such abilities demands the integration of cutting-edge artificial intelligence with domain-specific expertise to promote interdisciplinary thinking and innovation.

(3) Feedback-based optimization methods for scientific ideas: While large models can generate numerous innovative scientific ideas at minimal cost, the practicality and feasibility of these ideas may still be constrained. Given the high costs associated with scientific simulations or experiments, establishing effective feedback mechanisms that integrate generated scientific ideas with real-world data, domain expert evaluations, and existing research outcomes is an urgent research priority. This involves creating continuous experimental feedback loops to iteratively validate, refine, and enhance these ideas, thereby progressively enhancing their quality and feasibility.

(4) The scale of data and engineering efforts is huge, requiring close collaboration among scientific research institutions, enterprises, and universities to share data resources, technical experience, and talents. In addition, it may be necessary to organize and implement a large-scale scientific initiative "oriented to the independent discovery of scientific knowledge" to coordinate and manage research work in different fields and ensure the long-term sustainability and efficient operation of the platform.

(5) Supervision of scientific norms and ethics. As the foundation of human social progress, ensuring the quality of scientific knowledge discovery, compliance with scientific research norms, and compliance with basic human values are necessary requirements for the autonomous scientific knowledge discovery platform. Therefore, it is necessary to explore the role of human researchers in the platform (such as research goal definers, research results reviewers, etc.) to ensure that the knowledge discovered conforms to scientific research norms and human values.

Acknowledgements During the writing process of this article, we received great support from Dr. Ben Fei of the Chinese University of Hong Kong in data collection, drawing, etc., and we would like to express our gratitude.

References

1. Park M, Leahey E, Funk RJ. Papers and patents are becoming less disruptive over time. Nature. 2023 Jan 5;613(7942):138–44.
2. LI Guojie. AI4R: The fifth scientific research paradigm[J]. Bulletin of Chinese Academy of Sciences, 2024, 39(1):1–9.
3. Lu X, Huo F, Wang Z, et al. The Methods of Information Analysis in Data Intelligence Era: Data Driven, Knowledge Driven, and Fusion Driven by Data and Knowledge. Journal of Library Science in China, 2024, 50(1):29–44.

4. Wang H, Zhang L, Han J, et al. DeePMD-kit: A deep learning package for many-body potential energy representation and molecular dynamics[J]. Computer Physics Communications, 2018, 228: 178–184.
5. Cui T, Tang C, Su M, et al. Geometry-enhanced pretraining on interatomic potentials[J]. Nature Machine Intelligence, 2024, 6(4): 428–436.
6. Merchant A, Batzner S, Schoenholz S S, et al. Scaling deep learning for materials discovery[J]. Nature, 2023, 624(7990): 80–85.
7. Bi K, Xie L, Zhang H, et al. Accurate medium-range global weather forecasting with 3D neural networks[J]. Nature, 2023, 619(7970): 533–538.
8. Lam R, Sanchez-Gonzalez A, Willson M, et al. Learning skillful medium-range global weather forecasting[J]. Science, 2023, 382(6677): 1416–1421.
9. Chen K, Han T, Gong J, et al. Fengwu: Pushing the skillful global medium-range weather forecast beyond 10 days lead. arXiv:2304.02948, 2023.
10. Guo Z, Lyu P, Ling F, Luo JJ, Boers N, Ouyang W, Bai L. ORCA: A Global Ocean Emulator for Multi-year to Decadal Predictions. arXiv:2405.15412. 2024.
11. Jumper J, Evans R, Pritzel A, et al. Highly accurate protein structure prediction with AlphaFold[J]. nature, 2021, 596(7873): 583–589.
12. Krishna R, Wang J, Ahern W, et al. Generalized biomolecular modeling and design with RoseTTAFold All-Atom[J]. Science, 2024, 384(6693): eadl2528.
13. Huang B, Xu Y, Hu X, et al. A backbone-centred energy function of neural networks for protein design[J]. Nature, 2022, 602(7897): 523–528.
14. Nguyen E, Poli M, Durrant M G, et al. Sequence modeling and design from molecular to genome scale with Evo[J]. BioRxiv, 2024: 2024.02. 27.582234.
15. Degrave J, Felici F, Buchli J, et al. Magnetic control of tokamak plasmas through deep reinforcement learning[J]. Nature, 2022, 602(7897): 414–419.
16. Zheng, Yuxiang, Shichao Sun, Lin Qiu, Dongyu Ru, Cheng Jiayang, Xuefeng Li, Jifan Lin et al. OpenResearcher: Unleashing AI for Accelerated Scientific Research. arXiv:2408.06941, 2024.
17. Wang, Yidong, Qi Guo, Wenjin Yao, Hongbo Zhang, Xin Zhang, Zhen Wu, Meishan Zhang et al. AutoSurvey: Large Language Models Can Automatically Write Surveys. arXiv:2406.10252, 2024.
18. Zhang, Kaiyan, Sihang Zeng, Ermo Hua, Ning Ding, Zhang-Ren Chen, Zhiyuan Ma, Haoxin Li et al. Ultramedical: Building specialized generalists in biomedicine. arXiv:2406.03949, 2024.
19. Zhang, Di, Wei Liu, Qian Tan, Jingdan Chen, Hang Yan, Yuliang Yan, Jiatong Li et al. Chemllm: A chemical large language model. arXiv:2402.06852, 2024.
20. Qi, Biqing, Kaiyan Zhang, Kai Tian, Haoxiang Li, Zhang-Ren Chen, Sihang Zeng, Ermo Hua, Hu Jinfang, and Bowen Zhou. Large Language Models as Biomedical Hypothesis Generators: A Comprehensive Evaluation. arXiv:2407.08940, 2024.
21. Si, Chenglei, Diyi Yang, and Tatsunori Hashimoto. Can llms generate novel research ideas? a large-scale human study with 100+ nlp researchers. arXiv:2409.04109, 2024.
22. Ghafarollahi, Alireza, and Markus J. Buehler. SciAgents: Automating scientific discovery through multi-agent intelligent graph reasoning. arXiv:2409.05556, 2024.
23. Boiko D A, MacKnight R, Kline B, et al. Autonomous chemical research with large language models[J]. Nature, 2023, 624(7992): 570–578.
24. Lu C, Lu C, Lange R T, et al. The AI scientist: Towards fully automated open-ended scientific discovery. arXiv:2408.06292, 2024.
25. Huang, Qian, Jian Vora, Percy Liang, and Jure Leskovec. "MLAgentBench: Evaluating Language Agents on Machine Learning Experimentation." Forty-first International Conference on Machine Learning, 2024.
26. Qi, Biqing, Xinquan Chen, Junqi Gao,et al. "Interactive continual learning: Fast and slow thinking." *Proceedings of the IEEE/CVF Conference on Computer Vision and Pattern Recognition*, pp. 12882–12892. 2024.
27. Xu Y, Wang F, An Z, Wang Q, Zhang Z. Artificial intelligence for science—bridging data to wisdom. The Innovation. 2023 Nov 13;4(6).

28. Zhou J, Luo M, Chen L, Zhu Q, Jiang S, Zhang F, Shang W, Jiang J. A multi-robot-multi-task scheduling system for autonomous chemistry laboratories. ChemRxiv. 2024.

Bowen Zhou is the Director and Chief Scientist of Shanghai Artificial Intelligence Laboratory, Huiyan Chair Professor at Tsinghua University, and the Fellow of IEEE/CAAI. Formerly, he served as the Dean of IBM Research Artificial Intelligence Institute, Chief Scientist of IBM Watson Group, Senior Vice President of JD Group, Chairman of the Group Technical Committee, and the President of Cloud and AI. Professor Zhou has rich experience in international scientific research and management and has long been engaged in research on cutting-edge international theories, technological innovation, talent cultivation, and large-scale industrial applications of artificial intelligence. He has not only promoted the integration and linkage between the industry and academia, but also achieved seamless connection and efficient collaboration in the international field. He has won the highest award in the intelligence field of China, such as the "Wu Wenjun Outstanding Contribution Award for Artificial Intelligence". Prof. Zhou has published hundreds of papers in top international journals and academic conferences, with tens of thousands of citations, including several pioneering papers that have been cited thousands of times. He has achieved outstanding achievements in the core fields of artificial intelligence technology and large-scale industrial applications, and has a high international influence.

Lei Bai is a research scientist at the Shanghai Artificial Intelligence Laboratory, where he leads the AI for Science Center. His research interests lie in Scientific Foundation Models, Autonomous Discovery System, and their cross-disciplinary applications. Lei has published over 100 research papers in top-tier journals and prestigious AI conferences, including Nature Communications, Nature Machine Intelligence, Science Advances, IEEE TPAMI, NeurIPS, CVPR, KDD etc. Lei has won a set of awards, including the 2024 TCSVE Best Paper Award, 2022 Yunfan Award at the World Artificial Intelligence Conference, 2020 Engineering Research Excellence Award from the University of New South Wales, and the 2019 Google PhD Fellowship.

Ding Ning is a assistant professor at Tsinghua University. He has long been engaged in research regarding Large Language Models (LLMs) and related technologies.He has published over 50 papers in Nature partner journals and top-tier AI conferences and journals (such as NeurIPS, ACL, ICLR, etc.). He regularly serves as a reviewer or program committee member for relevant academic venues and has accumulated 4,000 citations on Google Scholar. His honors include the ACL Best System Demonstration Paper Award, the World Artificial Intelligence Conference (WAIC) Youth Outstanding Paper Award, the WAIC Yunfan Award, and the Baidu Scholarship.

Biqing Qi is a Young Researcher at the Shanghai Artificial Intelligence Laboratory. His research areas include robust and online learning theory, foundation models for knowledge composition, and applications for human-machine collaborative knowledge discovery.

Constructing Knowledge Base of Scientific and Technological Literature to Support AI for Science

Zhixiong Zhang, Yi Liu, Jie Li, Li Qian, Xin Lin, Mengting Zhang, Yang Li, and Yajiao Wang

Abstract Currently, artificial intelligence-driven scientific research (AI for Science, AI4S) has become the high ground of international scientific and technological (Sci&Tech) competition. The essence of artificial intelligence lies in knowledge acquisition and application. In the process of rapid development of Sci&Tech large models and the advancement of AI4S research paradigms, a critical issue faced by many scientific research teams is the lack of reliable high-quality data and specialized domain knowledge. As the main carrier of human knowledge, Sci&Tech literature is rich in rigorous, credible, systematic, and cutting-edge knowledge, which has important value for supporting the training of Sci&Tech large models and driving the AI4S scientific research paradigm transformation in China. In this paper, the concept of Sci&Tech literature knowledge base to support AI4S is proposed. This paper reviews

Z. Zhang (✉) · Y. Liu · J. Li · L. Qian · X. Lin · M. Zhang · Y. Li · Y. Wang
National Science Library, Chinese Academy of Sciences, Beijing, China
e-mail: zhangzhx@mail.las.ac.cn

Y. Liu
e-mail: liuyi@mail.las.ac.cn

J. Li
e-mail: lijie201909@mail.las.ac.cn

L. Qian
e-mail: qianl@mail.las.ac.cn

X. Lin
e-mail: linxin@mail.las.ac.cn

M. Zhang
e-mail: zhangmengting@mail.las.ac.cn

Y. Li
e-mail: liyang2022@mail.las.ac.cn

Y. Wang
e-mail: wangyajiao@mail.las.ac.cn

Z. Zhang · L. Qian · X. Lin · M. Zhang · Y. Li · Y. Wang
Department of Information Resources Management, School of Economics and Management, University of Chinese Academy of Sciences, Beijing, China

China's e-Science Blue Book 2025, https://doi.org/10.1007/978-981-96-8252-2_8

the requirements for constructing knowledge base of Sci&Tech literature supporting AI4S. The overall idea of constructing AI4S Sci&Tech literature knowledge base through Sci&Tech literature data mining to support Sci&Tech large models training and scientific research paradigm transformation is elaborated. The National Science Library, Chinese Academy of Sciences has been actively exploring the scientific knowledge and high-quality data embedded in the domain Sci&Tech literature and making efforts to construct the domain knowledge base and Sci&Tech literature large model to support AI4S. This paper concludes by proposing the future development direction and strategy.

Keywords AI4S · Scientific and technological literature · Knowledge base · Knowledge mining

1 The Demand for Building AI4S Knowledge Base for Scientific and Technological Literature

1.1 What is AI4S

In recent years, the development of Artificial Intelligence (AI) has been advancing by leaps and bounds, and all fields around the world are facing the key question of how to integrate and apply AI technology. In the new wave of AI technology, the in-depth application of AI in the field of science has already brought significant changes and new opportunities for basic scientific research such as materials science and drug development. With the advancement of this change, the deep integration between AI and scientific research has accelerated in the mode of "data-knowledge two-way driven", which profoundly changed the traditional scientific research pattern dominated by "academic experience" and led to the emergence of a new paradigm of scientific research-Artificial intelligence-driven scientific research (AI for Science, AI4S).

In 2018, Academician E Weinan of Peking University put forward the concept of "AI for Science" [1], emphasizing the use of AI to learn scientific principles and create scientific models to solve practical problems, and he believed that "AI for Science: an ongoing technological revolution". At the same time, Academician Li Guojie pointed out that the intelligence of scientific research "will bring opportunities and challenges that will determine the next 20 years, China's scientific and technological development may widen the gap with the international advanced level or work hard to catch up". Academician Li said [2], "'AI for Science' is essentially 'AI for Scientists'. AI scientists and engineers are not the protagonists of intelligent scientific research, but scientists from various disciplines are the true protagonists, because intelligent modeling of each field must be done mainly by scientists in their field". At present, AI4S has become the high ground of international scientific and technological competition. Many fields in China are actively building various types of AI4S scientific and technological large models, to promote the change of scientific research paradigms, such as the GeoGalactica geographic large model led by Academician Zhou Chenghu, the CAS MatChat materials field large model developed by the Institute of Physics of the Chinese Academy of Sciences, and the Baiyulan scientific large model of Shanghai Jiao Tong University, and so on.

Based on comprehensive analysis, it can be concluded that AI4S has the following characteristics: ① AI4S relies on reliable high-quality data and specialized domain knowledge. High-quality data provide key corpus support for scientific and technological large model training, while specialized domain knowledge supports AI4S intelligent models to make reasonable assumptions, achieve fact-based predictions, and carry out logical reasoning. ② AI technology is an important cornerstone for the transformation and development of scientific research into AI4S. Using the powerful predictive analysis and inductive deduction capabilities of AI technology, AI4S can effectively achieve the acquisition, value-added and analysis of data through literature mining and knowledge integration, assisting scientific researchers in discovering the key knowledge, potential patterns and distribution of features implied in massive data, thus accelerating the process of scientific discovery. Taking biomedicine as an example, AI technology has already played an important role in key aspects such as protein 3D structure prediction, new drug development [3], and high-throughput screening of candidate compounds [4]. ③ AI4S accomplishes intelligent modeling of scientific research mainly by scientists from various fields [2]. In the intelligent modeling process of AI4S, scientists in various fields play an important role. Relying on the rich experience and professional knowledge accumulated in scientific research, field scientists strongly promote the intelligent transformation of scientific research to ensure the scientific soundness and practicality of the AI4S intelligent model. ④ The new research paradigm of AI4S has significant transformative features. Through automated experimental design and simulation processes, AI4S has profoundly changed the ways of formulating scientific research hypotheses, assisting scientists in generating hypotheses, designing experiments, collecting and interpreting datasets, to gain insights that the traditional methods struggle to uncover, inspiring scientists to generate valuable scientific insights, and providing new perspectives for scientific research.

1.2 *Importance of AI4S*

At the end of 2022, ChatGPT, a generative AI chatbot launched by OpenAI, triggered a landmark technological revolution in the history of AI development, leading people to a profound realization that the era of artificial intelligence has arrived. The ability of AI to learn knowledge, master knowledge, and apply knowledge to solve problems has amazed the world. In the scientific community, various disciplines are actively utilizing AI to help scientific research and build an intelligence-driven scientific research model, and it is anticipated that AI4S will not only drive various fields to achieve scientific breakthroughs, but also emerge as a key development strategy for the country to remain competitive in future scientific and technological advancements.

In March 2023, in order to implement the national "Development Planning for a New Generation of Artificial Intelligence", the Ministry of Science and Technology of the People's Republic of China, together with the National Natural Science Foundation of China, initiated the special deployment of "Artificial Intelligence-driven Scientific Research" [5], which is focusing on the needs of key disciplines such as drug research and development (R&D), genetic research, bio-breeding and new materials R&D, formulating the cutting-edge science and technology R&D system of "Artificial Intelligence-driven Scientific Research (AI4S)", in order to promote the deep integration of AI technology within scientific research, and to advance AI4S as a new paradigm for solving the major scientific problems in basic disciplines.

On October 8, 2024, the Nobel Prize in Physics was awarded to AI scientists John J. Hopfield and Geoffrey Hinton for their "fundamental discoveries and inventions in artificial neural networks and machine learning". Hopfield and Hinton's work draws inspiration from physics, utilizing physics principles and methods to identify underlying patterns in data. Hopfield created the Hopfield network structure with nodes and connections to store and reconstruct information based on the principles of atomic spin systems and energy function minimization in physics, while Hinton further invented a method that can discover data attributes by autonomously adjusting the connection weights between nodes based on statistical physics. Awarding the Nobel Prize in Physics to two scientists is not only a recognition of the role physics plays in advancing the development of AI, but also reflects the physics community's affirmative attitude toward AI in advancing physics itself. Ellen Moons, chair of the Nobel Committee for Physics, said that, "In physics, we use artificial neural networks in a wide range of areas, such as developing new materials with special properties" [6]. In this regard, Academician Li Guojie stated in his article "Why the 2024 Nobel Prize in Physics and Chemistry Favors Artificial Intelligence" that "Regardless of how much more effort is required for AI and physics to achieve a deep convergence, it has already been seen that the paths of 'AI for Physics (Science)' and 'Physics (Science) for AI' will both be broader." [7].

On October 9, 2024, the Nobel Prize in Chemistry was announced, another award for AI scientists. The Royal Swedish Academy of Sciences announced that the 2024 Nobel Prize in Chemistry was awarded to scientists David Baker, Demis Hassabis and John M. Jumper for their outstanding contributions to "computational protein design" and "protein structure prediction". Baker used AI technology to create new proteins that can be used in a variety of fields, including drugs, vaccines, nanomaterials, and microsensors; AlphaFold2, an AI model developed by Hassabis and Chopp, can accurately predict the complex structure of almost all known proteins. This work has enabled scientists worldwide to conduct more in-depth research on antibiotic resistance, enzyme design, and more. Their groundbreaking results demonstrate once again the great potential and significant value of AI-enabled chemistry research.

1.3 AI4S's Need for Data and Knowledge

Knowledge reflects objective laws, can be used to solve real-world problems, and is a force for historical development [8]. Intelligence is essentially an ability to learn and apply knowledge. Currently, artificial intelligence has developed rapidly, which appears to be the integrated application of big data, large models (algorithms), and high computing power, while essentially, the breakthrough of AI capability lies in the significant improvement of the ability to learn from data to acquire knowledge and apply knowledge [9].

Accurate AI predictions require high-quality AI corpus. While focusing on the unprecedented innovation potential and development opportunities that AI4S brings to scientific research, greater emphasis should be placed on the significant demand for high-quality data and knowledge behind the development of AI4S. If AI technology is the main engine for transforming and developing scientific research to AI4S, then data and knowledge are the fuel. AI4S scientific research paradigm needs substantial data and reliable knowledge to support hypothesis formulation, prediction verification, and reasoning supervision. The current construction of various types of scientific and technological large models also requires large-scale and high-quality data and knowledge to realize knowledge enhancement, such as the foundation large models need to obtain reliable commonsense knowledge from large-scale and high-quality textual data, the domain large models need to be trained with domain-specialized knowledge, and structured knowledge is especially needed to achieve models' knowledge enhancement and supervision, in order to achieve the goal of knowledge-based reasoning.

Currently, in the process of training Sci&Tech large models and AI4S scientific research paradigm advancement, it has been observed that two important issues faced by many scientific research teams are: the lack of reliable and high-quality data to support the training of Sci&Tech large models; and the lack of specialized domain knowledge to support the AI4S intelligent models to put forward reasonable assumptions, make fact-based predictions, and carry out logical reasoning. Where and how to obtain reliable and high-quality data and knowledge, as well as methods for obtaining them, have become an important problem to be solved in the process of advancing the AI4S research paradigm.

1.4 The Value and Significance of Literature Data to AI4S

Sci&Tech literature contains human knowledge, expresses scientific mechanisms, reveals scientific research results, is the main carrier of human knowledge exchange, and is also recognized as a reliable source of human knowledge, characterized by its cutting-edge insights, as well as its rigor and systematicity. In the context of AI, Sci&Tech literature is considered a high-value corpus, serving as a key source of knowledge acquisition for artificial intelligence.

From the perspective of logical structure, a research paper can be regarded as a collection of various types of knowledge, research programs and research data for solving research questions. Specifically, a research paper contains at least the following four types of knowledge: ① Commonsense knowledge. For example, the research question, the research area to which the research question belongs, the basis for the research, the value, and significance of the research, and the consensus that has developed on the question and related areas. ② Advancing knowledge. For example, regarding a specific research question, it is necessary to systematically examine: which scholars have explored specific aspects of the problem, the theoretical frameworks and methodologies they employed (e.g., techniques, analytical tools), key advancements achieved through their research, and valuable insights generated that contribute to the field. ③ New knowledge obtained by the author through their research. For example, given a particular research question, targeted hypotheses, experiments, results, and research conclusions that the author put forward. ④ Knowledge of the authors' viewpoints and tendencies by combing through relevant research. This includes an analysis and evaluation of the current research field, the summary of future development trends, research focus, proposed ideas, methods, and programs. The knowledge in research papers will be valuable in supporting the training of Sci&Tech large models and AI4S scientific research paradigm revolution.

In the era of AI, the important value and role of Sci&Tech literature has become increasingly prominent, how to mine Sci&Tech literature resources accumulated by the literature institutions, and leverage the strengths of literature institutions in knowledge organization, knowledge graph construction, and domain ontology construction, is of great significance for literature institutions to keep pace with the era of AI. Some foreign researchers have actively explored effective methods for mining, sorting, organizing, and correlating of knowledge content in Sci&Tech literature, revealing the rich semantic knowledge embedded in Sci&Tech literature [10–12].

To address the bottleneck that intelligent scientific research urgently needs the support of domain knowledge and data, based on the concept that Sci&Tech literature serve as a carrier of human knowledge, the National Science Library, Chinese Academy of Sciences leverages its extensive collection of high-quality Sci&Tech literature data across various disciplines, and cooperates actively with domain scientists to propose transforming Sci&Tech literature data into knowledge base of Sci&Tech literature supporting AI4S scientific research paradigm, and subsequently facilitate intelligent scientific research solution.

2 The Approaches and Frameworks

In 2021, the National Science Library, Chinese Academy of Sciences, outlined two critical initiatives in its data platform construction plan under the "14th Five-Year Plan": evidence-based scientific strategy decision-making and AI-based novel scientific research.

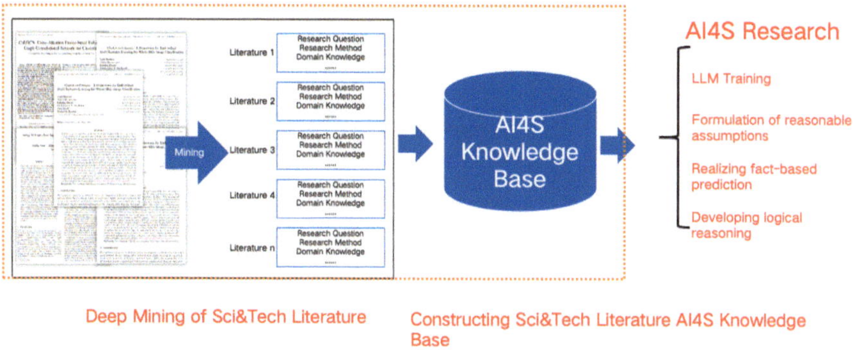

Fig. 1 General approach of deep mining Sci&Tech literature knowledge to construct AI4S knowledge base

By 2023, with the accelerated development of AI4S, a framework was further proposed to transform conventional Sci&Tech literature repositories into an AI4S knowledge base through deep mining of fine-grained knowledge embedded in Sci&Tech literature to support scientific large model training and research paradigm shifts (Fig. 1). The study proposed detailed approaches and workflows on two key processes of the framework: deep mining of Sci&Tech literature and construction of AI4S Sci&Tech literature knowledge base.

2.1 Approaches for Sci&Tech Literature Deep Mining

To enable deep mining of Sci&Tech literature, the author's team summarized eight approaches based on practical work to extract and reveal the knowledge content embedded in Sci&Tech literature. The key approaches are as follows.

(1) **Approach for Mining the IMRaD Structure**

IMRaD is a widely adopted structure for scientific writing in the international academic community [13], including Introduction, Methods, Results, and Discussion. The chapter structure extraction method is primarily based on the IMRaD paradigm to enable the chapter-level knowledge object extraction such as Introduction, Method, Result, Conclusion, and Discussion (Fig. 2). The extracted fragments can effectively support information retrieval, knowledge graph construction, and fine-grained knowledge mining.

(2) **Approach for Mining Multi-category Knowledge Centered on Research Question**

The research question is the core of the research paper. This approach centers on the research question, mining common knowledge, such as the value and significance of the research question, the research basis of the research question, and relevant

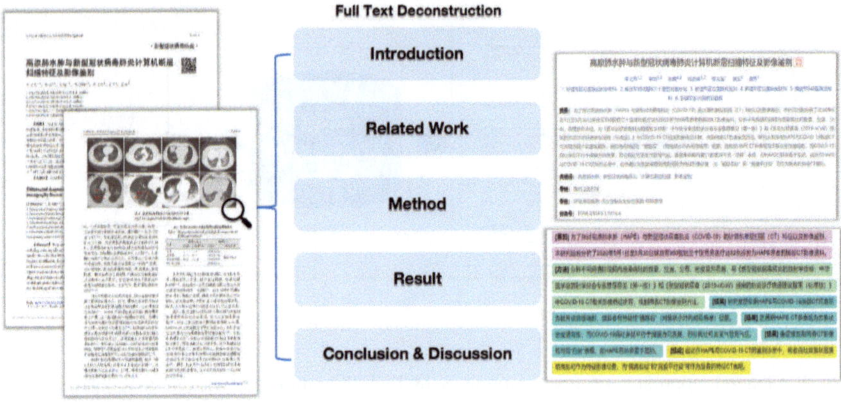

Fig. 2 Approach for mining the IMRaD structure

definitions. Mining advancing knowledge, such as the research method adopted in the relevant literature, the results obtained, and the limitations of the research specified in the relevant literature. Mining the new knowledge the author obtained, such as research hypotheses, research plans, experimental steps, and research innovation. Mining the viewpoint knowledge put forward by the authors, such as the author's judgment on the future research interests of the research question, and the proposed recommendations. (Fig. 3).

(3) **Approach for Mining Author's Viewpoints on Specific Research Question**

The authors' viewpoints on specific research questions are of great significance in reviewing the current research and predicting future trends. This approach focuses

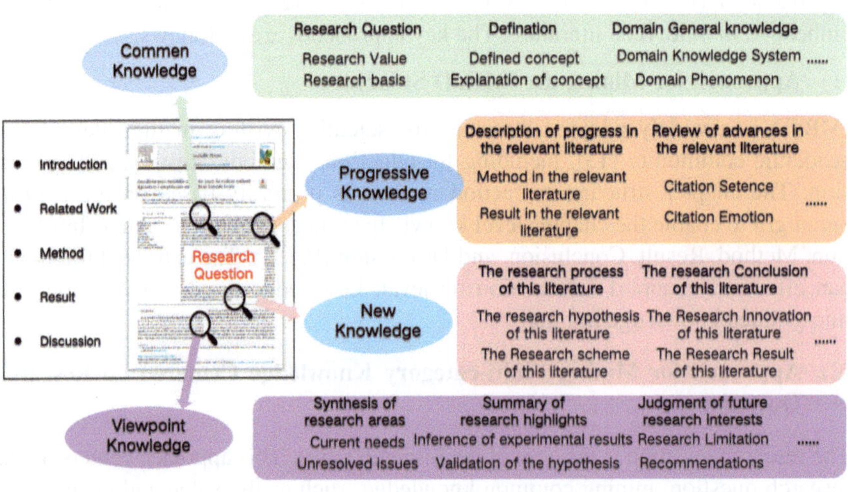

Fig. 3 Approach for extracting knowledge relevant to the research question

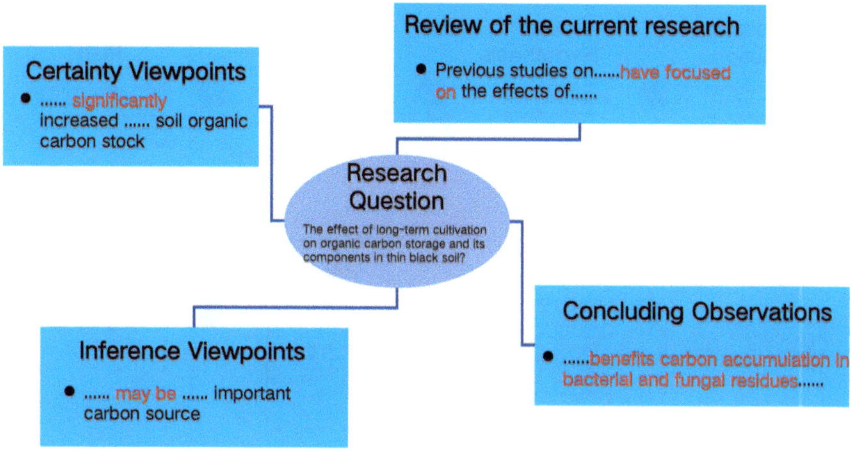

Fig. 4 Approach for mining author's viewpoints on the specific research question

on analyzing the types of research viewpoints, including review-based, summary-based, and judgment-based viewpoints. This approach also focuses on analyzing the confidence of the research viewpoints including the assertion viewpoints supported by clear data and the inferred viewpoints based on incomplete information (Fig. 4).

(4) **Approach for Mining the Research Logical Structure**

Hypothesis formulation and verification are the key processes of scientific research. This approach focuses on extracting research question, research method, research plan, experimental procedure, research results, research conclusion, and research prospect. These extracted knowledge elements are then correlated with the corresponding research hypothesis to reveal the logical structure of scientific research papers (Fig. 5). Unlike the IMRaD writing structure, explicitly mining the logical structure enhances readers' comprehension of the research process by illustrating how the authors conducted their research. Additionally, this approach provides critical evidence for verifying and reproducing research outcomes.

(5) **Approach for Mining General Knowledge and Domain Knowledge**

The content of Sci&Tech literature comprises both general knowledge which is common across research fields, and domain knowledge which is specific to particular scientific domains. This approach is designed to extract both types of knowledge (Fig. 6). In addition to identifying general knowledge objects such as methods and models, data, metrics and concept definition, it also focuses on extracting domain-specific knowledge objects. For example, in chemical domain, it extracts organic substances, inorganic substances, and reaction processes; in medical domain it extracts disease, drugs, and drug targets; in agricultural domain, it extracts crops, products, and pests.

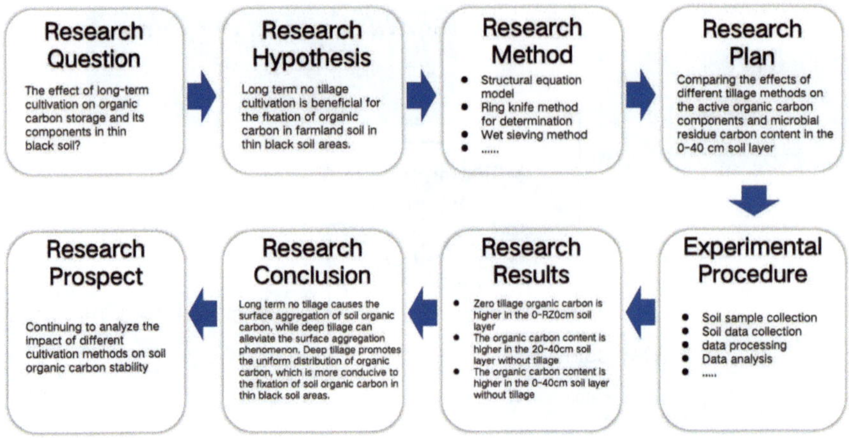

Fig. 5 Approach for mining research logical structure

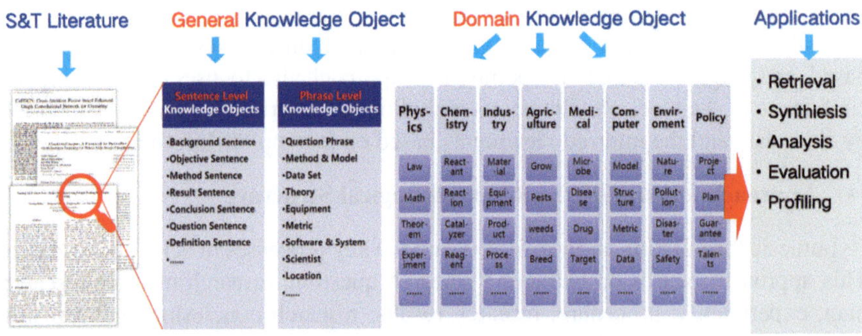

Fig. 6 Approach for extracting general knowledge and domain knowledge

(6) Approach for Mining Multimodal Knowledge Object Alignments Among Graphs, Tables, Formulas, and Text

Sci&Tech literature presents information in multiple modalities, including figures, tables, formulas, and text. While human readers can naturally recognize the relationship between these multimodal knowledge objects, enabling computers to automatically and explicitly identify these associations is highly valuable. This approach aims to achieve the automatic association of multimodal knowledge objects on Sci&Tech literature, ensuring they are aligned, complementary and accurately calibrated with one another (Fig. 7).

(7) Approach for Mining the Semantics of Associations between Citing and Cited Literatures

Citation relationship constitutes semantic connections between multiple pieces of Sci&Tech literature, while citation sentences convey the sematic features of these

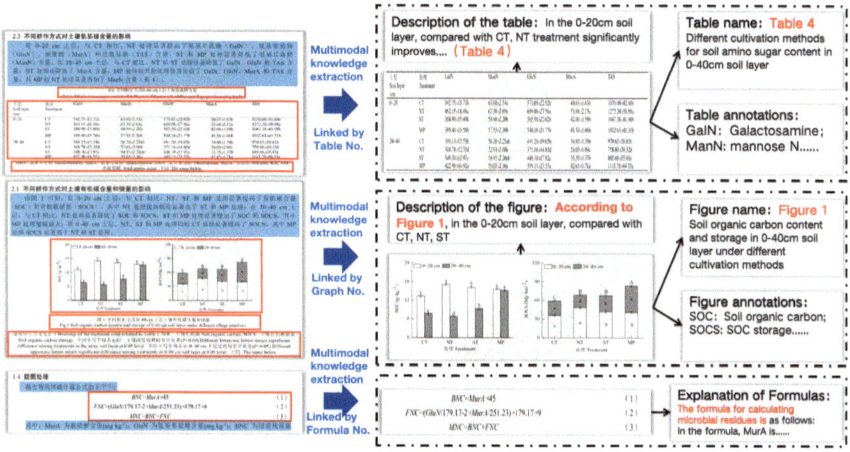

Fig. 7 Approach for achieving multimodal knowledge object association alignment among graphs, tables, formulas, and texts

connections. This approach extracts citation sentences, citation emotions, and citation intentions from Sci&Tech literature to analyze the citation relationship across multiple literatures. Citation emotion is categorized as neutral, positive, or negative. Citation intention is categorized as referencing the research background, research methods or research results for comparison. Each citation sentence, along with its emotion and intention, is associated with one or more cited literature, forming a structural citation association network. This network facilitates semantic citation analysis and enhances the understanding of citation dynamics in Sci&Tech literature (Fig. 8).

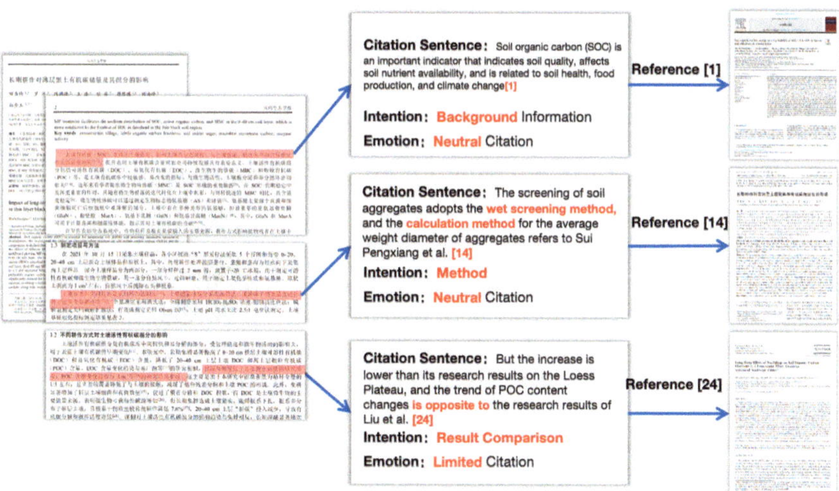

Fig. 8 Approach for mining semantic analysis of associations between citing and cited literature

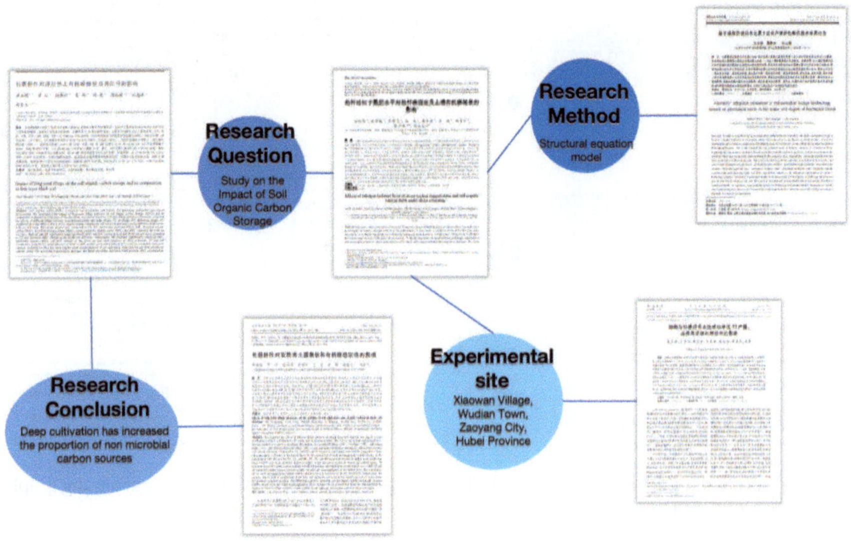

Fig. 9 Approach for mining multi-literature association structures mediated by knowledge objects

(8) **Approach for Mining Multi-literature Association Structures Mediated by Knowledge Objects**

Beyond mere citation correlations, this approach leverages knowledge objects to establish meaningful links between multiple pieces of Sci&Tech literature. Literature can be interlinked based on the occurrence of the same or similar research questions, research methods, research conclusions, or other fine-grained knowledge objects (Fig. 9). For example, papers that propose the same or similar research question are associated together; papers that employ the same research method and conduct experiments in the same location are linked.

2.2 The Process for Constructing an AI4S Knowledge Base

By summarizing the related work, the author's team proposes a process for constructing Sci&Tech literature knowledge base for AI4S (Fig. 10). The process begins with gathering high-quality Sci&Tech literature, followed by extracting fine-grained knowledge content such as domain knowledge, scientific data, and viewpoint tendencies. These extracted elements are then organized into specialized knowledge base, including AI-Ready scientific datasets, domain knowledge ontology database, viewpoint tendency database. Finally, the constructed knowledge bases are utilized to support intelligent scientific research applications.

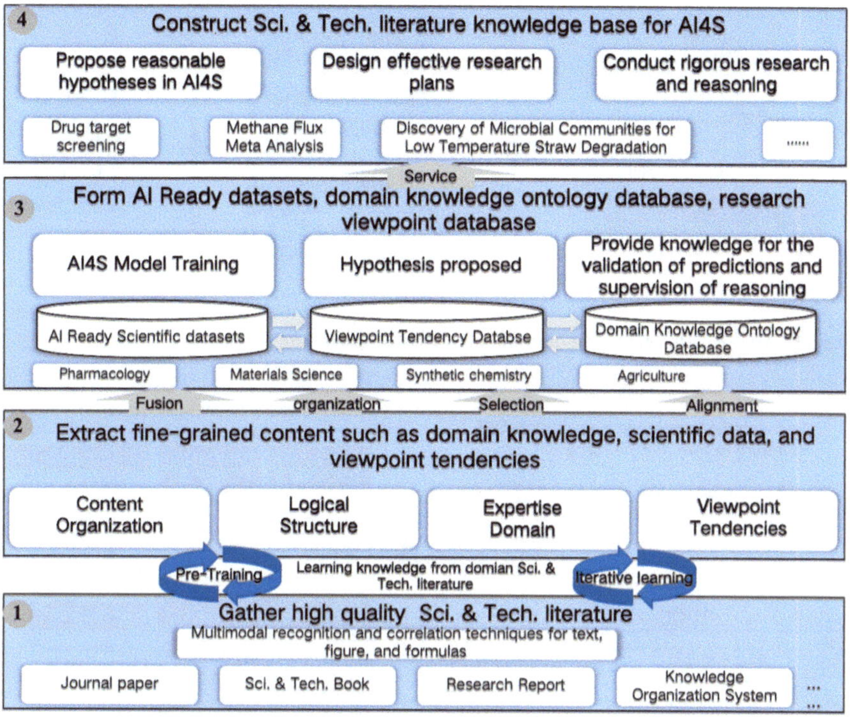

Fig. 10 The process for constructing an AI4S knowledge base

(1) Gathering High-quality Sci&Tech Literature

High-quality Sci&Tech literature is the basis for constructing the AI4S Sci&Tech literature knowledge base. This stage involves the gathering of high-quality and reliable journal papers, knowledge organization systems, Sci&Tech books, and so on. Through multimodal deconstruction and high-value data integration and cataloging of text, figures, formulas, etc. Within the full text of Sci&Tech literature, a Sci&Tech literature database is formed which contains traditional textual information, with non-textual elements such as charts and formulas (Fig. 11).

(2) Extracting Fine-grained Knowledge Content

Based on gathering high-quality Sci&Tech literature, we fully mine the knowledge of Sci&Tech literature, and construct an artificial intelligence engine (SciAIEngine) to support the in-depth mining of Sci&Tech literature content. Then we utilize SciAIEngine to extract sentence-level and phrase-level general knowledge, domain knowledge, scientific data, viewpoint tendencies, and other fine-grained knowledge contents in Sci&Tech literature (Fig. 12). SciAIEngine supports evidence-based traceability by establishing evidence-based relationships between fine-grained knowledge contents and sentences, paragraphs, and full texts.

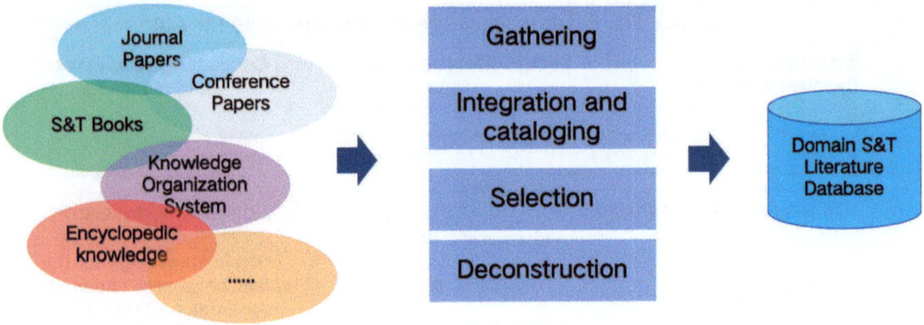

Fig. 11 Approach for gathering high-quality Sci&Tech literature

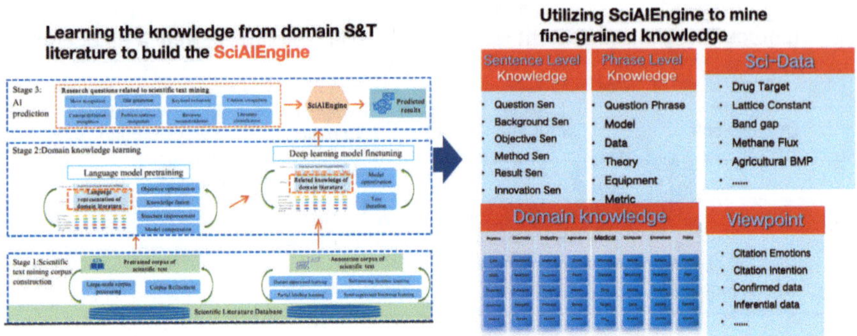

Fig. 12 Approach for extracting fine-grained knowledge content

(3) Forming Knowledge Bases

Based on fine-grained knowledge content mining, followed by further fusion, selection, alignment, and completion, we construct reliable, assured, and high-quality knowledge bases. Focusing on constructing AI Ready scientific datasets, domain knowledge ontology database, and viewpoint tendency database (Fig. 13).

(4) Construct Sci&Tech Literature Knowledge Base for AI4S

Eventually, constructing AI4S Sci&Tech literature knowledge base. Among them, the AI-Ready scientific dataset serves as the foundation for training AI4S intelligent models, the viewpoint tendency database supports the formulation of hypotheses for these models, and the domain knowledge ontology database provides essential knowledge for predictions validation and inference supervision (Fig. 14).

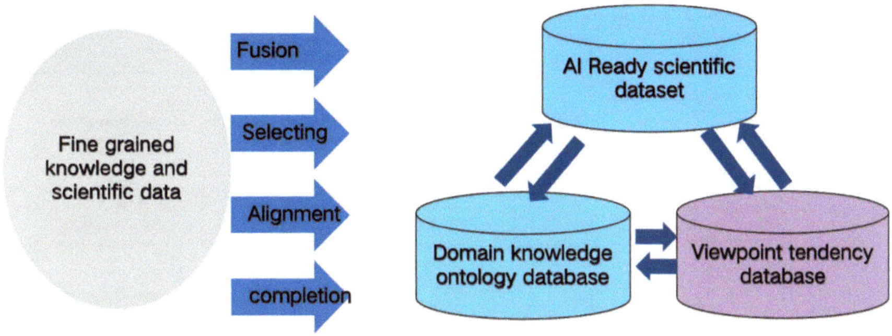

Fig. 13 Approach for transforming fine-grained knowledge into knowledge base

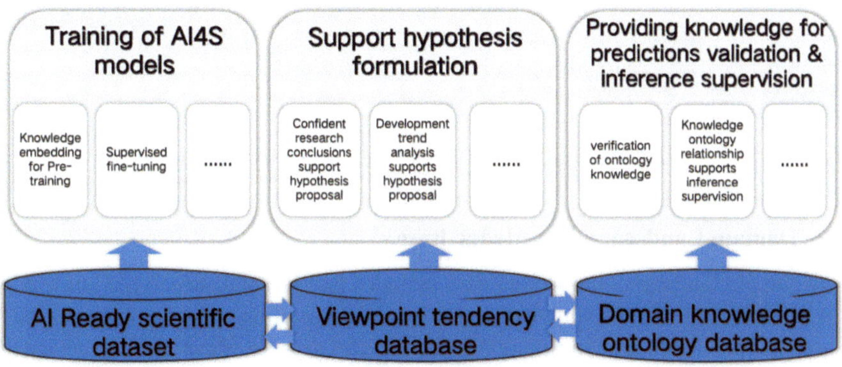

Fig. 14 Approach for constructing Sci&Tech literature knowledge base for AI4S

2.3 The Framework of Constructing AI4S Sci&Tech Literature Knowledge Base

Based on practical work, the author's team proposes the framework of constructing the AI4S Sci&Tech literature knowledge base, which contains four important components: domain primary knowledge base, domain landscape knowledge base, domain ontology knowledge base, and domain neural network knowledge base (Fig. 15). This framework provides solid data and a knowledge foundation for supporting AI for Knowledge Service and AI for Science.

(1) **Domain Primary Knowledge Base**

The domain primary knowledge base is the foundation of the AI4S knowledge base. Through the aggregation, selection, and cleaning of knowledge resources such as papers, patents, books, as well as the deconstruction and association of multimodal data such as text, figures, tables, and the semantic evaluation of Sci&Tech literature, a high-quality, reliable, and high-value domain primary knowledge base is constructed.

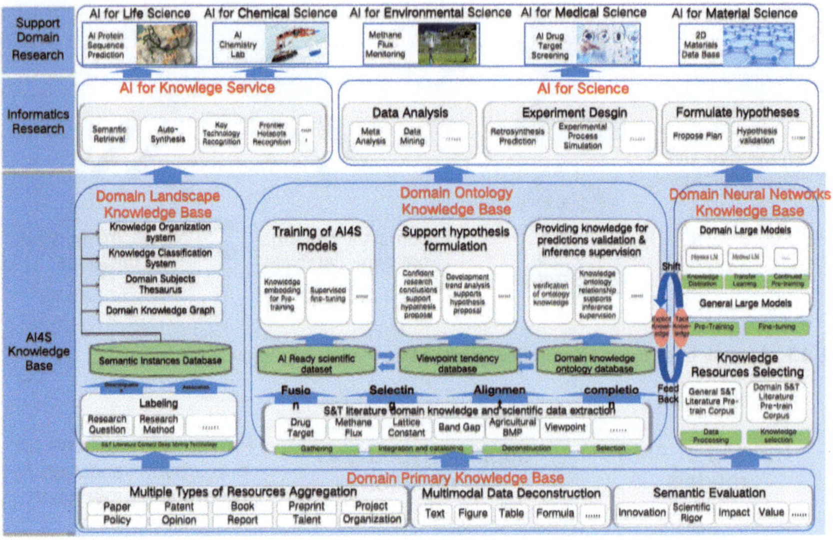

Fig. 15 The framework of constructing AI4S Sci&Tech literature knowledge base

(2) **Domain Landscape Knowledge Base**

The domain landscape knowledge base includes fine-grained, structured knowledge for Sci&Tech literature knowledge service. It is based on the primary knowledge base of a specific domain, utilizing AI technologies to mine, organize and manage the knowledge from Sci&Tech literature, supporting the intelligent knowledge service scenarios of semantic retrieval, reading assistance, correlation recommendation, and the information analysis scenarios of key technology identification, knowledge evolution analysis, frontier hotspot identification.

(3) **Domain Ontology Knowledge Base**

The domain ontology knowledge base includes convinced and explicit knowledge resources such as theoretical principles, scientific data, formulas, etc. oriented to scientific research. Knowledge extraction technology is utilized to extract domain knowledge and scientific data from Sci&Tech literature to support scenarios such as the formulation of scientific research hypotheses, inference supervision, and experimental design.

(4) **Domain Neural Networks Knowledge Base**

The domain neural network knowledge base includes high-quality tacit knowledge used for training and fine-tuning large models. High-value general and domain-specific Sci&Tech Corpus is derived from the primary knowledge resources, enabling the construction of general large model and several domain large models through continuous pre-training and fine-tuning processes. These models facilitate various scenarios such as the knowledge Q&A, scientific writing, and creative inspiration for scientific research.

3 Initiatives for Constructing the AI4S Scientific Literature Knowledge Base

Based on the aforementioned concepts and framework of the Knowledge Base, the author actively promotes the implementation of initiatives related to the AI4S Scientific Literature Knowledge Base by integrating literature resources, developing intelligent tools, and collaborating with enterprises and teams of domain scientists.

3.1 Actively Promoting the Construction of an Intelligent Data System

Based on compliance with intellectual property regulations and international standards, the author's team has fully leveraged the advantages of possessing vast literature resources and a well-established knowledge organization system. Concrete practices have been carried out in areas such as the integration of diverse knowledge resources, multimodal data deconstruction, and association.

By collecting and organizing vast amounts of research data, including scientific papers, patents, standards, scientific books, funding information, science and technology policy texts, talent information, research institution profiles, and intelligence resources, the team has implemented strategies for data integration, classification, association, and semantic enrichment. These efforts have transformed previously scattered and unstructured information into a semantically rich and structured intelligent data system. This system enables the aggregation of diverse knowledge resources, thereby providing a solid foundation for subsequent data analysis and knowledge discovery.

For multimodal data deconstruction and association, the author's team decomposes various modalities of knowledge in literature, such as paragraph texts, images, tables, and formulas, into fundamental data units. Through techniques like entity alignment and relationship mining, the team identifies and establishes the intrinsic connections among these data units, uncovering the complex relationships between them. This approach facilitates the construction of a multidimensional and multimodal knowledge network.

3.2 Advancing Key Technologies for Developing an AI Engine for Scientific Literature Knowledge

Advancing Key Technologies for Mining Scientific Literature Content. To enable knowledge extraction at the sentence and discourse levels, our team proposed a knowledge extraction method based on the Masked Sentence Model (MSM) [14] and a two-stage knowledge extraction approach [15]. For knowledge extraction at the phrase level, we introduced a keyword recognition method that integrates embedded dictionaries and part-of-speech features [16], combined with a semi-supervised iterative learning technique [17] for entity recognition using limited data. For the task of scientific literature classification, a hierarchical classifier ensemble method [18] was proposed to implement classification based on the Chinese Library Classification system across thousands of categories.

Developing an AI Engine for Scientific Literature Knowledge [19]. By fully leveraging scientific literature resources, our team has developed methods to automatically learn and extract significant knowledge embedded in scientific texts. Based on this extracted knowledge, core AI components have been built to form the Scientific Literature Knowledge AI Engine, SciAIEngine. This engine supports a range of capabilities, including multi-level and fine-grained automatic classification of scientific literature, as well as automated knowledge extraction at the sentence and phrase levels. The development and application of SciAIEngine are expected to advance the intelligence and automation of scientific literature mining, empowering researchers to access the latest research findings and technological trends in their fields more efficiently.

3.3 Collaborating with Leading AI Companies to Develop Large Models for Scientific Literature

The National Science Library, Chinese Academy of Sciences has collaborated with leading AI companies to jointly develop large-scale scientific literature models and research assistants. These large-scale models significantly enhance the efficiency of information extraction and intelligent processing of scientific literature, demonstrating superior content understanding and knowledge generation capabilities in comparison with general large language models.

The research assistant, developed based on large-scale scientific literature models, employs retrieval-augmented generation (RAG) strategies grounded in a paper knowledge base to ensure that the generated results are reliable and well-supported. It integrates functions such as research review, paper reading, and academic writing support: the research review function can automatically summarize and analyze the content of scientific literature search results and generate review reports based on user-selected papers; the paper reading function can answer users' questions about papers with sources and supporting evidence, as well as provide translation for selected texts and recommend related papers; and the academic writing function assists users with Chinese-English translation and offers editing and polishing services for English papers.

3.4 *Actively Collaborating with Domain Experts to Support Domain-Specific AI4S*

The author's team actively collaborates with domain experts from domestic research institutions and universities to build domain-specific scientific literature knowledge bases through the integration of domain knowledge and scientific data mining. This effort supports research in domain-specific informatics and AI4S, including the following aspects:

(1) **Constructing a Drug Target Knowledge Base**

In collaboration with relevant research institutions, the team has conducted research on mining drug target knowledge from medical literature to build a drug target knowledge base that supports new drug development. Based on core scientific literature in related fields, the team proposed a "two-stage approach" for extracting drug target knowledge. In the first stage, drug target sentences are extracted from medical literature through text classification. In the second stage, sequence labeling is used to identify the embedded knowledge elements—such as drug targets, diseases, and genes—within these sentences. This approach leads to the construction of a drug target knowledge base that encapsulates knowledge relationships in the form of "target sentence-disease-gene-target".

(2) **Constructing a Knowledge Base for Two-Dimensional Materials**

In collaboration with relevant universities, the team has conducted research on mining knowledge of two-dimensional materials from physics literature to build a knowledge base that supports research and applications in this field. The team first identifies knowledge-rich regions for two-dimensional materials and then extracts knowledge entities and relationships, capturing information such as band gaps, crystal structures, lattice constants, and space groups. Additionally, the team extracts opinion sentences from authors regarding two-dimensional materials and establishes evidence-based relationships between domain knowledge, opinion sentences, knowledge-rich regions, paragraphs, and literature. This approach ultimately leads to the construction of a comprehensive knowledge base for two-dimensional materials.

(3) **Constructing a Knowledge Base for Green Hydrogen Production**

In collaboration with relevant universities, the team has conducted research on mining knowledge related to green hydrogen production from scientific literature. The team focuses on extracting and analyzing core elements such as hydrogen production materials, catalysts, process flows, and energy conversion efficiency. By constructing a knowledge base for green hydrogen production, this effort aims to provide comprehensive background knowledge and literature evidence to support the optimization of hydrogen production technologies and the reduction of application costs.

4 Case Study on Knowledge Base Construction for Lignocellulose Biodegradation from Scientific Literature

The National Science Library, Chinese Academy of Sciences, leverages its extensive collection of high-quality scientific literature to collaborate with relevant research institutions in conducting knowledge mining and database construction for lignocellulose biodegradation. This effort aims to establish a robust knowledge foundation supported by scientific literature, facilitating AI for Science (AI4S) in the field of lignocellulose biodegradation. Such advancements are critical for promoting the efficient utilization of biomass in black soil regions.

Black soil, renowned for its high organic matter content, is a vital resource for ensuring China's food security. The degradation and return of straw to fields play a crucial role in black soil conservation and sustainable utilization, as effective straw decomposition enhances soil organic carbon content and fertility. However, in cold regions, low temperatures significantly inhibit the decomposition of organic matter, posing a major challenge to sustainable agricultural development. Lignocellulose, as the primary component of straw, has become a focal point in research on novel straw incorporation methods. In particular, the study of "lignocellulose-degrading microorganisms adapted to perennial low-temperature environments" has emerged as a critical area of scientific investigation.

To address the significant research questions and knowledge objects of interest to domain scientists, this paper followed the fundamental process for constructing the AI4S scientific literature knowledge base (Fig. 10). We gathered technological literature in the field of lignocellulose biodegradation, extracting general knowledge, domain-specific knowledge, and opinion tendencies to form a semantically enriched database, an ontological knowledge base, and a repository of research opinion tendencies. This supports applications such as meta-analysis in lignocellulose biodegradation and microbial discovery. Regarding the extraction of knowledge entities and research opinions in this field, we propose a technical pipeline encompassing the aggregation and representation of domain scientific literature, knowledge-rich paragraph localization, extraction of domain knowledge and scientific data, and the construction of a domain database (Fig. 16).

4.1 Domain Literature Aggregation and Representation

Our research team conducted a systematic literature search focusing on research subjects and degradation processes. The search was performed using targeted keywords and supplemented by identifying leading scholars, core journals, and research institutions in the field of lignocellulose biodegradation. The retrieved literature was then processed through multimodal full-text parsing, converting PDF files into structured JSON format. This process extracted and linked key elements such

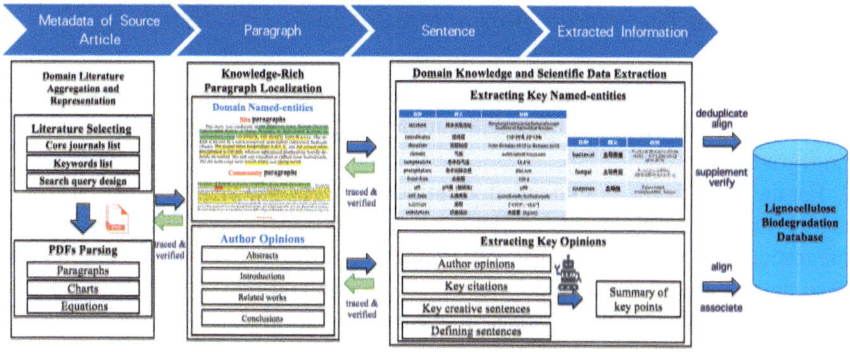

Fig. 16 The pipeline of automatic knowledge base construction for lignocellulose biodegradation

as text paragraphs (e.g., titles, abstracts, and descriptions), figures, and formulas, ultimately forming a comprehensive scientific literature database.

4.2 *Knowledge-Rich Paragraph Localization*

Knowledge-rich paragraphs refer to segments or regions within a text where target entities or key information appear with high frequency. These paragraphs typically encompass core knowledge points and associated information on specific themes. To accurately identify the enriched areas of knowledge within the field of lignocellulosic biodegradation, our research team has proposed distinct localization methods for domain knowledge entities and viewpoints.

For domain knowledge entities, we categorized target entities into classes such as 'Site' and 'Community', and developed corresponding algorithms for identifying entity-rich areas. For instance, paragraphs that include entities such as locations, geographical coordinates, annual precipitation, or unique units (e.g., mm, °C, %) are classified as 'Site'. Similarly, segments containing terms ending in "-ase" or specific names of microorganisms and enzymes are designated as 'Community'.

Regarding domain knowledge viewpoints, we designed an algorithm for locating viewpoint-rich areas. This approach focuses on critical sections such as literature abstracts, backgrounds, related research, and conclusions. These parts typically serve as key regions where authors summarize, discuss, and conclude their research topics, containing a wealth of perspectives on the subjects of study.

4.3 Domain Knowledge and Scientific Data Extraction

Building on the identification of knowledge-rich areas, our research team further extracted key knowledge entities and author opinions in the field of lignocellulose biodegradation.

For the identification of knowledge entities in this domain, we proposed a prompt engineering approach based on four fundamental principles: instructional clarity, integration of domain knowledge, structured output, and avoidance of hallucination. This method guides large models to extract precise and relevant information from the identified knowledge-rich areas. To ensure the traceability of all knowledge and data, we constructed a traceability evidence chain that follows the extracted entities back through their source sentences, source paragraphs, and the article's metadata.

In terms of opinion identification for lignocellulose biodegradation, our team developed a relevant annotated dataset and trained models on this dataset to automatically identify opinion sentences. For the identified opinion sentences, we further utilized APIs from large models to summarize these sentences, refining the author's viewpoints while maintaining fidelity to the original intent, and assessing the credibility of these opinions.

4.4 Construction of Domain Knowledge Base and Knowledge Discovery Platform

Building on the extraction of domain knowledge and scientific data, our team developed an evidence-based and reliable knowledge base for lignocellulose biodegradation through processes of deduplication, alignment, completion, validation, and association (Fig. 17).

Specifically, this knowledge base establishes a network of interaction relationships among key knowledge entities in the field, including degradation substrates,

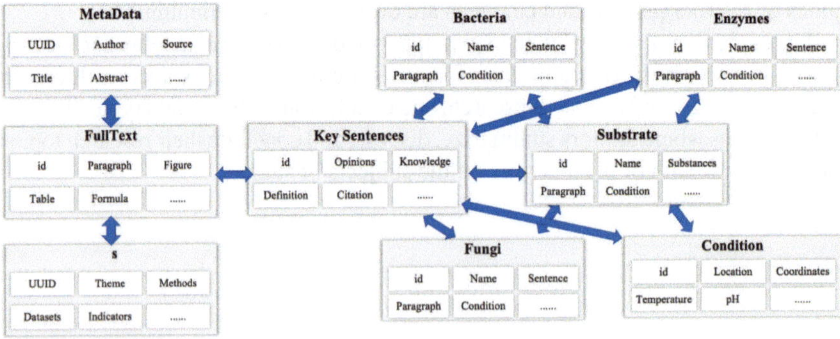

Fig. 17 Structure of the lignocellulose biodegradation knowledge database

bacteria, fungi, enzymes, and experimental conditions. It also creates a relational network connecting domain knowledge, source sentences, source paragraphs, and source literature, enabling traceable evidence of knowledge. Furthermore, we analyze the co-occurrence relationships between sentences, paragraphs, and entire texts to reveal the close connections among research questions, methodologies, and experimental data. Ultimately, this results in a comprehensive knowledge base for the field of lignocellulose biodegradation, which encompasses relevant literature, general knowledge objects, domain-specific knowledge objects, and research opinion trends.

The knowledge base not only includes full-text literature data and general domain knowledge objects extracted at the sentence and phrase levels but also comprises information on 205 bacteria, 898 fungi, 3,656 degradation enzymes, and 11 degradation experiments related to lignin, cellulose, and hemicellulose, along with over 30,000 research opinion sentences related to lignocellulose biodegradation.

Based on the constructed knowledge base of lignocellulose biodegradation, the Lignocellulose Biodegradation Knowledge Discovery (LBKD) Platform was developed. This platform allows for literature retrieval across two dimensions: foundational knowledge and domain-specific knowledge (Fig. 18). Foundational knowledge encompasses entities such as research questions, methodologies, innovations, and concept definitions, which are applicable across various types of literature. In contrast, domain-specific knowledge includes relevant entities specific to the field of lignocellulose biodegradation. For individual documents, the platform provides knowledge structure navigation, displaying sentence-level knowledge objects (including opinion sentences, citation sentences, definition sentences, and innovation sentences) as well as phrase-level knowledge objects (comprising both foundational knowledge entities and domain-specific entities).

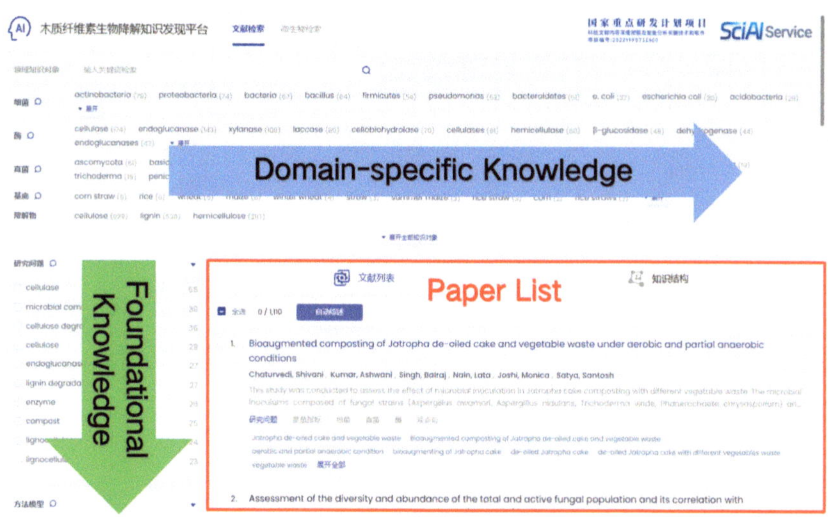

Fig. 18 Search page of LBKD platform

The platform constructs a knowledge structure diagram of relevant literature, systematically associating and visualizing research themes such as different microbial communities, types of degradation products, and soil types (Fig. 19). Through data visualization and knowledge integration, it not only achieves effective consolidation of literature resources but also provides researchers with a multidimensional analytical perspective, allowing them to observe the synergistic effects of specific strains during the degradation process, such as the mutual promotion with symbiotic bacteria or the collaborative degradation capabilities with certain fungi.

Compared to traditional retrieval systems, the platform offers more granular information, such as the specific degradation abilities of different strains and their interactions with other microorganisms or enzymes, supporting analyses of differences in degradation efficiency, enzyme activity, and environmental adaptability among various strains. It can also reveal previously overlooked potential knowledge relationships, such as the synergistic effects between specific strains and certain fungi or new roles of enzymes discovered during cellulose degradation. This helps provide a scientific basis for optimizing the lignocellulose degradation process, thereby offering new ideas and strategies for the efficient utilization of biomass energy.

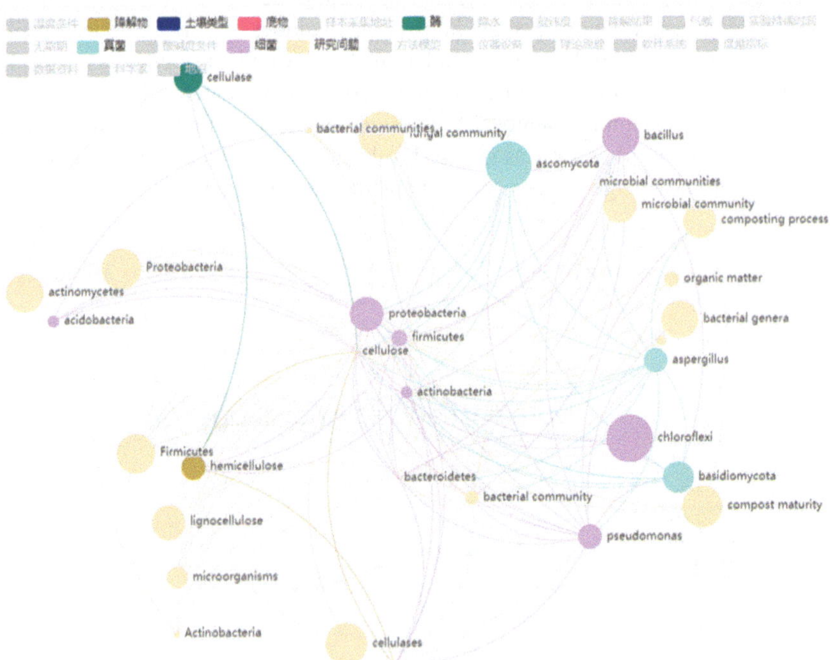

Fig. 19 Literature knowledge structure diagram display of LBKD platform

5 Conclusion and Outlook

The new paradigm of artificial intelligence-driven scientific research is opening a new era of intelligent scientific research. AI4S has penetrated into all fields of scientific research and profoundly affected the process and outcome of scientific innovation. In the process of advancing the scientific research paradigm of AI4S, the various kinds of high-quality data and specialized domain knowledge contained in Sci&Tech literature are of great value in supporting the training of sci&tech large models and the change of the scientific research paradigm of AI4S in China.

It is a milestone that the 2024 Nobel Prize in Physics and the Nobel Prize in Chemistry have awarded for AI-related scientific research, respectively. It not only reflects that AI technology models are highly recognized on the Nobel Prize stage, but also signifies that AI4S has played an important impact and positive role in promoting and accelerating research in basic science fields. At present, the traditional paradigm of scientific research has changed. In the future, China's scientific community will continue to promote the advancement of intelligent scientific research by focusing on the major strategic needs of the country. Constructing AI4S knowledge base of Sci&Tech literature can fully explore the knowledge content hidden in Sci&Tech literature, so that AI4S, the new scientific research paradigm, can be built on a solid knowledge foundation, thus supporting scientists to carry out more efficient scientific research work in related scientific research fields.

Acknowledgements This work is supported by the project "Chinese Academy of Sciences for the special consulting research project on network security and informatization" (Grant No. CAS-WX2023ZX01-0703), and the National Key Research and Development Program of China "Key Technology and Software for Deep Mining and Intelligent Analysis of Sci-Tech Literature" (Grant No. 2022YFF0711900).

References

1. Xinhua News Agency. AI for Science: New paradigm of scientific research. [R/OL] [2024-9-29].
2. LI G J. AI4R: The fifth scientific research paradigm[J]. Bulletin of Chinese Academy of Sciences, 2024, 39(1): 1–9.
3. Chan H C S, Shan H, Dahoun T, et al. Advancing Drug Discovery via Artificial Intelligence[J]. Trends in Pharmacological Sciences, 2019, 40(8): 592–604.
4. YOU Y J, LAI X, PAN Y, et al. Artificial intelligence in cancer target identification and drug discovery[J]. Signal Transduction and Targeted Therapy, 2022,7(1): 156–179.
5. The Central People's Government of the People's Republic of China. The Ministry of Science and Technology has launched a special deployment for "Artificial Intelligence Driven Scientific Research" [EB/OL] [2024-10-24].
6. The Nobel Prize. They trained artificial neural networks using physics [EB/OL] [2024-11-11].
7. LI G J. Why does 2024 Nobel Prize favor artificial intelligence[J]. Science & Technology Review, 2024, 42(19): 6–9.
8. Internet Encyclopedia of Philosophy and its Authors. Francis Bacon (1561–1626) [EB/OL] [2024-06-06].

9. ZHANG Z X, Zeng J X, XIA C J, et al. Information Resource Management Researchers' Thinking about the Opportunities and Challenges of AIGC[J]. Journal of Library and Information Science in Agriculture, 2023, 35(1): 4–28.
10. Ribaupierre H, Falquet G. An Automated Annotation Process for the SciDocAnnot Scientific Document Model[C]//Proceedings of the 5th International Workshop on Semantic Digital Archives. 2015: 30–41.
11. Liakat M, S T, A S, et al. Corpora for the conceptualization and zoning of scientific papers [C]//Proceedings of the Seventh International Conference on Language Resources and Evaluation (LREC'10). 2010: 2054–2061.
12. Gomez C, Guardia A, Mantari JL, et al. A contemporary approach to the MSE paradigm powered by Artificial Intelligence from a review focused on Polymer Matrix Composites[J]. Mechanics of Advanced Materials and Structures, 2021, 29 (21): 3076–96.
13. WU, J. Improving the writing of research papers: IMRAD and beyond[J]. Landscape Ecol, 2011, 26: 1345–1349.
14. YU G H, ZHANG Z X, LIU H, et al. Masked Sentence Model Based on BERT for Move Recognition in Medical Scientific Abstracts[J]. Journal of Data and Information Science, 2019 (4): 42–55.
15. LI X S, DING L P, ZHANG Z X. Drug Target Extraction from Biomedical Articles Based on a Two-Stage Cascading Framework[C]//2023 ACM/IEEE Joint Conference on Digital Libraries (JCDL). IEEE, 2023: 245–246.
16. DING L P, ZHANG Z X, LIU H, et al. Automatic keyphrase extraction from scientific Chinese medical abstracts based on character-level sequence labeling[J]. Journal of Data and Information Science, 2021, 6(3): 35–57.
17. DING L P. Named Entity Recognition Based on Partial Annotation Learning[D]. Beijing, China: University of Chinese Academy of Sciences, 2023.
18. ZHAO Y, ZHANG Z X, LIU H. A Research on Automatic Classification of Chinese Medical Literature Based on Hierarchical Classification[J]. Research on Library Science, 2021(21): 49–55, 61.
19. ZHANG Z X, LIU H, YU G H. Building an Artificial Intelligence Engine Based on Scientific and Technological Literature Knowledge[J]. Journal of Library and Information Science in Agriculture, 2021, 33(1): 17–31.

Zhixiong Zhang Doctor, Professorate Senior Engineer (Level II), doctoral supervisor, Deputy Director of the National Science Library, Chinese Academy of Sciences. Winners of the National Hundred and Ten Thousand Talents Project, the "Distinguished Researcher Program of the Chinese Academy of Sciences", and he granted by the State Council Special Allowance, which is for experts, scholars and technical personnel who make prominent contributions. He is the director of national major projects, such as the National Key Research and Development Program of China "Key Technology and Software for Deep Mining and Intelligent Analysis of Sci-Tech Literature". His main research fields are semantic intelligence, Sci&Tech literature mining, knowledge organization, etc. Hosted over 60 national and provincial-level projects, and published over 200 papers. Led his team to develop multiple in-use Sci&Tech literature intelligent mining tools and academic communication platforms, such as the AI engine of Sci&Tech literature knowledge, the Chinese Academy of Sciences Sci&Tech preprint platform (ChinaXiv) and the public academic platform (PubScholar), etc.

Yi Liu, Doctor, Senior Engineer at the National Science Library, Chinese Academy of Sciences, and Young Innovative Researcher. His main research fields include fine-grained knowledge content mining, automatic structured synthesis and semantic retrieval of scientific literature.

Jie Li, Doctor, Engineer at the National Science Library, Chinese Academy of Sciences, and Postdoctoral Researcher. Her main research fields include deep mining of scientific literature, deep clustering of scientific texts, and refined corpus construction.

Li Qian, Director of the Department of Data Resources at the National Science Library, Chinese Academy of Sciences, Doctoral Supervisor, and Professorate Senior Engineer. He also serves as Director of the Key Laboratory of New Publishing and Knowledge Services for Scholarly Journals, National Press and Publication Administration, and is a member and Distinguished Researcher of the Youth Innovation Promotion Association of CAS. Recognized as a Leading Young Information Scientist in China, his research focuses on big data analytics and knowledge mining of scientific literature.

Xin Lin, Ph.D. candidate at the National Science Library, Chinese Academy of Sciences. Her main research interests include plausibility assessment of scientific claims and move recognition in scientific literature.

Mengting Zhang, Ph.D. candidate at the National Science Library, Chinese Academy of Sciences. Her main research interests include academic opinion mining and scientific entity recognition in scientific literature.

Yang Li, Ph.D. candidate at the National Science Library, Chinese Academy of Sciences. His main research interests include knowledge extraction from scientific literature in vertical domains and multimodal data mining of scientific literature.

Yajiao Wang, Ph.D. candidate at the National Science Library, Chinese Academy of Sciences. Her main research interests include domain knowledge extraction and scientific data extraction from scientific literature.

Construction Achievements and Innovative Application Exploration of the CSTR Identification Platform

Yuanchun Zhou, Shu Wang, Jia Liu, Xiaolei Xia, Lijuan Wang, and Xuefeng Lv

Abstract With the rapid development of technology, the management and utilization of science and technology resources have become key to enhancing national innovation capacity and driving industrial upgrades. The Common Science and Technology Resource Identification Platform (CSTR identification platform), based on China's national standard GB/T 32843-2016 "*Science and Technology Resource Identification*", was established by the Computer Network Information Center, Chinese Academy of Sciences. This platform provides unique identification service for 11 resources types, including academic papers, dissertations, scientific data, and preprints, etc. It promotes the establishment of a globally interconnected system for science and technology resources across disciplines, regions, and platforms, enabling the tracking of global impacts and facilitating rapid access to resources. This article analyzes international policies on science and technology resources identification, summarizes the platform's construction achievements, and explores its innovative applications across multiple disciplines. These efforts aim to enhance the management, sharing efficiency, and utilization benefits of science and technology resources. Looking forward, the platform will play a pivotal role in driving digital transformation and intelligent development, providing robust support for technological innovation and socio-economic progress.

Keywords Science and technology resource identification (CSTR) · Identification services · Citation tracking · Data certification

1 Introduction

In the context of increasingly fierce global technological competition, the efficient management and open sharing of science and technology resources have become critical elements in driving national science and technology innovation and industrial upgrading. China is currently at a pivotal stage in its efforts to build a world-class

Y. Zhou (✉) · S. Wang · J. Liu · X. Xia · L. Wang · X. Lv
Computer Network Information Center, Chinese Academy of Sciences, Beijing, China
e-mail: zyc@cnic.cn

China's e-Science Blue Book 2025, https://doi.org/10.1007/978-981-96-8252-2_9

science and technology power and promote innovative development. In 2016, China issued the national standard GB/T 32843-2016 "*Science and Technology Resource Identification*," providing a unified standard for the identification of various types of science and technology resources and serving as the foundation for related services and applications in China. In 2018, the Chinese Ministry of Science and Technology and Ministry of Finance jointly issued the "*Administrative Measures for the National Science and Technology Resource Sharing Service Platforms*," promoting the open sharing of science and technology resources to society. These measures state: All science and technology resources published on national platforms must be labeled in accordance with the national standard; when users utilize science and technology resources from national platforms to produce works, papers, or other publications, they must clearly indicate the identifiers of the science and technology resources and explicitly disclose the circumstances of their use.

To implement national standards, in 2017, the Computer Network Information Center, Chinese Academy of Sciences constructed and launched the Common Science and Technology Resources Identification Service Platform (CSTR identification platform). The platform applies CSTR through heterogeneous identification resolution technologies and system tools, enabling unique identification, location, open access, and sharing of various science and technology resources. This breaks down information silos, promotes the interconnection and sharing of science and technology resources information, enhances the standardization and normalization of resource management, and provides researchers with more convenient and efficient access to resources. Additionally, it accelerates the conversion and application of science and technology achievements.

This article introduces the current development status of international policies on science and technology resource identification, analyzes the challenges faced by the development of science and technology resources identification in China, and summarizes the construction achievements of the CSTR identification platform. Additionally, it showcases the platform's innovative applications across multiple disciplinary areas. The aim is to further improve the management level, sharing efficiency, and utilization benefits of science and technology resources in China, drive digital transformation and intelligent development, and provide robust support for science and technology innovation and socio-economic development.

2 Development Status of Science and Technology Resource Identification

2.1 Policy Status

The identification of science and technology resources spans the entire lifecycle of these resources—from their generation, utilization, publication, dissemination, to evaluation. Functioning as the "digital ID" of science and technology resources, its crucial role has become a widely recognized consensus both domestically and internationally.

In 2021, the United Nations Educational, Scientific and Cultural Organization (UNESCO) released the *Recommendation on Open Science*, which, for the first time, designated Persistent Identifiers (PIDs) as a key component of open science infrastructure. The recommendation emphasizes the need for the continuous and stable operation of PID systems and affirms the significant role of PIDs in promoting the open sharing of science and technology resources. Following this, countries such as the United States, the United Kingdom, Germany, the Netherlands, and Switzerland, etc., have successively published national strategies and roadmaps for PIDs.

The issuance of these policies clearly demonstrates that science and technology resources identification, along with other types of PIDs, have become essential tools in the global governance of science and technology resources.

International policies on PIDs primarily focus on three aspects:

(1) **Encouraging Researchers to Adopt PIDs for Enhanced Research Interoperability**

Leveraging the uniqueness and permanence of PIDs, researchers are encouraged to use PIDs throughout the entire research lifecycle. This makes research outcomes more easily discoverable and sharable, enhances the interoperability of research results, and benefits researchers in multiple ways. By utilizing PIDs, researchers can improve the efficiency of maintaining their publication lists, reduce the time and effort spent on routine administrative tasks, alleviate the burden of administrative work, and better protect the intellectual property rights of their research outcomes. Overall, the adoption of PIDs supports a more streamlined, interconnected, and secure research environment.

(2) **Promoting Institutional Use of PIDs to Improve Research Tracking**

Research institutions are encouraged to incorporate PIDs into personnel and project management processes. By leveraging PIDs, these institutions can enable seamless connectivity between diverse science and technology resources, such as researchers, research outputs, research institutions, and research equipment. This approach simplifies the tracking processes, effectively promoting the monitoring and evaluation of research outcomes. Additionally, it reduces the complexity and costs associated with resource tracking while advancing the implementation of open science principles.

(3) **Building a PID Ecosystem to Boost National Innovation Competitiveness**

By leveraging PIDs to support national research and innovation ecosystems, countries can better assess research quality and influence, improve the efficiency of their research and innovation systems, strengthen their competitiveness in global research and innovation, promote cross-departmental collaboration, and further enhance their position and influence in open science activities.

Against this global backdrop, China has also placed great emphasis on the standardized management, open sharing, and secure utilization of science and technology resources in recent years. Additionally, China has achieved notable progress in the construction of its science and technology resources identification system. The country has issued national standards and a series of management measures related to science and technology resources identification. It has been gradually improving the science and technology resources identification system and has introduced policies to support the use of unique identifiers for science and technology resources. Furthermore, China plans to implement related projects focusing on science and technology resources identification and drive technological breakthroughs in key areas.

(1) **CSTR Standard Development and Policy Implementation**

In August 2016, China issued the national standard GB/T 32843-2016 "*Science and Technology Resource Identification*" [1], establishing Common science and technology resource Identification (CSTR). The standard provided detailed specifications for the objects and generation methods of technology resource identification, the structure and writing rules of identifiers, as well as the management and application of technology resource identification. It created a standardized framework for subsequent cataloging, registration, publication, querying, maintenance and management of technology resources.

(2) **Subsequent Policy Development and Resource Management**

To further strengthen and regulate the management of technology resources, ensure their security, and enhance their level of openness and sharing, the state sequentially issued the "*Administration Measures for Science Data*" and the "*Administration Measures for the National Science and Technology Resource Sharing Service Platform*" [2]. These measures clarified the division of responsibilities for technology resource management among different departments and regions, covering aspects including scientific data collection, submission and preservation, sharing and utilization, as well as confidentiality and security. From a policy perspective, these measures further reinforced the role of researchers using technology resource identifiers [3, 4]. These provisions provided clear guidance for subsequent management departments and researchers [5, 6], aiding in better understanding, mastering and implementing relevant regulations, and thereby providing strong support for national innovation, socio-economic development and national security.

Looking to the future, with the continuous improvement and implementation of relevant policies, China will further strengthen the integration and utilization of technology resources, providing powerful support for national technological progress and security assurance.

2.2 Science and Technology Resource Identification Application Status

Currently, major international PIDs for science and technology resources include the Digital Object Identifier (DOI), the Open Researcher and Contributor ID (ORCID), and the Research Organization Registry (ROR), among others.

(1) Usage and Application Scenarios of DOI

The DOI was initially developed based on a joint initiative by three professional associations within the publishing industry: the International Publishers Association, the International Association of Scientific, Technical & Medical Publishers, and the American publishers Association [7]. In May 2012, the DOI was formally certified by the International Organization for Standardization (ISO) and the standard "ISO 26324:2012 *Information and documentation—Digital Object Identifier system*" was released [8]. Currently, the DOI is developed and managed by the International DOI Foundation (IDF) [9]. Although the DOI was originally designed for the publishing industry, it has now been widely adopted in multiple domains, including data management and preservation, image processing, and intellectual property protection [10]. DOI registration is handled by 12 global registration agencies authorized by the IDF. As of December 2024, the cumulative registration volume has exceeded 300 million entries, with resolution instances exceeding 949 billion.

(2) Usage and Application Scenarios of ORCID

ORCID is a global identifier for individuals, designed to uniquely identify scientists, academic authors, and contributors, addressing the challenges of identity verification and name ambiguity [11]. Initially known as the "Open Researcher and Contributor Identification Initiative," its software prototype was adapted from the ResearcherID system used by Thomson Reuters. ORCID, Inc., was officially established and began operations in the United States in August 2010 [12]. As of September 4, 2024, ORCID has registered 8.68 million active researcher accounts, integrated with 5,757 active systems, and serves 1,405 institutional users, making it one of the most widely used for researcher.

(3) Usage and Application Scenarios of ROR

ROR is a community-driven identifier for research organizations, providing unique organization identifiers (ROR IDs) for institutions related to research. It builds upon the GRID initiative's seed data and is jointly managed by the California Digital Library, CrossRef, and DataCite. With a community-driven governance model, ROR determines its future direction and strategic decisions. As of now, ROR has assigned ROR IDs and related metadata to over 111,392 research organizations. It is supported as an organizational identifier in CrossRef DOI metadata, DataCite DOI metadata, and ORCID. ROR finds extensive application in journal publishing systems, data repositories, funding projects, and grant management platforms.

Currently, international PIDs such as DOI, ORCID, and ROR continue to dominate the market [13, 14], and the role of science and technology resource identifiers in science and technology development and innovation is increasingly significant [15].

3 CSTR Identification Platform Effectiveness

To implement the national standard GB/T 32843-2016 "*Science and Technology Resource Identification*," the Computing Network Information Center of the Chinese Academy of Sciences developed the science and technology resource identification Platform in 2017 [16] (see Fig. 1). The platform provides unique resource identification services to global science and technology resources, including datasets, research papers, preprints, and patents, etc. It serves as a crucial component of China's national standard in addressing global science and technology resource management and is a vital foundational platform for promoting international recognition and interoperability of China's standards.

Currently, the CSTR identification platform has integrated 11 resources types, including scientific datasets, academic papers, theses, patents, preprints, and species, etc. [16], and provides services such as identification resolution and multi-dimensional citation tracking. These offerings support the enhancement of China's governance of science and technology resources and facilitate international recognition and open sharing. As of December 2024, the platform has cumulatively registered over 5.57 million CSTRs, covering 11 resource types and serving more than 262 institutional users.

Fig. 1 CSTR identification platform homepage

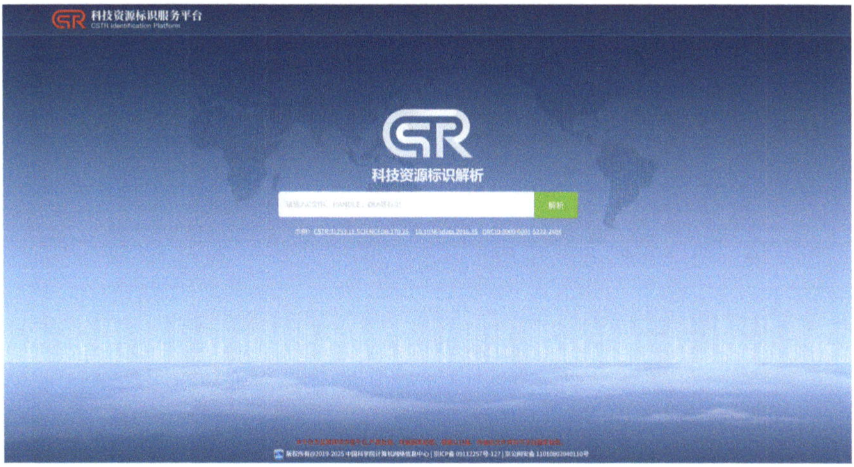

Fig. 2 Identifier resolution service interface

3.1 Service and Application Scenarios

3.1.1 CSTR Resolution Service

Persistent access to science and technology resources is a crucial foundation for their open sharing. The CSTR resolution service supports researchers in accessing resources through stable resolution addresses, thus avoiding disruptions due to changes in resource URLs. To enhance the reliability and stability of the resolution service, the CSTR identification platform ensures high availability through the Same-City Dual-Active with Cross-Region Backup deployment strategy. Additionally, the CSTR identification platform provides resolution services for mainstream international and domestic identifiers, including Handle, DOI, ORCID, and ROR, and among others (see Fig. 2). Researchers can initiate resolution either by using the "CSTR identification platform address + identifier" format or by entering the identifier directly in the CSTR identification platform's online interface.

3.1.2 Citation Tracking Service

The CSTR identification platform provides citation tracking services based on registered science and technology resources' CSTR and DOI, allowing precise tracking of the usage and citation of various resources such as datasets, journal articles, and patents, etc. This service comprehensively analyzes the changes in their influence and impact. Through functions like citation analysis and new citation email notifications (as shown in Fig. 3), the platform assists research institutions and researchers in effectively evaluating the academic influence and social value of science and

Fig. 3 Platform analysis of science and technology resource citation

technology resources. Additionally, the platform supports analysis of citation distributions and journal impact factors, providing precise and reliable evidence for the management, evaluation, and optimization of resources. This further promotes the efficient development of global scientific innovation and collaboration.

3.1.3 Institutional User Report Service

The CSTR identification platform provides regular user reports for institutional users (see Fig. 4). By statistically analyzing key metrics such as the registrations, resolutions, and citations, combined with multi-dimensional indicators such as citation rankings, resolution rankings, and rankings of countries accessing resources, the platform comprehensively evaluates the usage and influence changes of resources. The platform presents these insights in an intuitive and easy-to-understand manner, aiding institutional users in maintaining the accessibility of registered resources and ensuring their long-term stability. Additionally, by providing detailed statistical analyses of various indicators, the platform helps institutional users understand the attention levels of researchers from different countries toward the resources, offering valuable evidence for resource optimization and research collaboration.

3.1.4 Data Certification Service

One of the key challenges facing researchers in the process of open sharing is the issue of asserting rights over scientific data. As the demand for openness and sharing of science and technology resources grows, clearly defining the ownership of scientific data and safeguarding the rights of researchers have become pressing needs for

Fig. 4 Institutional user report

resolution. Collaborating with the Scientific Data Chain, the CSTR identification platform has introduced a scientific data asset attestation service (see Fig. 5). By providing a unique and permanent identifier for attested scientific data assets and their associated proofs, and combining this with blockchain technology, the platform ensures the immutability of the data and the exclusive rights to intellectual property. This effectively protects researchers' rights during the registration and sharing of science and technology resources, preventing disputes over data property rights.

科学数据链 Science Data Chain | 科学数据银行 Science Data Bank

科学数据资产存证证明

本证书予以证明用户通过科学数据链（Science Data Chain）平台进行了证书内文件的存证，证明文件自存证时起即客观存在且内容完整，未被篡改。

数据集：Spatial transcriptomics data of mouse thymus

作者：Jingwei Ma,Liang Tang,Jingxuan Xiao,Bo Huang,Junwei Liu

CSTR：31253.11.sciencedb.18291

DOI：10.57760/sciencedb.18291

数据量：136075966912 字节

文件数： 8

版本：V1

上链日期：2024-12-13 23:33:28 GMT+0

时间戳（中国科学院国家授时中心）：1734104008889

TXID：c0c00417ae8b7e68df85d53a9ba3f9050749ee7adcf118c574c01650fd16b57f

数据Hash：7c0d05ad5fdd7cd2cce79a4e7a8b70cae82c3fe15a65f0c61fbd409819595ea7

数据链接： https://cstr.cn/31253.11.sciencedb.18291

科学数据链技术支持： https://www.scichain.cn/

32126.41.SCICHAIN.274126083.CN.V1

Fig. 5 Scientific data assets archival proof

3.2 *Platform Achievements*

3.2.1 Supporting Cross-Border Flow of Species Resources in "The Chinese Species Catalogue"

With the rapid growth of biodiversity data and the deepening of global research collaborations, the standardized management of science and technology resources has become a critical foundation for sharing and utilizing biodiversity resources. Collaborating with the Species 2000 China Node, the CSTR identification platform provides unique identification services for over 200,000 species in the 2024 edition of "The Chinese Species Catalogue" (see Fig. 6). This collaboration integrates species data from various countries and standards, enhancing the traceability of species data, and driving the in-depth development of biodiversity conservation and ecological research. Furthermore, it provides technical support for the continuous discovery and application of global biodiversity resources.

The CSTR identification platform assigns CSTR identifiers to species resources, helping to break down data silos between domestic and international platforms in the biodiversity field. By adhering to unified standards, it promotes the coordinated development of global species resources, establishing a standardized framework for global science and technology resource sharing. This ensures the interoperability and uniformity of cross-domain species data, providing a robust technical foundation for collaborative research among different countries and regions under a consistent data standard.

Fig. 6 Details page of "The Chinese Species Catalogue" 2024 edition

Fig. 7 Example of intellectual property registration for scientific data

3.2.2 Supporting and Safeguarding Intellectual Property Rights for Scientific Data

The CSTR identification platform collaborates with the Zhejiang Intellectual Property Bureau based on the "Opinions of the Zhejiang Provincial People's Government Office on Deepening Data-Related Intellectual Property Reform for High-Quality Development." This partnership promotes the registration of intellectual property for open science and technology resources by key domestic universities, research institutions, laboratories, and other research organizations using CSTR. The aim is to clarify ownership of intellectual property, boost researchers' enthusiasm, and enhance the operational efficiency of research institutions.

The CSTR identification platform facilitates the integration and fusion of science data platforms and intellectual property registration platforms (see Fig. 7). This promotes the registration, circulation, and sharing of intellectual property related to science and technology resources across fields such as natural sciences and engineering sciences. It also advances the development of key technologies for trusted circulation of science and technology resources, aids in tracking the influence of management institutions, and fosters seamless connectivity of science and technology resources throughout the research process.

4 CSTR Innovative Applications

4.1 CSTR Biodiversity Application

Biodiversity serves as the foundation for human survival and societal development, and it is also crucial for ecological civilization construction and the sustainable development of nations. However, with the intensification of climate change and human activities, biodiversity now faces unprecedented threats. Therefore, strengthening biodiversity conservation and research has become particularly important. Against this backdrop, exploring the application of science and technology resource identification in the biodiversity field holds significant meaning. By fully utilizing its unique advantages, CSTR can drive the standardization, regularization, and sharing of biodiversity data, providing strong support for biodiversity conservation and research.

The CSTR identification platform collaborates with the Tashkent Botanical Garden of the F. N. Rusanov Botanical Institute, part of the Academy of Sciences of the Republic of Uzbekistan (see Fig. 8). The collaboration registers species resource data from Uzbekistan based on the CSTR species nomenclature coding standards, constructs a global network of species interconnections, and offers related application services.

(1) Data Standardization and Normalization

CSTR assigns unique identifiers to biodiversity data, promoting standardization and normalization. This addresses issues such as inconsistent formats and standards during data collection, storage, processing, analysis, and sharing. As a result, data comparability and reusability are enhanced. The South China Botanical Garden of the Chinese Academy of Sciences has utilized CSTR to precisely locate and retrieve botanical species and herbarium resources, ensuring data accuracy and reliability.

(2) Data Sharing and Integration

The CSTR-based biodiversity data sharing platform enables cross-institutional and cross-domain data sharing and integration. This collaborative mechanism helps break down information silos, optimize resource allocation, and improve the efficiency of

Fig. 8 Species resource introduction page of Uzbekistan

data utilization. Researchers can access required biodiversity data quickly, fostering interdisciplinary and cross-sectoral collaboration.

(3) **Intelligent Analysis and Decision Support**

The integration of CSTR with cutting-edge technologies such as big data and artificial intelligence enables the intelligent analysis of biodiversity data. Through in-depth mining and analysis of vast biodiversity data, underlying patterns and trends are revealed, thus providing robust decision-making support for biodiversity conservation and research.

(4) **Germplasm Resources Conservation and Utilization**

In biodiversity conservation, the protection and utilization of germplasm resources are paramount. CSTR facilitates comprehensive management and protection of such resources. By assigning unique identifiers to germplasm resources, precise tracking and management are ensured, safeguarding their security and integrity. Moreover, germplasm resource repositories built on CSTR enable long-term preservation and effective utilization, providing critical support for biodiversity conservation and sustainable use.

(4) **Public Education and Engagement**

CSTR can also be applied in biodiversity education and public engagement. By assigning easily understandable and memorable identifiers to biodiversity resources, the public can access and learn about biodiversity knowledge more conveniently. Additionally, CSTR-based educational platforms offer diverse content, including images, videos, and animations, enhancing public awareness and participation in biodiversity conservation.

The application of CSTR in the biodiversity field is broad and far-reaching. It not only promotes the standardization, normalization, and sharing of biodiversity data but also improves resource utilization efficiency, providing robust support for biodiversity conservation and research. With advancing technology and deepening applications, the future of CSTR in this field is poised to grow even further.

5 Future

With the increasing application of science and technology resource identifiers in various fields, the CSTR identification platform will further promote technological integration and intelligent development under the impetus of data and intelligence. By deeply integrating with artificial intelligence (AI), the platform will undergo a transformative shift toward advanced intelligent, realizing automatic identification, classification, and management of science and technology resources. Leveraging AI technology, CSTR will be widely applied in areas such as automatic annotation, intelligent retrieval, and personalized recommendations, thereby enhancing the efficiency of resource utilization and user experience. Addressing the challenges in tracking and analyzing the entire lifecycle of science and technology resources, the CSTR identification platform will adopt digital solutions to achieve comprehensive resource management and optimized utilization, further improving the open sharing and interoperability of resources.

To promote the sharing and exchange of science and technology resources, the standardization of CSTR will become a significant trend. By establishing unified identifier standards and specifications, the platform ensures interoperability between different systems, thereby enhancing resource utilization efficiency and accuracy. Moving forward, CSTR will advance in the directions of intelligent, digitization, and standardization, providing users with more convenient and efficient services for science and technology resource identification.

Acknowledgements Special thanks to the National Key Research and Development Program "*Key Technology Research and Application of Trusted Services for Typical Science and Technology Resource Identification*" (2023YFF0616900) for its support.

References

1. National Standardization Technical Committees. Science and Technology Resource Identification. GB/T 32843-2016 [S]. Beijing: Standards Press of China, 2016: 1.
2. XING W, HONG C. Opening is Normal—Interpretation of Items about Scientific Data Sharing and Utilization in The Scientific Data Management Rules [J]. Library Tribune, 2019, 39(01): 117–124.
3. DU P, LI Y, ZHANG X. Research on Conceptual Domain and Management Boundary of Data Set in China's Measures for Scientific Data Management[J]. Library Journal, 2022, 41(04): 96–105.114. https://doi.org/10.13663/j.cnki.lj.2022.04.013
4. ZHU Y, HU L, KONG L, et al. Research and Popularization of Scientific Data Citation for National Standard of the People's Republic of China[J]. Frontiers of Data & Computing, 2018, 9(06): 25–30.
5. WU Y. The Construction Trend of Digital Archives in the Internet Environment: Digital Object Unique Identification Technology[J]. China Management Informationization, 2017, 20(24): 187–188.
6. ZHANG J. Research on the Construction of a Cooperative Sharing Platform for Scientific and Technological Information Resources[J]. Science and Technology Innovation Herald, 2012, (13): 41–46. https://doi.org/10.15941/j.cnki.issn1001-0424.2012.13.016.
7. GU Q. The Application Research of DOI[J]. Journal of Modern Information, 2009, 29(05): 73–76.
8. DAI X. The Role of Establishing Chinese Identifiers in the Publication of Scientific Journals[J]. New Media Research, 2016, 2(18): 55–56. https://doi.org/10.16604/j.cnki.issn2096-0360.2016.18.026.
9. REN R, LIU L, PU D, MI J. The Path for Chinese Exploring the Development Direction of Chinese Digital Object Unique Identifiers through the Relationship between Bibliographic Citations and DOI[J]. Journal of Library Science in China, 2010(2): 115–121.
10. TU Y, PENG J. Research on the Application of Digital Object Identifier in China's Scientific Data[J]. Digital Library Forum, 2013, (08): 31–36.
11. Let the World Understand a Better You: Building a Complete Author Academic ID-ORCID [J]. Chinese Journal of Tissue Engineering Research, 2025, 29(17): 3519.

12. ORCID as an International Tool for Academic Communication [J]. Journal of Practical Stomatology, 2023, 39(06):813.
13. HE D ZHANG X. Study of the Applications of DOI in Scientific and Technical Information Sharing[J]. Data Analysis and Knowledge Discovery, 2007, (08): 26–29.
14. ZHANG T, CHEN J. Restraints and Countermeasures for the Sharing of Science and Technology Resources in China[J]. Academic Exchange, 2010, (07): 131–134.
15. Chinese Association for Science and Technology. The Blue Book on the Development of China's Science Journals 2021 (Special Topic on Academic Publishing under Open Science Environment)[M]. Beijing: Science Press, 2021: 115–116.
16. LIU J, XIA X, WANG S, WANG L, et al. Science and Technology Resource Identification Service System and Its Innovative Applications [J]. Frontiers of Data and Computation Development, 2020, 2(6): 62–73.

Yuanchun Zhou Ph.D. supervisor, researcher, deputy director and academic committee chair of the Computer Network Information Center, Chinese academy of sciences, director of the Scientific Data Center of CAS, and director of the special committee on scientific data of the Chinese information association. He has been engaged in research on scientific data, data intelligence, and data identification, and has undertaking multiple national-level projects including national natural science foundation key projects, National Natural Science Foundation of China major research plan key support projects, and National Key Research and Development Program of China. He has been awarded 7 national standards; authored 4 monographs; secured over 40 authorized patents; and published more than 100 high-quality papers in international renowned journals and conferences such as KDD, WWW, AAAI, VLDB, SIGMOD, IEEE TKDE, AIJ, and cell research. He has received multiple first and second prizes at provincial and ministerial levels.

Shu Wang PhD, is a Senior Engineer at the Lab of Identification Technology and Application Services, Computer Network Information Center, Chinese Academy of Sciences. She specializes in Identification technology research and application services. She oversees the National Key R&D Program subproject on Secure and Efficient Hierarchical Management and Cross-domain Secure Resolution of Heterogeneous Identification under the Cybersecurity Special Project. She is also involved in the 2019 Industrial Internet Development Project and the Phase III Scientific Big Data Project on Heterogeneous Identification Management and Service Platforms. Previously, she contributed to the NDRC Industrialization Special Projects, including Research and Formulation of IoT Foundation Standards and the IoT Identifier Public Service Platform. She authored the book Key Technologies and Applications of IoT Identifiers, participated in the development of 9 national standards and 2 international proposals, published over 10 papers, and holds 3 patents.

Jia Liu with a master's degree, is a member of the Chinese Academy of Sciences' Youth Promotion Association. She serves as the Director and Senior Engineer of the Identification Technology and Application Services Lab at the Computer Network Information Center, Chinese Academy of Sciences. She is a member of the State Council's Food Safety Special Committee, an expert in the International Organization for Standardization's Subcommittee on Identification and Description

(ISO/TC46/SC9/WG8), and a member of the Persistent Identifiers for eResearch (ePIC) Management Committee. Her research focuses on the policy, standards, and practices of data identifiers, distributed data resolution technology and systems, and big data management and analysis. She leads the Ministry of Science and Technology's National Basic Conditions Platform Center project, Research and Development of Generic Technical Standards for the Construction of the Science and Technology Resource Identification System. As a principal investigator, she has undertaken over 20 research projects, including the National Key R&D Program subproject Dual Identification Fusion and Resolution Technology for Scientific Data, the National Natural Science Foundation of China, and the Ministry of Industry and Information Technology's Industrial Internet Innovation and Development Project. She has developed 17 standards (including 7 national standards and 1 international proposal) and holds 3 authorized patents.

Xiaolei Xia with a master's degree, is an Engineer at the Identification Technology and Application Services Lab of the Computer Network Information Center, Chinese Academy of Sciences. Her main work involves researching international standard developments. She has participated in national key research and development projects, including Key Technologies and Applications for Trusted Services of Typical Science and Technology Resource Identification and Big Data Platform for Smart Ecological Livestock Based on Blockchain Technology. She has contributed to 2 published national standards, authored 3 papers, and holds 1 authorized patent.

Lijuan Wang with a master's degree, is an Engineer at the Identification Technology and Application Laboratory within the Big Data Technology and Application Development Division at the Computer Network Information Center, Chinese Academy of Sciences. Her primary work involves research on Identification resolution technology and the design of platform products. She has participated in the following national key research and development projects:- Key Technologies for Trustworthy Services of Typical Science and Technology Resource Identification and Identification Applications – 2019 Industrial Internet Innovation and Development Project: Public Service Support Platform for Industrial Internet Identification Resolution – National Import Cold Chain Food Traceability Management Platform.

Xuefeng Lv is an Engineer at the Identification Technology and Application Services Lab of the Computer Network Information Center, Chinese Academy of Sciences. His primary research focuses on the technology and application of scientific data Identification. He has served as the principal investigator for the National Key R&D Program subproject, Key Technology Research on Evidence Testing and Source Tracing for Endangered Animal Crimes. Additionally, he has participated in other research projects, including the National Industrial Internet Big Data Center Construction Project and the Imported Cold Chain Food Traceability Management Platform.

Practical Applications

Artificial Intelligence-Driven Civil Aircraft Design

Guanghui Wu, Miao Zhang, Jing Wang, Hairun Xie, Dayong Dong, Zhaoguang Tan, and Junqiang Bai

Abstract Modern civil aircraft design is a complex system engineering task that involves multiple disciplines and requires design teams to possess a rigorous knowledge base and innovative capabilities to ensure precision and foresight. However, with the intensifying competition in the civil aircraft market, the design and development cycles are continually being shortened, leading to significant bottlenecks in existing design methods that struggle to adapt to rapidly changing market demands. This paper focuses on the empowering strategies of artificial intelligence technology in civil aircraft design, delving into the concept of intelligent design and its common challenges. First, it discusses the potential of intelligent design in enhancing design efficiency, accuracy, and innovation. Next, it introduces key technologies related to intelligent civil aircraft design, including machine learning, optimization algorithms, and simulation technologies. Finally, it elaborates on China's explorations and practical applications in overall design, aerodynamic optimization, structural analysis, and cockpit design, demonstrating how artificial intelligence technology plays a vital role in various aspects of civil aircraft design.

Keywords Civil aircraft design · Artificial intelligence · Intelligent design · System engineering

G. Wu · M. Zhang (✉) · H. Xie · D. Dong · Z. Tan
Shanghai Aircraft Design and Research Institute, Shanghai, China
e-mail: zhangmiao@comac.cc

J. Wang
School of Aeronautics and Astronautics, Shanghai Jiao Tong University, Shanghai, China

J. Bai
School of Aeronautics, Northwestern Polytechnical University, Xi'an, China

China's e-Science Blue Book 2025, https://doi.org/10.1007/978-981-96-8252-2_10

1 Introduction

In the development of modern civil aircraft (hereinafter referred to as civil airplanes), while continuously pursuing safety, environmental friendliness, and comfort, there is an increasingly higher demand for economic efficiency. Boeing's B737 and Airbus's A320 are classic works in the history of aircraft design and occupy a large market share in the field of civil airplanes. The C919 is a 150-seat passenger aircraft with completely independent intellectual property rights in China. It has been successfully put into safe operation by China Eastern Airlines, China Southern Airlines, and Air China, marking that China's civil airplanes are competing with models of the same class as the B737 and A320 in the market. At the same time, China is carrying out the design of various derivative models of the C919 and is committed to researching the long-haul wide-body C929 passenger aircraft with a higher flight speed, a longer range, and a larger commercial load [1]. However, with the increasingly stringent competitive demands of large passenger aircraft in terms of economic efficiency, safety, and environmental friendliness indicators, the design and manufacturing capabilities of China's civil airplanes face higher challenges.

The design of modern civil aircraft is a highly integrated and complex systems engineering involving multiple disciplines. It requires a close combination of a rigorous knowledge system and innovative capabilities to ensure the accuracy and forward-looking nature of the design. Usually, it starts from market demands to propose an initial concept, and through continuous iteration and optimization, the final design goal is gradually achieved. At each stage of civil aircraft design, professional teams need to conduct multi-level analysis and optimization of the design plan from preliminary to detailed levels. This includes overall layout, aerodynamic performance, structural strength, handling and stability characteristics, performance evaluation, safety and economic analysis, etc. The plan is refined through continuous iteration, ultimately forming an overall optimal design [2, 3]. Since the design of civil aircraft highly depends on rich engineering experience, a large amount of historical data, and complex design tools, foreign countries have long imposed strict technological blockades in this regard. Therefore, there is an urgent need for us to independently develop core technologies and establish design standards, processes, models, and tools that meet our own needs, so as to break through technological barriers and enhance the international competitiveness of China's civil aircraft.

In terms of design methods and means, aviation-powerhouses such as the United States, European Union countries, and Russia, relying on profound design experience and technological iterations, have established mature rapid-design methods, tools, and rich engineering databases. In China, numerical simulation methods are mainly used for design. For example, CFD (Computational Fluid Dynamics) simulation is used to simulate the aerodynamic characteristics of the aircraft's flow field, and FEM (Finite Element Method) simulation is used to evaluate the stress, strain, vibration and other responses of the structure. Practice has proved that these technical means can meet the urgent design schedule requirements and achieve relatively higher design goals [4].

However, with the intensification of competition in the civil aircraft market, the design and development cycle of aircraft needs to be continuously shortened, and the existing design methods and means still face the following bottlenecks: First, the traditional aircraft design process is complex, involving numerous professional departments. There is a large amount of manual coordination and data transfer in the analysis and evaluation, which significantly reduces the design efficiency. Second, although numerical simulation methods can provide relatively accurate single-discipline results in the performance evaluation of each discipline, the overall efficiency is still insufficient, restricting the effective shortening of the development cycle. Finally, when facing multiple sets of multi-level indicators, the existing design methods are difficult to meet all requirements at once. Usually, multiple iterations and adjustments in each discipline are required, resulting in a reduction in design efficiency. Moreover, multi-discipline plan conflicts may occur in the later stage of the design, forcing the team to make important trade-offs.

In response to the above-mentioned challenges faced by China's civil aircraft design, it is necessary to increase research and development efforts in aspects such as the theorization of design methods, the systematization of tools and experience, the systematization of multi-disciplinary coordination, and the standardization of professional organization management. Especially in terms of the continuous in-depth development of basic scientific research capabilities and the accumulation of systematic innovation, Artificial Intelligence (AI), as a disruptive technology, is profoundly influencing the transformation of various engineering fields. At present, large AI models have begun to surpass local human thinking abilities, and their reasoning abilities are increasingly improving. This progress will bring profound changes to the field of aircraft design. Firstly, AI can significantly improve the efficiency of numerical simulation, making multi-disciplinary performance evaluation faster and more accurate, thus effectively shortening the iteration cycle. Secondly, AI technology can comprehensively reconstruct the existing engineering knowledge, experience system, and design process. This not only helps to quickly adjust the design plan and improve the efficiency of data transfer among various professional departments, but also enables engineers to more easily access knowledge from other specialties, allowing them to focus more on plan evaluation and innovative thinking. Finally, the improvement of AI's reasoning ability is conducive to better handling multiple sets of multi-level complex design index requirements. Combined with big data analysis and optimization algorithms, it can identify better design plans in the initial stage of design, reduce multiple iterations among various disciplines, and ensure the improvement of overall design quality and efficiency.

This paper is based on the enabling strategy of artificial intelligence technology in the field of civil aircraft design and explores the intelligence of civil aircraft design. Firstly, it discusses the concept of intelligent civil aircraft design and the common problems it faces. Then, it introduces the key technologies related to intelligent civil aircraft design. Finally, it specifically elaborates on the exploration and practical applications of China in the design of civil aircraft overall layout, aerodynamics, structure, and cockpit, etc., providing a reference for the further application of artificial intelligence technology in the whole life cycle of civil aircraft design.

2 Definition and Problem Analysis of Intelligent Civil Aircraft Design

2.1 Intelligent Design Concepts for Modern Civil Aircraft

The core objective of modern intelligent design for civil aircraft is to enhance the efficiency and performance of aircraft design, reduce R&D costs, and support the sustainable development of the aviation industry through the in-depth application of artificial intelligence technology. Based on this, an intelligent design framework as shown in Fig. 1 is proposed, aiming to achieve the digital and intelligent transformation of all aspects of the aircraft's entire life cycle, including design, manufacturing, operation, and maintenance. The concept of intelligent design not only requires the full sharing of data and knowledge in all stages and disciplines but also needs to integrate physical and virtual simulations to promote the iteration and self-evolution of design plans. Combining the key engineering problems in civil aircraft design practice and relevant breakthroughs in basic research, the intelligentization of modern civil aircraft design is expected to play an important role in the following three fields:

1. Multidisciplinary Collaborative Intelligent Design: By introducing the concept of multidisciplinary collaborative design, AI technology is utilized to extract and integrate massive data from different disciplines, identifying the mutual coupling among various disciplines, thus optimizing the design process. Machine learning technology can assist experts in extracting useful information from experimental data, standards, specifications, and knowledge bases, providing

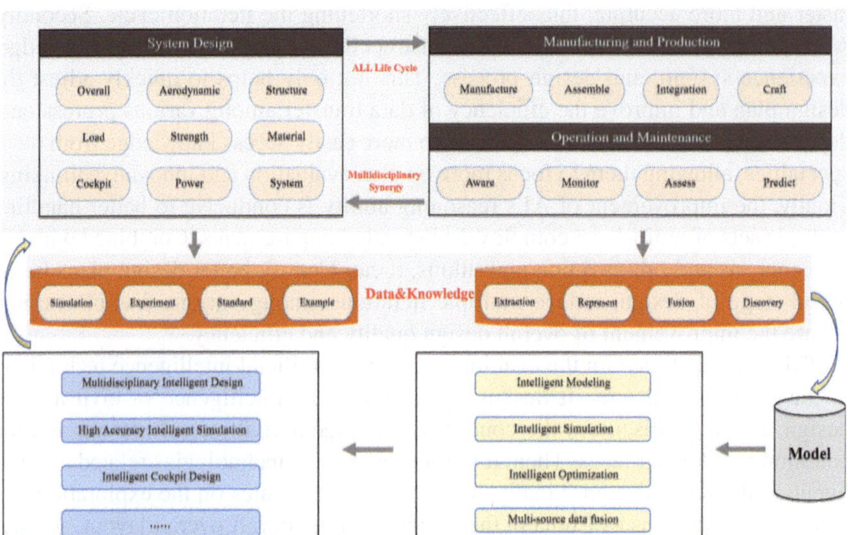

Fig. 1 Overall framework of modern civil aircraft intelligent design

support for decision-making. This ensures that a global optimal solution can be found under multiple sets of multi-level design indicators, thereby enhancing the overall design performance.

2. Intelligent Simulation Models and Intelligent Optimization Techniques: The introduction of intelligent simulation technology makes performance evaluation and optimization more accurate, reduces the number of iterations, and accelerates the design cycle. Through multi-scale, multi-disciplinary, and multi-physical-field modeling, AI can efficiently integrate sensor data with the simulation results in the design phase, achieving an accurate match between the simulation model and the actual aircraft state. At the same time, AI can also intelligently optimize complex design indicators, search for the global optimal solution, and ensure that the design scheme meets the performance requirements in various aspects.
3. Intelligent Management throughout the Life Cycle: Intelligent design should not be confined to the initial design stage of the aircraft but should span the entire product life cycle. By integrating the design, manufacturing, and operation data generated at each stage, big data technology is applied for condition assessment, life prediction, and mission execution assessment. The feedback information is then used to adjust and optimize the design in real-time, making the design anticipatory and forward-looking. Meanwhile, AI technology can combine off-line assessment with actual operations, enabling the aircraft design scheme to have the ability of self-evolution. With the help of a digital simulation platform, decision-makers can quickly grasp the current state of the system, providing more accurate guidance for future maintenance and improvement.

Intelligent design of civil aircraft, as an emerging field, combines cutting-edge AI technology with century-old classic aviation engineering technologies, demonstrating a trend of interdisciplinary integration. However, during this development process, it still faces some common problems and challenges.

2.2 *Common Challenges in Intelligent Civil Aircraft Design*

2.2.1 High-Quality Data Acquisition

Data lies at the core of modern artificial intelligence models, and model training relies on a large amount of high-quality data. It is precisely the emergence of massive data such as Internet big data and industrial big data that has propelled the development and maturation of large-scale models, attracting widespread attention across various industries. However, in the field of civil aircraft design, due to the involvement of numerous technical and commercial secrets, obtaining high-quality open-source or public data resources is extremely difficult. The directly collected data often contains uncertainties caused by simulation accuracy, experimental errors, measurement noise, etc. Therefore, the screening, cleaning, and calibration of data are of crucial importance to ensure the final acquisition of a high-quality dataset. In addition, the multi-disciplinary performance analysis and evaluation of civil aircraft design require the integration of heterogeneous data sources from different fields, such as flight data, structural data, material data, and environmental data. These heterogeneous data significantly differ in format and accuracy, and their validity also needs further verification. At the same time, although methods such as simulation and experimentation can be used to generate high-precision and high-fidelity data, this process is often costly, severely restricting the scale and diversity of the data, thus limiting data acquisition and application.

2.2.2 Reliability of Models

In recent years, data-driven artificial intelligence models have witnessed rapid development and have been widely applied and studied in various fields. However, civil aircraft design, as a high-risk and high-precision-demanding field, places additional requirements on the reliability of intelligent models. First of all, the prediction results and design schemes of the model should possess reasonable and consistent accuracy or fidelity. This not only requires the model's accuracy to match the needs of different design processes but also demands that the model can provide sufficiently accurate results under all possible scenarios or offer the confidence level of the results for designers to make judgments. Secondly, merely improving the model's accuracy is insufficient. Only models that can be understood can gain the trust of designers. This is partly because the data used to train the model may have fidelity issues. Even high-precision simulation results and experimental data may contain certain errors. In addition, the decisions of "black-box" models are difficult to trust. The model should be able to provide the basis for its predictions and design decisions. In terms of usability, intelligent models should be easy to express design intentions, support manual control and data analysis, thus providing effective support for engineering design.

2.2.3 Generalizability of Models

Different civil aircraft exhibit significant design differences, including but not limited to aerodynamic layout, airframe structure, engine type, and avionics systems. These differences often make AI models trained for a specific aircraft model inapplicable to others, requiring retraining or adjustment for new aircraft types. The design data of different aircraft models possess unique characteristics and patterns, such as variations in data dimensions, data range, and data precision. When transferring AI models across different aircraft types, inconsistencies in data must be addressed, which may result in poor training outcomes on new data or even render the model nonfunctional. Therefore, a major challenge in introducing intelligent methods into civil aircraft design lies in effectively leveraging shared knowledge across different models while developing intelligent models that integrate both general and specific characteristics.

3 Key Technologies in Intelligent Civil Aircraft Design

3.1 Intelligent Modeling

Digital modeling provides a precisely defined digital design space through computer-aided methods, serving as a fundamental cornerstone for transitioning from conceptual design to evaluation and final production [5]. Simulations of aerodynamics and structural performance rely on accurate and detailed three-dimensional digital models. Conventional modeling tasks typically involve repetitive modeling processes, requiring designers to iteratively refine models based on past experience, which incurs significant computational and time costs. Although traditional parametric modeling methods attempt to alleviate this issue, they still face challenges such as difficulties in parameter definition and complex logical structures. These limitations result in a strong dependence on user experience and a lack of flexibility in method adaptation during practical applications. In aerodynamic shape design, the modeling process must also address surface combination and stitching, discrete meshing for computational evaluation and optimization, and final surface reconstruction. Consequently, in experience-based traditional modeling workflows, a vast amount of accumulated modeling data and samples over long design cycles often remain underutilized, limiting their potential for reuse.

In recent years, with the rapid advancement of deep neural networks and generative artificial intelligence technologies, intelligent modeling has achieved breakthroughs in academia and garnered widespread attention in industry. By leveraging deep learning to process 3D scanned point clouds, mesh surfaces, and implicit function representations, intelligent tools for generating 3D models from text and images have been widely adopted in the entertainment and art sectors. Meanwhile, the adoption of generative AI in professional Computer-Aided Design (CAD) is revolutionizing design workflows. For instance, Autodesk's Project Bernini enables the generation of 3D models from text, images, point clouds, or voxels [6]. Intelligent modeling technology learns from existing design data and utilizes techniques such as Generative Adversarial Networks (GANs), Variational Autoencoders (VAEs), and diffusion models to reduce complex design spaces into lower-dimensional feature spaces. This process can even distill efficient, compact, and interpretable design representations, significantly reducing the number of parameters required for geometric modeling. Additionally, intelligent modeling allows for direct manipulation within the feature space based on design requirements, enabling the rapid generation of surface geometries. This approach simplifies traditional modeling tasks such as stitching, stacking, and inverse fitting, dramatically accelerating the modeling process.

However, the application of intelligent modeling technology in engineering design still faces a series of challenges that need to be addressed. Currently, both academia and industry have primarily focused on intelligent modeling for the art and entertainment sectors, where the geometric precision requirements for model surfaces are relatively low. These fields typically employ point cloud-based surface mesh reconstruction, which differs significantly from the continuous surface models used in industrial applications. As a result, intelligent modeling techniques based on continuous surface definitions are still under active development.

Furthermore, aerodynamic shape design requires enhanced flexibility and precision in surface definition. Intelligent modeling must tackle challenges such as high coupling in shape deformations, redundant information caused by repeated fitting, and ensuring high-order surface continuity in design definitions. Additionally, for complex multi-component surface structures, it is essential to maintain strong design correlations among different components to construct an effective full-aircraft primary geometry model and ultimately accomplish the generation of large-scale and complex digital prototypes. Finally, due to fundamental differences between deep neural network-based generative modeling and traditional experience-driven modeling workflows, it is crucial to integrate human-in-the-loop design principles. This involves developing interactive, visual, and intuitive modification tools to bridge the gap between AI-generated models and human design expertise.

3.2 *Intelligent Simulation*

In civil aircraft design, simulation technology is employed to model physical processes, system behaviors, and complex scenarios to predict system performance and facilitate optimized design. However, traditional numerical simulations often require substantial computational resources and time, potentially leading to extended design cycles and increased costs. This challenge is particularly pronounced in multi-physics coupling scenarios, where traditional simulation techniques struggle to handle large-scale input data and intricate computational processes, making data transmission and integration significant bottlenecks. Additionally, when dealing with nonlinear complex systems, conventional methods face difficulties in maintaining accuracy and stability, necessitating higher efficiency and precision in simulations.

Intelligent simulation technology integrates artificial intelligence, big data, and traditional simulation models, achieving advancements in the following aspects of complex system simulation, accelerated simulation computation, and optimized simulation workflows:

1. Intelligent Model Correction: By integrating data-driven methods, traditional physical models can be calibrated to enhance their predictive capability for real-world physical phenomena. For instance, deep learning-based model correction techniques can learn from large-scale experimental data to automatically adjust parameters in turbulence models. This enables real-time model updates and optimizations, improving the accuracy and reliability of simulations for more precise turbulence modeling [7].
2. Simulation Substitution: Convolutional Neural Networks (CNNs) effectively process image and mesh data, Graph Neural Networks (GNNs) are well-suited for handling complex structures and relational networks, and Transformer models

excel in processing time-series data and dynamic system simulations. By developing data-driven models to directly replace traditional simulation methods, computational accuracy can be maintained while significantly improving efficiency. For example, Airbus has developed a simulation surrogate model for supercritical airfoil flow fields based on implicit neural representations [8]. Compared to the traditional Reynolds-Averaged Navier-Stokes (RANS) simulation method, this approach achieves a fivefold acceleration, providing a reliable solution for rapid simulations in complex aerodynamic design scenarios.

3. Simulation Acceleration: AI techniques can replace or accelerate iterative solvers in certain simulation methods without compromising accuracy. For example, researchers at Google have employed data-driven discretization to interpolate differential operators onto coarse grids with high precision, achieving a 40–80× speedup in computation [9]. This approach maintains high result accuracy while significantly enhancing simulation speed, offering a more efficient solution for engineering design.

Despite the immense potential of intelligent simulation technology, its widespread application in engineering design still faces numerous challenges. First, while intelligent simulations can significantly speed up the simulation process, their accuracy depends heavily on the quality and quantity of the training data. In the future, achieving a balance between high-fidelity simulation and intelligent simulation will require integrating data-driven methods with physical models to enhance both simulation accuracy and efficiency. Second, for complex and highly nonlinear physical flows such as turbulence and shockwaves, the challenge lies in enabling intelligent simulations to precisely capture the nonlinear characteristics of dynamic systems without relying on extensive computational resources. This remains a key difficulty in applying intelligent simulation technology to engineering design. Additionally, many machine learning models are considered "black boxes," with their internal decision-making mechanisms being difficult to interpret. In the future, it will be crucial to introduce explainability techniques that help designers understand simulation results and effectively apply them in real-world designs. Finally, the future trend of intelligent simulation is to achieve deep automation and intelligence. Future research needs to explore how to deeply integrate automation technologies with intelligent algorithms to realize fully automated simulation processes, addressing potential anomalies and uncertainties that may arise during automation.

3.3 Intelligent Optimization

In the field of engineering design, optimization has always been a core step to enhance product performance, efficiency, and market competitiveness. Optimization design involves a complex decision-making process, relying on accurate computational models, comprehensive performance testing, and cost-benefit analysis. In traditional optimization methods, engineers define objectives, directions, and constraints based on their experience and prior knowledge. While this approach ensures design reliability, it also brings challenges related to knowledge transfer, low automation, and interdisciplinary validation. As the complexity of system design increases and the parameter space becomes vast, traditional methods consume significant resources and struggle to keep up with the ever-evolving design requirements and technological advancements.

Intelligent optimization technology leverages AI algorithms to quickly find optimal solutions within large datasets, while satisfying various design requirements and constraints. It offers advantages such as efficiency, accuracy, and adaptability. Through parallel computing and efficient search techniques, it significantly enhances simulation efficiency and shortens the optimization cycle. Additionally, by learning from historical design data, it acquires digital prior knowledge, accelerating the design process and driving the generation of innovative solutions. Below are several key aspects of current intelligent optimization technologies:

1. Dimensionality Reduction and Optimization of High-Dimensional Design Spaces: Aircraft design often involves extremely high-dimensional design parameters. Artificial intelligence technologies can be employed to reduce dimensionality and optimize the design space. By using traditional dimensionality reduction methods, manifold learning, and other techniques, the high-dimensional design space can be mapped to a lower dimension. This significantly reduces the number of function evaluations and parameter magnitudes, thereby effectively improving the quality of the solution space and increasing the reliability and interpretability of the optimization path.
2. Intelligent Optimization Based on Reverse Design: In cases where large amounts of design data are available, intelligent optimization can use regression models to predict target design parameters, enabling reverse design. Existing applications include the AI-sizer model for composite laminate optimization based on neural networks, as well as a reverse design method for supercritical wings that generates designs from pressure distribution using generative AI.
3. Optimization Strategy Reuse Based on Deep Reinforcement Learning: Currently, the reuse of optimization strategies is limited. Deep reinforcement learning enables agents to learn prior knowledge from past optimization problems, simulating the decision-making process of engineers and accelerating the tuning and convergence of optimization tasks. For example, in the optimization design of supercritical wing in buffet, exploratory studies have used reinforcement learning methods to derive different optimization strategies. Through repeated trials and learning, the agent can quickly apply the acquired optimization experience to new problems [10].

Despite the tremendous potential of intelligent optimization technology in enhancing design efficiency and solving complex problems, its future development still faces several challenges. First, current optimization problems often involve complex design integration, coupling, and multi-objective, multi-disciplinary issues. The complexity of multi-disciplinary, multi-objective optimization makes it difficult for practical optimization to converge to the theoretical global optimum. Second, in data-driven intelligent optimization, surrogate models may introduce accuracy and information loss. Future research needs to focus on optimizing the construction of dimensionality reduction models to ensure the integrity of reconstructed data, the precision of the design space, and the interpretability of latent space features. Finally, although intelligent optimization reduces computational costs in design, its development process still requires significant amounts of data and computational resources. Therefore, how to design more efficient algorithm architectures to reduce the computational burden and data requirements of model training will be an important research topic in the future.

3.4 *Multi-source Data Fusion*

Currently, the aviation design field faces increasingly complex design demands, and traditional single-source data is no longer sufficient to meet the precision requirements of optimization and design tasks. The primary data acquisition methods for aircraft design include numerical simulations, wind tunnel testing, and flight testing. However, the data obtained through these methods often exhibit significant discrepancies. For example, numerical simulation data is dependent on the accuracy of the models, wind tunnel test data is limited by experimental conditions, and flight test data is constrained by uncontrollable external environmental factors. Therefore, how to effectively integrate these diverse data resources to enhance the consistency and comprehensiveness of design data is a key issue that needs to be addressed.

Currently, data fusion primarily relies on techniques such as data estimation, modeling, and acquisition management. It involves processing multi-source data through redundancy or complementary methods to obtain more accurate and consistent information. Data fusion techniques integrate data from different sources over time or space based on specific criteria, thereby improving the credibility and consistency of the data. In aerospace design, research has been conducted on the consistency between ground and space data, as well as the exploration of methods for fusing high- and low-precision data. Fusion methods for data from different states are also gradually being applied. The use of these technologies is expected to significantly enhance the overall accuracy of databases, reduce data acquisition costs, and obtain more authoritative design data through the integration of redundant information. During the multi-source data collection process, modeling and analysis are used to ensure that various types of data complement and validate each other, thereby improving the comprehensiveness of the design.

Although significant progress has been made in existing data fusion technologies, many challenges remain. First, the criteria and algorithms for data fusion still require further optimization. Specifically, when dealing with data from different methods and states, how to reasonably integrate this information and identify effective fusion methods remains an area that requires in-depth research. Second, combining intelligent data fusion with error source theory is one of the key development directions for the future. By using intelligent algorithms to analyze the sources of errors in greater depth, it is possible to identify ways to reduce inconsistencies and improve the accuracy of data fusion, thereby further enhancing the reliability and practicality of the data.

4 Applications of Intelligent Civil Aircraft Design

In the practical application of intelligent design, concepts such as generative intelligence and AI for Science are gradually penetrating various stages of the design process, greatly driving the intelligence and automation of design procedures. Generative artificial intelligence, especially in design tasks, has become a crucial tool for solving complex design problems through deep exploration and optimization of design spaces using generative models. The AI for Science concept combines artificial intelligence with scientific research, particularly in fields like fluid dynamics and materials science, using intelligent algorithms to accelerate traditional computational processes. This concept is especially well-suited for multidisciplinary collaborative design and optimization processes. It can automatically analyze data and create models, uncovering patterns hidden within vast datasets, and use this foundation for optimization design.

In the international aviation industry, leading companies such as Boeing and Airbus are actively exploring the application of artificial intelligence (AI) technologies in the aviation sector, implementing "AI + Aviation" integration through initiatives such as aviation data analysis laboratories, smart sky platforms, and the AnalytX data platform. NASA (National Aeronautics and Space Administration) [11, 12] and EASA (European Union Aviation Safety Agency) [13] have both made detailed plans for AI-assisted flight assistance. They have collaborated with research institutions and businesses to conduct research in areas such as AI-driven design and airworthiness certification.

4.1 Multidisciplinary Collaborative Intelligent Design

4.1.1 Multidisciplinary Objective Collaborative Overall Design

The overall design of commercial aircraft is a complex and critical process that involves knowledge across multiple disciplines, including aerodynamics, structures, and materials. Traditional design methods often rely heavily on extensive experimentation and experience, which is not only time-consuming and labor-intensive but also challenging to achieve optimal design outcomes. With the rapid advancement of AI technology, its application in optimization design has become increasingly widespread, bringing revolutionary changes to aircraft design. Leading institutions such as NASA, Boeing, and Airbus have made significant progress in this field, providing strong support for multidisciplinary optimization in aircraft design.

Researchers abroad have developed multidisciplinary collaborative optimization frameworks that integrate design variables and constraints from different disciplines. By leveraging AI-driven optimization methods, these frameworks enable the realization of globally optimal designs.

Ensuring the precise definition and control of key aircraft metrics is central to achieving the goal of "designing exactly what is intended." However, the current workflow remains relatively traditional, with low efficiency and common issues such as inaccurate metric definitions, limited control methods, challenges in multi-department collaboration, and delays in data transmission. To address these challenges, the design team at the Shanghai Aircraft Design and Research Institute of COMAC is exploring the application of intelligent design technologies in the field of "collaborative design and data analysis of top-level metrics for commercial aircraft." Their research focuses on the computational definition and deviation assessment of critical requirement metrics, key design parameters, and essential process variables. By leveraging intelligent technologies, the team aims to establish a more scientific and rational approach to defining top-level aircraft metrics.

The design team has developed an automated design tool that enables the generation of two-dimensional design solutions based on design specifications while managing various constraints through generative design methods. For example, as illustrated in Fig. 2, by leveraging a generative model trained on existing design data, multiple geometric shapes and flow field distributions that meet specific performance criteria (such as lift and drag) can be generated within just one second, without the need for lengthy optimization iterations [14, 15]. However, generative models may sometimes produce designs that do not comply with physical laws or engineering constraints. To address this issue, the research team has incorporated physical constraints into the generative model and applied a filtering mechanism to improve the physical validity and engineering feasibility of the generated designs. This approach is particularly crucial in the early design phase, as it helps designers efficiently identify promising design candidates. Traditional design methods often rely on the experience and expertise of designers, making it challenging to explore a broad design space within a limited timeframe. Generative design overcomes this limitation by leveraging the powerful generative capabilities of deep learning models, enabling the creation of innovative design concepts and solutions.

In the future, this automated design tool will be extended to three-dimensional design solutions, encompassing aerodynamic configuration, structural layout, and propulsion systems. The next step for the research team is to automatically generate alternative design solutions based on aircraft design requirements and employ multi-objective optimization algorithms to balance multiple design goals. This approach aims to achieve overall performance optimization by minimizing weight, cost, and fuel consumption while maximizing passenger comfort and safety. Compared to traditional manual iterations and single-objective optimization methods, AI-based optimization is more efficient and comprehensive. In future trade-off processes, the team will leverage AI technology for automatic evaluation, scoring, and ranking to rapidly identify the optimal design solutions, thereby improving the efficiency and accuracy of decision-making. Additionally, the research team plans to allocate key

Fig. 2 Conditioned generative design method based on diffusion models

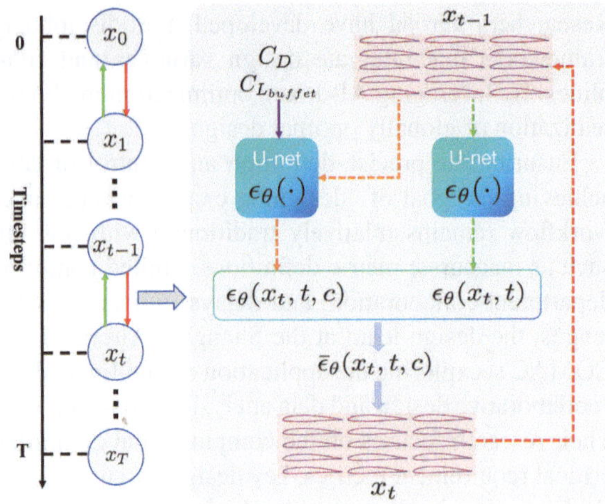

performance indicators to relevant disciplines and users, enabling collaborative workflows to ensure timely and precise control of these indicators throughout the aircraft's lifecycle. These measures are expected to enhance the coordination of key technical indicators across the entire aircraft, improve the accuracy of critical design parameters, accelerate deviation assessments, and facilitate timely and efficient design adjustments. Moreover, these efforts will contribute to the establishment of a comprehensive database for intelligent design and evaluation of key aircraft performance indicators, providing a robust data foundation for future AI-driven design processes.

4.1.2 Interactive Intelligent Modeling

The Intelligent Aerodynamic Design Team at COMAC Shanghai Aircraft Design and Research Institute, in collaboration with the Shanghai Artificial Intelligence Laboratory, has developed the "Shusheng·Yifei" interactive design model (see Fig. 3). This model enables designers to efficiently generate and edit two-dimensional airfoils using three different input modalities: text, voice, and graphical interaction [16, 17]. On one hand, as the first parametric approach to supercritical airfoil design that balances interpretability and representation accuracy, this model allows designers to precisely control airfoil generation while ensuring that the generated results meet design requirements. On the other hand, airfoils can be edited through simple input descriptions or graphical manipulations, achieving a control precision of up to the 10^{-3} scale. This interactive process facilitates seamless designer-system interaction, significantly enhancing the efficiency and convenience of aerodynamic design. Practical application has demonstrated that interactive design not only accelerates design iterations but also helps designers gain deeper insights into the complex relationships within the design space, enabling more informed design decisions. The advantage of interactive design lies in its ability to leverage both the expertise of designers

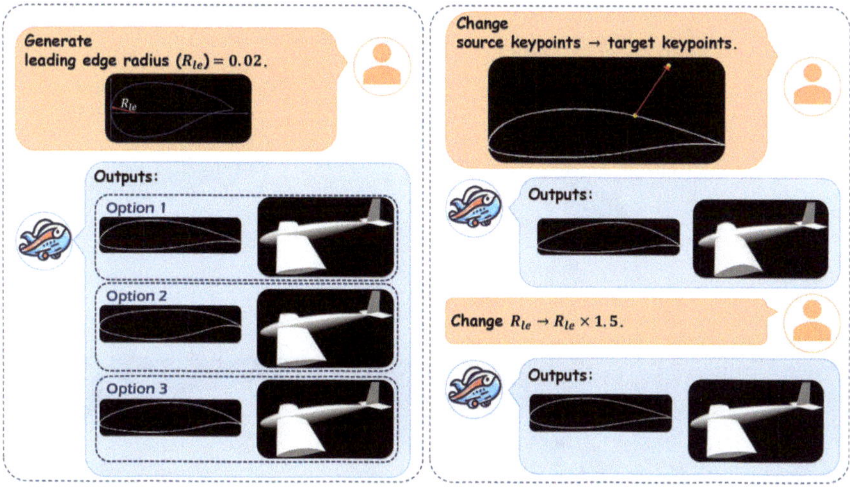

Fig. 3 "Intern·Wing-Wing" interactive design large model

and the computational power of AI, providing a more flexible and efficient design solution. Traditional design methods typically follow rigid optimization workflows, limiting the extent to which designers can apply their expertise during the process. In contrast, interactive design fosters human-machine collaboration, allowing designers to adjust their models at each iteration based on real-time feedback, thus significantly improving both design efficiency and overall outcomes.

For intelligent modeling tasks of three-dimensional aerodynamic shapes, the research team directly learned from Initialization Graphics Exchange Specification (IGES) file samples and applied generative methods to achieve dimensionality reduction in the encoded latent space and decoding-based generation of 3D wing and fuselage shapes (see Fig. 4). Currently, this technique compresses the hundreds of parameters used to describe surface geometry into a reduced latent space of 7–15 dimensions, while precisely capturing variations in the 3D wing and fuselage shapes through a generative network. The constructed latent space maintains continuity in feature variations and exhibits a certain degree of disentanglement. By manipulating feature space vectors, the approach enables adaptable transfer learning across different latent space dimensions and types. Additionally, by interfacing with CAD modeling software, the system allows millisecond-level generation of Standard for the Exchange of Product Model Data (STEP) and IGES modeling files while ensuring curvature continuity across sections. Subsequently, the generated STEP/IGES models are input into a mesh generator based on generative models for meshing, followed by simulation analysis. In terms of geometric representation, this technology enables direct manipulation of 3D shape variations, bypassing the traditional 2D stacking modeling process. By using parametric surfaces as the foundation for modifications, the approach eliminates the iterative conversions between

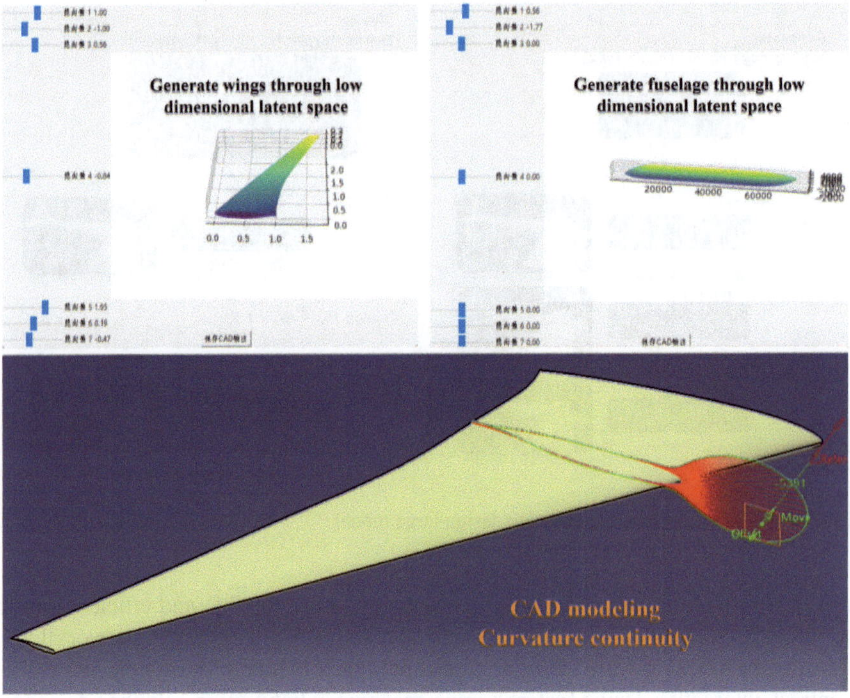

Fig. 4 Intelligent modeling of 3D wings and fuselage

meshes, discrete point stacking, and surface fitting, thereby significantly accelerating the design and modeling process.

4.2 High-Precision Intelligent Simulation

4.2.1 Knowledge-Assisted Design Data Sampling

Different problems and models have varying requirements for sample selection. However, collected samples should simultaneously exhibit rationality, relevance, and diversity, meaning:

1. Samples must satisfy fundamental engineering constraints, such as avoiding negative volumes or discontinuous geometries.
2. Samples should specifically address the intended research problem while eliminating irrelevant variables, ensuring that they adhere to necessary physical characteristics.
3. Within the scope of the study, samples should be as diverse as possible, covering different flow conditions, geometries, physical characteristics, and performance

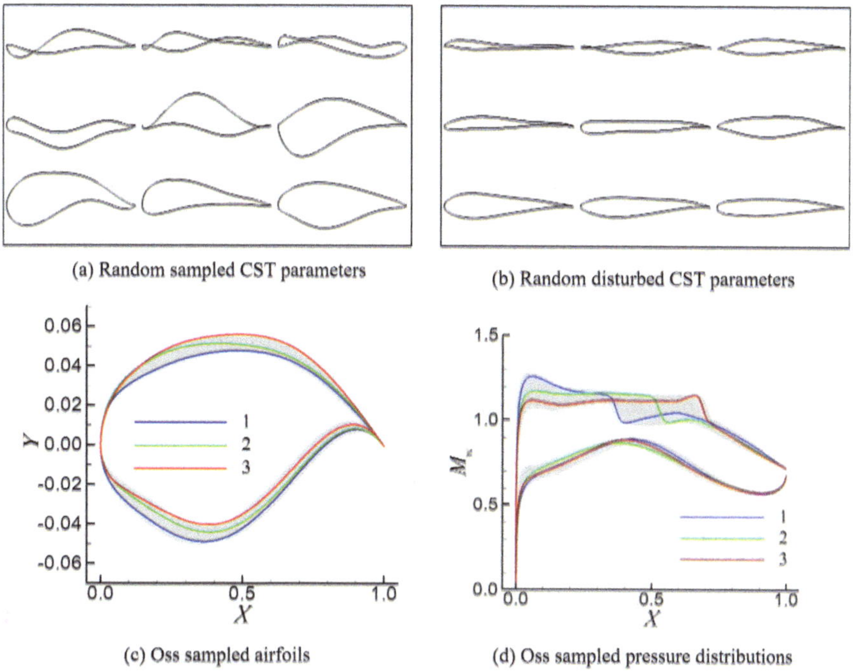

(a) Random sampled CST parameters

(b) Random disturbed CST parameters

(c) Oss sampled airfoils

(d) Oss sampled pressure distributions

Fig. 5 Supercritical airfoil sampling method

metrics. Existing static sampling methods yield a low proportion of effective samples, and classical adaptive sampling techniques are not always applicable. The research team proposed an Output Space Sampling (OSS) method based on Kullback-Leibler divergence for flow field structure-oriented adaptive sampling [18]. This method enables constrained sample acquisition within any feature space, providing an effective tool for generating a well-distributed and diverse high-quality sample set. Taking airfoil design as an example, compared to traditional design variable sampling methods, the OSS approach enables the collection of supercritical airfoil samples that maintain a similar lower surface pressure distribution while ensuring a uniform variation in the upper surface shockwave position (as shown in Fig. 5). This demonstrates that OSS significantly improves the specificity and diversity of sampling, effectively meeting the demands of complex aerodynamic design tasks.

4.2.2 End-to-End Intelligent Simulation Large Model

In response to the urgent need for complex flow simulation scenarios of supercritical airfoils for large passenger aircraft, the Intelligent Aerodynamic Design Team at COMAC Shanghai Aircraft Design and Research Institute is dedicated to solving the issues of long time consumption and high costs in current aerodynamic design methods. Based on the Ascend and Ascend Soft-Hardware platform, they developed the industrial-grade 2D fluid simulation large model, "Dongfang·Yufeng" (as shown in Fig. 6) [19, 20]. This model can replace traditional Computational Fluid Dynamics (CFD) solvers based on the Navier-Stokes equations, enabling efficient and high-precision simulations of the flow field around the 2D airfoil of large passenger aircraft. When the airfoil geometry, inflow Mach number, and angle of attack change, the model can generalize and infer the flow field around the airfoil in real-time. The error across the entire flow field reaches the order of 1/10,000, with a single simulation speedup of over 24 times.

Building on the 2D model, the research team overcame a series of challenges brought about by transitioning from 2 to 3D problems and further developed the industrial-grade 3D fluid simulation model, "Dongfang·Yifeng" (see Fig. 6c). This model enables real-time inference of millions of grid flow fields in seconds when changes occur in geometric configuration, inflow Mach number, and angle of attack. Compared to traditional simulation methods that require exascale computing resources, the simulation error is controlled within the thousandth range, with a speedup of over 1,000 times for each simulation. The 3D wing model uses a complex grid topology with scales reaching millions, posing significant challenges for both data formatting and hardware memory. The research team, leveraging existing hardware conditions and combining multi-grid fusion technology, developed a componentized model architecture and distributed parallel training approach under big data samples, significantly enhancing the research and development efficiency of the new model. Moreover, 3D supercritical wings typically exhibit complex flow phenomena, leading to strong nonlinear and discontinuous flow features such as shock waves and separation vortices in transonic regions. Even with deep learning technologies, these issues present substantial technical difficulties. The research team used prior physical knowledge as a breakthrough point, developing a geometric information encoding method that integrates physical information. Through multi-level wavelet transformation techniques and dynamic weight distribution balancing error loss techniques, they improved the prediction accuracy in complex flow regions (such as shock waves), enabling precise capture of complex flow characteristics. This technology can significantly shorten the development cycle of large passenger aircraft, reduce design and research costs, and is expected to be applied in related fields such as automobiles, high-speed trains, drones, and aircraft engine blades.

In addition to fluid simulation models, other discipline teams are actively advancing intelligent simulation models. For example, high-precision load prediction models have been developed to simulate the load conditions of an aircraft under different flight conditions, providing a basis for strength design. Fatigue prediction models are established by learning historical data and simulation results, enabling predictions of the fatigue life of aircraft structures under various operational conditions. Machine learning algorithms are used to analyze large volumes of materials data, uncovering the complex relationships between material performance, composition, and structure, guiding the design of new aerospace materials with higher

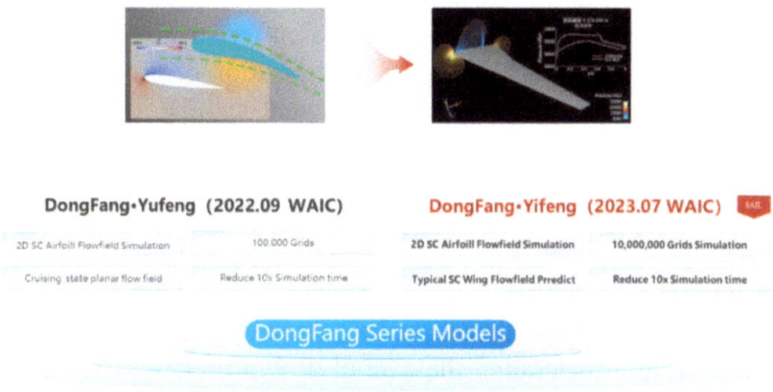

a) Progress in AI Fluid Simulation Fundamental Models

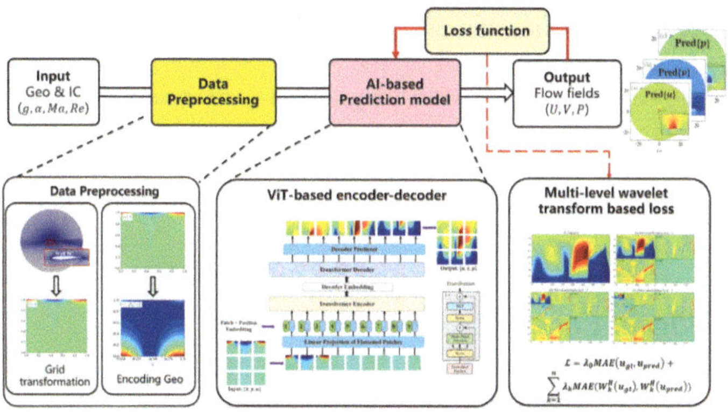

b) Dongfang. Yufeng Large Model

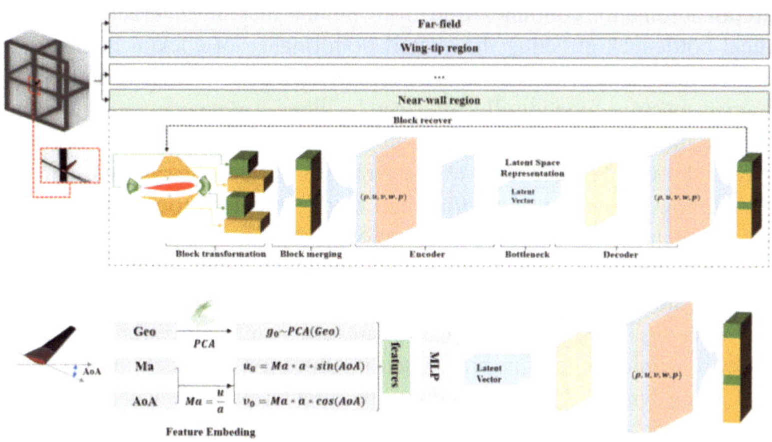

c) Dongfang Yifeng Large Model

Fig. 6 End-to-end intelligent simulation large model

strength, lower density, and corrosion resistance. Moreover, intelligent optimization algorithms are applied to automatically search for structural topology, dimensions, and other parameters, achieving the minimization of structural weight while satisfying constraints such as strength, stiffness, and stability.

The development of high-efficiency and high-precision simulation technology for large commercial aircraft based on artificial intelligence holds significant potential. On one hand, it is expected to overcome bottlenecks in the design and development of domestically produced large aircraft, enhancing the reliability and efficiency of existing aerodynamic design tools. This advancement will significantly shorten the design cycle of domestic large aircraft, reduce research and development costs, and play a crucial role in accelerating the overall design process. On the other hand, it contributes to breaking the monopoly of foreign technologies in large aircraft design and development by integrating the latest AI research achievements with high-end equipment manufacturing. This integration will facilitate the creation of proprietary design software based on the AI for Science paradigm, significantly advancing the development of wide-body aircraft and future commercial aircraft models. Furthermore, this technology is expected to be widely applied in other fields such as automobiles, high-speed rail, ships, unmanned aerial vehicles, and aero-engine blades, fostering technological advancements across related industries.

4.3 Intelligent Cockpit Design

4.3.1 Intelligent Cockpit Human-Machine Collaboration Design

As aircraft automation continues to advance, human-machine interaction has become a critical bottleneck affecting the overall performance of cockpit human-machine systems. While humans and automated systems each have distinct advantages, effective human-machine function allocation can facilitate system integration and enhance overall reliability. In an intelligent cockpit, more traditional pilot-executed functions or tasks will be delegated to the aircraft's automated systems, shifting the pilot's role toward system monitoring and management. Consequently, a reassessment of human-machine function allocation within intelligent cockpits is necessary. To maintain pilot workload at an optimal level—preventing excessive workload that could compromise flight safety and avoiding underload conditions that might reduce vigilance and situational awareness—an adaptive function allocation (AFA) or dynamic human-machine function allocation approach must be adopted. Implementing dynamic function allocation requires an assessment of multiple dimensions, including situational demands, the relative strengths of human and automated systems, emergency scenarios, crew workload, pilot capability evaluation, and human-machine system performance forecasting. By integrating multiple modes into an automation-level triggering mechanism, an engineering-feasible, scenario-based multimodal cockpit adaptive function allocation method can be developed. This approach involves defining human-machine collaboration models tailored to different

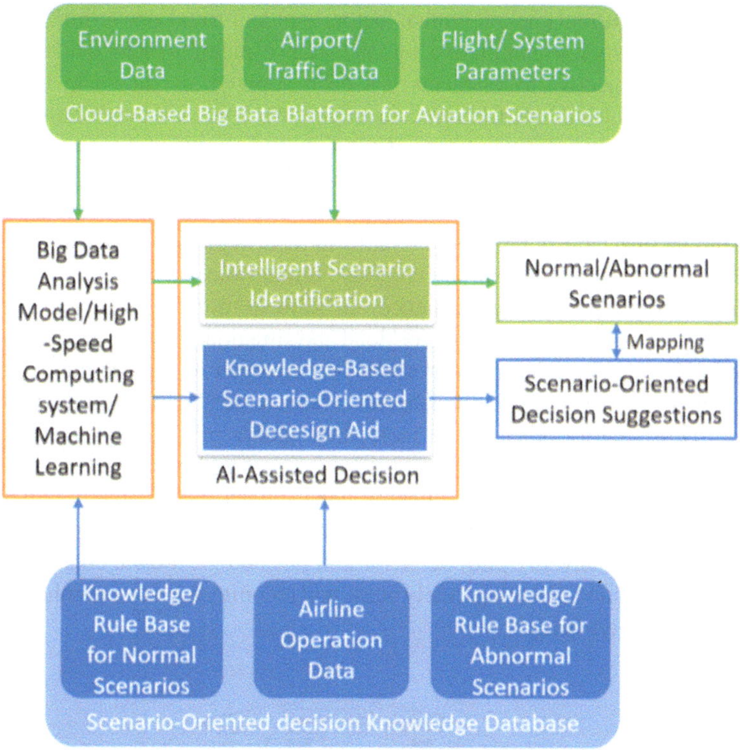

Fig. 7 Intelligent decision support technology

scenarios based on human cognitive models, system architecture and functional models, and overarching human-machine function allocation strategies. Additionally, a large-scale aviation operations data platform and a flight operation response knowledge/rules database will be established. Utilizing generative artificial intelligence, a virtual co-pilot flight assistance system can be developed. As illustrated in Fig. 7, the integration of flight assistance systems, intelligent interaction interfaces, pilot status monitoring, and flight data analysis will enhance piloting performance and flight safety.

4.3.2 Intelligent Cockpit Human Factors Evaluation

Cockpit human factors evaluation has long faced challenges such as pilot resource scarcity, high costs, difficulties in data acquisition, heavy reliance on subjective pilot judgment, and lengthy assessment cycles. However, with the advancement of artificial intelligence and big data processing technologies, it is now possible to conduct large-scale, quantitative, and objective cockpit human factors evaluations.

In COMAC's current cockpit human factors evaluation and airworthiness compliance verification work, a quantitative human factors evaluation method that combines subjective and objective assessments has been adopted, as shown in Fig. 8. By utilizing devices such as eye trackers, electrocardiographs, and depth cameras, pilot eye movement, electrocardiographic (ECG) data, and operational actions are monitored. These physiological parameters and operational data are then integrated with flight parameters extracted from flight simulators and a parameterized flight performance model to establish a mapping relationship between physiological data, operational metrics, flight data, and key cognitive dimensions such as pilot perception, cognition, effort level, and performance. To achieve more accurate and real-time monitoring and assessment of crew workload and situational awareness in the future, deep learning and Bayesian networks will be employed to explore the underlying physiological and psychological mechanisms represented by monitored parameters. This will help identify objective variables that anchor pilot cognitive characteristics and workload, improving the automation of cockpit human factors evaluation experiments and data processing, thereby enhancing test efficiency. Furthermore, real-time monitoring of pilot fatigue, attention levels, and emotional states will be implemented. By analyzing pilot behavior and physiological data, changes in pilot states can be identified, enabling timely alerts and recommendations to enhance flight safety and pilot well-being.

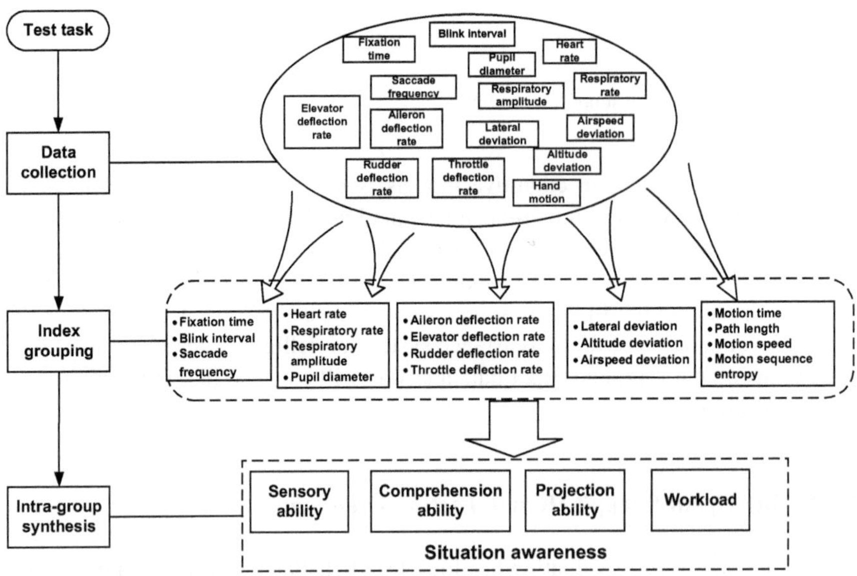

Fig. 8 Comprehensive situation awareness measurement

5 Conclusion and Outlook

To promote the practical application of artificial intelligence in domestic civil aircraft design, it is essential to enhance the reliability, generalizability, interpretability, and usability of intelligent models based on the characteristics of multidisciplinary design trade-offs and multi-source data fusion in engineering design. While AI has made significant breakthroughs in intelligent modeling and design, laying the foundation for future development, its application in aircraft design is not merely a straightforward extension of data-driven methods. Instead, it requires a deep integration of physical knowledge into intelligent models. Physics serves as the fundamental source of data, and the true value of data-driven models lies in their ability to learn, extract, and reveal underlying physical laws. Relying solely on data-driven models in engineering design neither achieves ideal cost control nor provides sufficient reliability and interpretability. Therefore, embedding knowledge directly into models to guide learning and knowledge discovery or constraining models based on physical principles can effectively address the shortcomings of purely data-driven approaches. By constructing a data-and-knowledge-driven intelligent design framework, as illustrated in Fig. 9, the practical applicability of AI in engineering can be significantly enhanced. Moreover, in civil aircraft design, a substantial amount of "precise" and "trustworthy" design knowledge is proprietary to manufacturing entities and cannot be publicly shared for large-scale model training, which limits the capabilities of purely data-driven models. To overcome this, AI models must leverage their reasoning abilities while integrating proprietary knowledge, internal data, and domain expertise from research institutions. This approach enhances the accuracy and relevance of model predictions, providing engineers with more precise design assistance and improving the scientific rigor and efficiency of decision-making. Ultimately, this integration will drive the intelligent transformation of domestic civil aircraft design, strengthening its competitiveness and innovation.

In integrating artificial intelligence technology with aircraft design, it is crucial not only to consider the application of the technology itself but also to clarify its relationship with human expertise. Currently, human creativity and comprehensive analytical capabilities still surpass AI models, and designers remain the fundamental safeguard

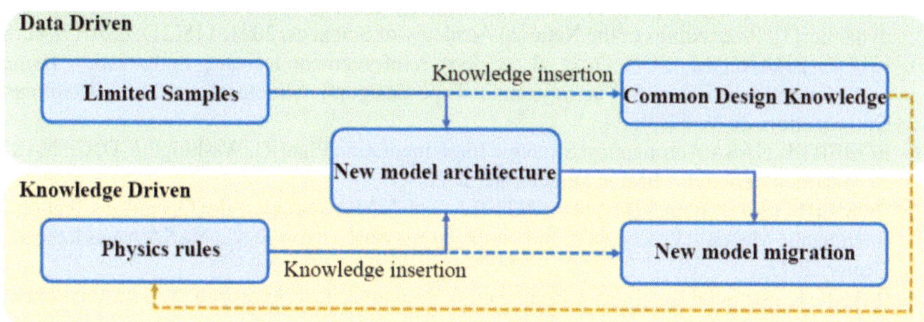

Fig. 9 Data and knowledge-driven intelligent design framework

for the reliability and safety of engineering design. However, AI's demonstrated capabilities in aerodynamic design highlight its potential as a powerful assistant to designers. At the same time, the new knowledge discovered by AI from data should form a feedback loop with designers, enhancing their cognitive abilities and design skills. As AI continues to evolve rapidly, it is essential to maintain a prudent and steady approach to prevent overhyped adoption or unrealistic expectations. The transformation of intelligent methods in engineering design should proceed in a measured manner, with a clear understanding of AI's capability boundaries. The goal should be a human-centered approach, rooted in knowledge, driven by real-world needs, and ultimately focused on practical implementation. By fostering next-generation foundational design tools for civil aircraft, AI-driven enhancements can effectively empower aircraft design and development, ensuring the successful application of AI research in industrial settings. This approach will lead to a tangible improvement in China's civil aircraft design capabilities, reinforcing innovation and competitiveness in the aviation sector.

References

1. WU G. A trio of commercial aircraft developments in china[J]. Engineering, 2021, 7(4): 424–426.
2. CHEN Y C, SONG W B, LIU H. Civil Aircraft Design[M]. Shanghai Jiaotong University Press, 2022.
3. Zhang X J. Aircraft Design Handbook Volume 6, Aerodynamic Design[M]. Beijing Aviation Industry Press, 2002.
4. CHEN Y C, ZHANG M H, ZHANG M, et al. Review of large civil aircraft aerodynamic design [J]. Acta Aeronautica et Astronautica Sinica, 2019, 40(1): 17.
5. WU G H, LIU H. Framework of Digital Design Support System-of-systems for Large Airliners[J]. Acta Aeronautica et Astronautica Sinica, 2008(5): 1386–1394.
6. Xu X, Lambourne J, Jayaraman P, et al. Brepgen: A b-rep generative diffusion model with structured latent geometry[J]. ACM Transactions on Graphics (TOG), 2024, 43(4): 1–14.
7. Duraisamy K, Iaccarino G, Xiao H. Turbulence modeling in the age of data[J]. Annual review of fluid mechanics, 2019, 51(1): 357–377.
8. Catalani G, Agarwal S, Bertrand X, et al. Neural fields for rapid aircraft aerodynamics simulations[J]. Scientific Reports, 2024, 14(1): 25496.
9. Kochkov D, Smith J A, Alieva A, et al. Machine learning–accelerated computational fluid dynamics[J]. Proceedings of the National Academy of Sciences, 2021, 118(21): e2101784118.
10. LIU Z, ZHANG M, SUN D, et al. A deep reinforcement learning optimization framework for supercritical airfoil aerodynamic shape design[J]. Structural and Multidisciplinary Optimization, 2024, 67(3): 34.
11. ROBERT P. NASA Aeronautics: Strategic Implementation Plan[R]. Washington, D. C.: NASA Aeronautics Research Mission Directorate, 2020.
12. COMERFORD D, BRANDT L, LACHTER J, et al. NASA's Single-Pilot Operations Technical Interchange Meeting: Proceedings and Findings[R]. Washington, D. C.: NASA Ames Research Center, 2013.
13. PATRIC K. Artificial Intelligence Roadmap: A Human-Centric Approach to AI in Aviation[R]. Cologne: EASA, 2020.
14. WANG J, LI R, HE C, et al. An inverse design method for supercritical airfoil based on conditional generative models[J]. Chinese Journal of Aeronautics, 2022, 35(3): 62–74.

15. SUN K, WANG W, CHENG R, et al. Evolutionary generative design of supercritical airfoils: an automated approach driven by small data[J]. Complex & Intelligent Systems, 2023: 1–17.
16. LIU J, WU J, XIE H, et al. AFBench: a large-scale benchmark for airfoil design[C]//Advances in Neural Information Processing Systems, Vancouver, Canada. 2024.
17. XIE H, WANG J, ZHANG M. Parametric generative schemes with geometric constraints for encoding and synthesizing airfoils[J]. Engineering Applications of Artificial Intelligence, 2024, 128: 107505.
18. LI R, ZHANG Y, CHEN H. Pressure distribution feature-oriented sampling for statistical analysis of supercritical airfoil aerodynamics[J]. Chinese Journal of Aeronautics, 2022, 35(4): 134–147.
19. DENG Z, WANG J, LIU H, et al. Prediction of transonic flow over supercritical airfoils using geometric-encoding and deep-learning strategies[J]. Physics of Fluids, 2023, 35(7): 075146.
20. WANG J, HE C, LI R, et al. Flow field prediction of supercritical airfoils via variational autoencoder based deep learning framework[J]. Physics of Fluids, 2021, 33(8): 086108.

Guanghui Wu Ph.D., Aircraft Design Expert, Academician of the Chinese Academy of Engineering. He previously served as the Chief Designer of a national key project aircraft and the ARJ21 aircraft. Currently, he is the Chief Scientist of the Commercial Aircraft Corporation of China (COMAC) and the Chief Designer of the National Major Project C919 Large Passenger Aircraft Series. During the development of the C919 aircraft, he proposed the top-level design philosophy and guiding principles, formulated the overall plan, and overcame numerous technical challenges restricted from export to China, leading the team to forge a path of fully independent intellectual property rights in the forward design and development of civil aircraft. He is a recipient of the Special Government Allowance from the State Council and was a member of the expert group in the field of modern transportation technology for the "12th Five-Year Plan" 863 Program. He served as the Chief Scientist for the National Basic Research Program (973 Program) project on drag reduction mechanisms and methods for large passenger aircraft. In 2007, he was personally awarded the "Major Contribution Award for a Certain Project" and a gold medal by the Central Committee of the Communist Party of China, the State Council, and the Central Military Commission. In 2010, he received the Special Prize of the National Science and Technology Progress Award (ranked first). In 2011, he was awarded the National May 1st Labor Medal, in 2013 he received the Second "Feng Ru Aviation Science and Technology Elite Award," and in 2014 he was honored as a National Outstanding Science and Technology Worker.

Miao Zhang Research Fellow at the Shanghai Aircraft Design and Research Institute. He has long been engaged in applied research on supercritical wings, integrated airframe–engine design, overall aircraft aerodynamic design, computational fluid dynamics, and intelligent aerodynamics. He has led more than 20 projects funded by the Ministry of Industry and Information Technology, the Ministry of Science and Technology, the National Natural Science Foundation of China, and the National Intellectual Property Administration. He has received the First Prize of the Science and Technology Award of the Chinese Society of Aeronautics and Astronautics, the First Prize of the Science and Technology Award of the Chinese Aerodynamics Research Society, and the First Prize of the Shanghai Science and Technology Award. He has published over 20 journal and conference papers and holds more than 20 patents.

Jing Wang Ph.D., Assistant Research Fellow. She completed her postdoctoral research at the Shanghai Aircraft Design and Research Institute of COMAC in 2022. Her research focuses on interdisciplinary research spanning aerodynamics, computational fluid dynamics, and artificial intelligence.

Hairun Xie Ph.D., Associate Research Fellow. He completed his postdoctoral research at the Shanghai Aircraft Design and Research Institute of COMAC in 2024. His research focuses on interdisciplinary research spanning aerodynamics, computational fluid dynamics, and artificial intelligence.

Dayong Dong Ph.D., Research Fellow, and Member of the National Human Factors Engineering Standards Committee. His research interests include cockpit integrated design, human–machine interface simulation and evaluation, and crew workload prediction and assessment. He has received the Second Prize of the Shanghai Science and Technology Progress Award.

Zhaoguang Tan M.S., Research Fellow, and Level-III Chief Designer at the Shanghai Aircraft Design and Research Institute. His research focuses on the overall aerodynamic layout design of civil aircraft. He has been responsible for several key development tasks for major national aircraft programs, including the C919, C929, and C909.

Junqiang Bai Ph.D., Professor and Director of the Unmanned Systems Technology Research Institute at Northwestern Polytechnical University. In recent years, he has led more than 40 national-level projects and contributed to the aerodynamic design of major aircraft such as the Y-20 large transport aircraft, the C919 large passenger aircraft, the C929 long-range wide-body aircraft, and the MA700 advanced turboprop regional aircraft. He has published more than 200 academic papers, holds over 20 granted invention patents, and has authored two monographs.

Climate-Smart Forestry: Paradigm Shift in Forest Science Research Driven by Digitization and Intelligence

Jiaojun Zhu, Tian Gao, Huaiqing Zhang, Yangang Wang, Zongguo Wang, Yirong Sun, Jinxin Zhang, Deliang Lu, Tingdong Yang, Dexiong Teng, Shuailing Hao, and Fengyuan Yu

Abstract Forest Science is the science of afforestation, management, utilization, conservation and restoration, as well as the related theories and technologies of forest ecology and sustainable utilization of forest resources. Forests, the main subjects of Forest Science, are characterized by long life periods, high spatial heterogeneity, and extremely complex structure and composition. In the context of climate change, forests play a key role as significant carbon sinks in mitigating climate change, while also being affected by climate change. Traditional studies of Forest Science involve a labor-intensive fieldwork, high trial-and-error costs, and a low decision-making efficiency, which makes it challenging to address the multi-faceted situations of the present and affects the sustainable management of forest ecosystems. Climate-smart Forestry (CSF) integrates the next-generation information and artificial intelligence (AI) technology to shift the traditional researches and management paradigms of Forest Science from three aspects: data acquisition, simulation analysis, and management decision-making, providing a new methodological framework for Forest Science research. The integration of AI and Forest Science knowledge can drive innovation in the field, shift traditional research paradigms, and promote

J. Zhu · T. Gao (✉) · Y. Sun · J. Zhang · D. Lu · D. Teng · F. Yu
CAS Key Laboratory of Forest Ecology and Silviculture, Institute of Applied Ecology, Chinese Academy of Sciences, Shenyang, China
e-mail: tiangao@iae.ac.cn

Qingyuan Forest CERN, National Observation and Research Station, Liaoning Province, Shenyang, China

J. Zhu
e-mail: jiaojunzhu@iae.ac.cn

H. Zhang · T. Yang
Research Institute of Forest Resource Information Techniques, Chinese Academy of Forestry, Beijing, China

Y. Wang · Z. Wang
Computer Network Information Center, Chinese Academy of Sciences, Beijing, China

S. Hao
Shenyang Artificial Intelligence Computing Center, Shenyang, China

China's e-Science Blue Book 2025, https://doi.org/10.1007/978-981-96-8252-2_11

the innovative development of CSF. Based on the prospect in the published paper titled "On the Research of Climate-smart Forestry", here we outline the concepts and research framework of CSF, introduce key technologies for data acquisition, simulation analysis, and management decision-making, and explore pathways for shifting in research paradigms. The study includes research cases such as intelligent data collection/IoT transmission from the Ker Towers of the Qingyuan Forest CERN, digital twin analysis in the Saihanba Forestry Center, and AI expert decision-making. Based on these cases, there is a new vision in three fields of paradigm shift: forest IoT devices and sensors, multi-process coupling models for ecosystem simulation, and decision systems for ecosystem services that integrate human-AI interaction.

Keywords Artificial intelligence · Climate-smart forestry · Forest science · Digital twin · Ecosystem services

1 Introduction

Since the Industrial Revolution, the rapid increase in global greenhouse gas emissions has surged, driving significant climate changes, characterized by rising temperatures and increased frequency of extreme weather events. These changes have fundamentally disrupted ecosystem structures and functions, endangering both ecological stability and human habitat security. In response to these challenges, 197 countries ratified the Paris Agreement in 2015, establishing a multilateral framework for mitigating greenhouse gas emissions. Notably, China proclamed its carbon peak and neutrality targets, pledging to achieve peak carbon emissions by 2030 and carbon neutrality by 2060. These targets not only respond to the global imperative for climate governance, environmental sustainability, and fostering a community with a shared future for mankind, but also align with China's long-term strategic objectives for high-quality development and industrial transformation, with far-reaching implications for sustainable economic growth and ecological resilience [1].

Achieving the carbon peak and neutrality targets requires tripartite strategic initiatives, with Nature-based Climate Solutions (NbCS) fundamentally enhancing carbon sequestration capacity through scientifically guided conservation, sustainable management, and optimized rutilization following natural principles, effectively offsetting anthropogenic CO_2 emissions [2–5]. Forest ecosystems dominate terrestrial carbon sinks, contributing over 80% of global biospheric carbon sequestration [6–9]. Silvicultural practices, particularly afforestation and reforestation, serve as primary NbCS strategies for increasing carbon sequestration capacity [3]. Crucially, while forests can significantly mitigate climate change through carbon sinks [10, 11], climate-driven disturbances such as wildfires, pests, and drought, are concurrently altering forest structure, function and dynamics. The inherent complexity of forest ecosystems, characterized by multi-decadal developmental cycles [12], extensive spatial heterogeneity, and intricate biophysical-biogeochemical couplings, creates systemic challenges in predicting ecological responses to environmental change

[13]. Climate change uncertainties further compound these challenges, necessitating the integration of large-scale, multidisciplinary and multimodal datasets spanning geoscience, biology, ecology and socioeconomics [14]. Traditional forestry research paradigms, particularly in experimental design, data collection protocols, and observational frameworks, demonstrate increasing limitations in addressing these multidimensional complexities [15, 16]. Advances in next-generation information technologies provide opportunities to improve forestry research. Climate-smart Forestry (CSF) leverages data-driven artificial intelligence (AI) approaches to revolutionize three core aspects: multi-source data acquisition, process-based ecosystem modeling, and adaptive management strategies. This technological transformation establishes novel methodological frameworks for forest research, accelerates knowledge generation, and enables climate-resilient forest management [17].

This study synthesizes the concepts of CSF through critical analysis of its technical frameworks and research priorities [15, 16]. A systematical characterization of core principles identifies the paradigm shift in forestry research and management: digital informatization, networking systematization, and intelligent automation. Representative case analyses across operational scales reveal how these technological synergies enhance forest ecosystems' dual capacity for adapting to and mitigating global climate change, promoting the sustainable forest management and effective utilization of forest resources.

2 Overview of Climate-Smart Forestry

2.1 *Key Scientific Issues in Climate-Smart Forestry*

In the context of climate change, CSF integrates next-generation information technologies, particularly AI, to improve the understanding of interaction mechanisms between forest ecosystems and climate change. This approach enables adaptive intelligent management strategies for optimizing forest resource utilization while maintaining ecological integrity, positioning CSF as an advanced paradigm in forestry research and development [15, 16]. With the advancement of digital informatization, networking systematization, and intelligence automation, CSF has further refined innovative and forward-looking approaches to forest ecosystem management and decision-making. This framework enhances the role of forests in mitigating climate change, ensuring their efficient and stable functioning, supporting the achievement of sustainability targets, and ultimately fostering harmonious interactions between humans and nature. Similar to previous forestry management paradigms, CSF emphasizes scientific forest management to better meet societal demands for forest resources and ecosystem services. However, it places greater focus on enhancing the resilience of forest ecosystems under changing climatic conditions, enabling forests to both adapt to and mitigate climate-induced disturbances while strengthening their carbon sequestration capacity. Accordingly, CSF research focuses on elucidating the mechanisms underlying interactions between forest ecosystems and climate change, with particular emphasis on ecosystem resilience and carbon sequestration functions [15, 16].

2.2 *Scientific Framework and Key Technologies for Climate-Smart Forestry*

Enabled by next-generation information technologies, the research framework and key technologies of CSF encompass the following (Fig. 1) [15, 16]:

(1) Multi-dimensional intelligent sensing technology for high-efficiency data acquisition: Utilizing next-generation intelligent forestry equipment and integrating digital information, networking systematization, and intelligence automation provides robust data support for forest resource inventories, monitoring, management, protection, restoration, and ecological service function assessments.
(2) Forest ecosystem digital twin technology: By constructing a digital twin of forest ecosystems, this technology advances three-dimensional precision modeling of forest components, enabling spatially high-resolution visualization and simulation of ecological processes under climate change scenarios [18, 19].
(3) Simulation and AI-driven analysis technology for forest ecosystem succession and management processes: Based on forest growth and dynamics models, this technology integrates ecological process models, physiological mechanism

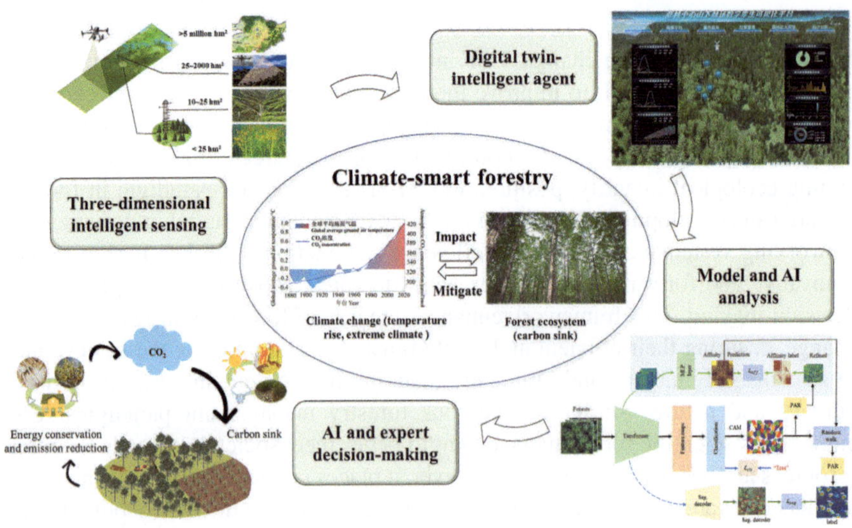

Fig. 1 Research and technical framework of climate-smart forestry [15]

models, and disturbance models to facilitate intelligent simulation and prediction of natural succession in forest ecosystems. It enables the analysis of biogeochemical cycles (e.g. carbon, nitrogen, and water) and the developmental and successional processes of these ecosystems.

(4) AI-expert hybrid management decision support system: Establishing a sustainable management decision-making system that integrates AI with expert knowledge, this system facilitates human-AI collaboration to optimize forest resource allocation, maintain and enhance forest service functions, protect and restore forest ecosystems, and achieve sustainable forest management in response to climate change.

3 Approaches to Transforming Research Paradigms in Climate-Smart Forestry

A scientific research paradigm constitutes the conceptual framework and system of shared methodologies employed by the scientific community to explore fundamental unknowns, expand cognitive boundaries, and advance knowledge innovation [20]. The multiscale structural complexity of forests, their long lifecycles, and the coupling of multiple ecological processes present substantial challenges to traditional research methods, including experimental design, in-situ observations, and field surveys. CSF facilitates interdisciplinary integration across forestry, global change ecology, computer science, data science, and engineering analytics, creating novel intellectual spaces for addressing cross-disciplinary challenges through hybridized epistemologies. Enabled by advances in digital informatization, networking systematization, and intelligent automation technologies, the evolution of knowledge within the new research paradigm is driven by developments in data, algorithms, and computational power, reflecting the co-evolution of cognitive frameworks, analytical approaches, and knowledge generation [21]. The transformation of the research paradigm in CSF encompasses the following three aspects.

(1) Advancing forest data acquisition methods

The next generation of intelligent forestry instrumentation provides comprehensive technical solutions and hardware support for forest resource inventory, precision silviculture, ecological restoration, risk assessment, and ecosystem service evaluation, facilitating comprehensive digital workflows [15]. Through automated networked monitoring systems incorporating specialized intelligent devices (e.g. such as automatic sensors for tree structure, physiological monitors for traits, and microenvironment loggers), these advancements significantly enhance forest data acquisition. These technologies also enable autonomous monitoring of forest structural dynamics, physiological processes, and ecological functions, including three-dimensional canopy structure, leaf functional traits, vegetation canopy photosynthesis, and transpiration. By improving data collection efficiency and reducing human-induced errors, such advancements enhance the accuracy and reliability of long-term ecological observations. Advances in multi-dimensional ecological sensing and multi-network fusion convergence have enabled the development of forestry resource and ecological environment data acquisition and updating technologies, allowing efficient and high-resolution retrieval of forest ecosystem parameters across spatio-temporal multi-scales [22]. The integration of forestry resource data with AI-driven data mining techniques facilitates multi-scale analysis, multi-modeling, and predictive simulations. Large-scale ecological models based on AI-ensembled algorithms improve the accuracy of forest structural parameter extraction and enhance the inversion of ecological service functions, supporting unprecedented accuracy in forest carbon sequestration estimation, hydrological regulation assessment, and biodiversity conservation planning [23, 24].

(2) Advancing forest data analysis

Applying AI technologies, particularly machine learning (ML) algorithms, enhances the processing and analysis of large and complex forest ecological datasets, enabling the rapid identification of patterns, trends, and anomalies. ML and deep learning (DL) models facilitate information extraction, text classification, and the construction of forest knowledge graphs. Limited labeled data are addressable through two compensatory strategies: ML implementations using coarse-grained manual features, or DL models employing low-dimensional dense vectors. Large volumes of easily accessible unlabeled forest data, including remote sensing images, soil properties and meteorological time series, and ground survey data, can be automatically labeled to develop datasets for extracting and interpreting forest information. AI and digital twin technologies provide new opportunities to explore the complex interactions and feedback mechanisms within forest ecosystems, supporting data-driven solutions for forest management under diverse scenarios and informing effective forest protection strategies. Leveraging extensive historical observational archives, literature, and real-time datasets, AI and digital twin technologies enable the development of more accurate ecological prediction models, reducing the need for repeated experimental plots and optimizing resource allocation in terms of funding, personnel, and time. These models also improve the accuracy of climate change impact assessments on forest ecosystem structure and function.

(3) Advancing forest management paradigms

Developing an AI-expert collaborative decision-making system enhances precision in forest ecosystem management. This system supports three critical pathways: 1) carbon sequestration through expanding forest area and improving quality, 2) emission reduction via mitigating disturbances (e.g. wildfires, pest outbreaks), and 3) the provision of diverse ecosystem services [13]. By integrating AI with expert knowledge, the sustainable forest ecosystem management framework enables multi-model synthesis, incorporating biodiversity, forest disturbances, species distribution, and ecological responses under different climate scenarios. This system facilitates intelligent and high-resolution monitoring and assessment of biotic components (e.g. forest structures, communities) and abiotic factors (e.g. soil, climate, and hydrology) within forest ecosystems [15, 16]. A decision-making system tailored to forest ecosystem functions and biodiversity conservation leverages ecosystem pattern optimization technology, structural and quality enhancement strategies, and carbon sequestration management simulations [25, 26]. This system provides intelligent decision support for forest operations, ecosystem service maintenance and enhancement, conservation and restoration efforts, and sustainable management strategies for adapting to climate change. Additionally, a complementary timber management and utilization decision-making platform, powered by big data analytics and AI technologies, supports the sustainable wood production for long-term carbon storage, the development of forestry biochar-based industries, and the provision of high-quality economic and ecological products. By integrating forest ecosystem management with sustainable timber utilization, this approach enhances decision-making to promote the efficient, stable, and sustainable functioning of forest ecosystems, ultimately fostering a balanced and harmonious relationship between humans and nature.

4 Scientific and Practical Applications of Climate-Smart Forestry

Through systematic investigation of scientific paradigm shift mechanisms in CSF, the research team has conducted practical implementations across multiple application domains, including intelligent data acquisition, digital twin analysis, and AI-expert decision-making.

4.1 Intelligent Data Collection and IoT Transmission

The Internet of Things (IoT) represents an advanced information integration technology that interconnects physical objects with digital networks through information-sensing devices based on agreed-upon protocols. This technology facilitates intelligent recognition, real-time control, precise management, and scientifically informed decision-making. In June 2016, China's State Forestry Administration promulgated the *Guiding Opinions on Promoting the Development of Forestry IoT in China*, outlining a strategic initiative to modernize forestry practices through IoT integration. The policy emphasizes three objectives: transforming forestry development, improving efficiency and quality, and accelerating comprehensive integration of IoT technologies with core forestry operations.

The IoT infrastructure development serves as a fundamental pillar of CSF, providing essential capabilities in forest information sensing and data acquisition. It has become a primary contributor of forestry big data resources. The IoT system architecture consists of three layers: the sensing layer, the network layer, and the application layer (Fig. 2). The Ker Towers at the Qingyuan Forest CERN, National Observation and Research Station, Liaoning Province (hereafter referred

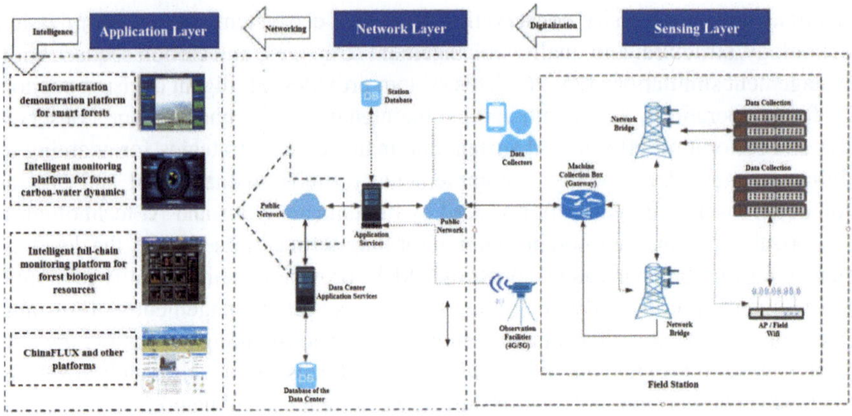

Fig. 2 Schematic diagram of the "digital-network-intelligent" architecture of the IoT system at Qingyuan Forest CERN

to as Qingyuan Forest CERN) demonstrate practical implementation of the IoT system, where the research team deployed numerous sensor nodes throughout the forest. These sensors collect data periodically and transmit it to the monitoring platform for storage and analysis, enabling automated, real-time, and intelligent forest resource management. The following section provides a detailed overview of the design, configuration, functions, and database development and application of the "digital-network-intelligent" architecture of the Qingyuan Forest CERN's IoT system.

4.1.1 Overall Design of the IoT System at Qingyuan Forest CERN

Qingyuan Forest CERN is located in the core area of the Liaodong mountainous region, where the complex terrain and climatic conditions present significant challenges for IoT deployment in forest ecosystems. In developing the IoT architecture for this site, the research team systematically addressed critical technical considerations: sustainable energy harvesting, low power operations, waterproof and cryogenic engineering, fire prevention with self-ignition suppression, secure early-warning signal feedback, and system robustness. By integrating wireless sensor networks, mobile communication technology, and edge gateway technology, the team developed a cloud-network-edge-terminal collaborative information solution engineered for complex terrains. This integrated system enables fully automated acquisition and real-time telemetry of field observations at Qingyuan Forest CERN.

The IoT infrastructure at Qingyuan Forest CERN is designed to systematically integrate comprehensive monitoring components, intelligent gateways, and cloud-edge collaboration systems, ensuring stable and reliable data acquisition. At the sensing and network layers, the research team developed low-power wide-area wireless networking transceivers with relay modules to address the connectivity challenges of ecological monitoring IoT systems in remote areas lacking base stations. Additionally, fiber optic deployment and other wired transmission methods enable the seamless integration of multi-source heterogeneous data from field observation and experimental facilities. This setup allows carbon flux and other critical datasets to be transmitted via wired connections to the data center and the China Ecosystem Research Network (CERN) data center, supporting terabyte-scale data convergence (Fig. 3): (1) Implementation of stable wired fiber and backbone network bridges ensures the reliable transmission of data collected by the observation towers to central repositories. (2) Deployment of edge network bridges concentrates data from key field plots surrounding the towers, and wireless access points (APs) are applied to unify sensor networks within the flux source area. (3) Installation of 4G/5G cellular infrastructure provides supplementary network coverage in areas without wireless signal availability, ensuring seamless and comprehensive data transmission.

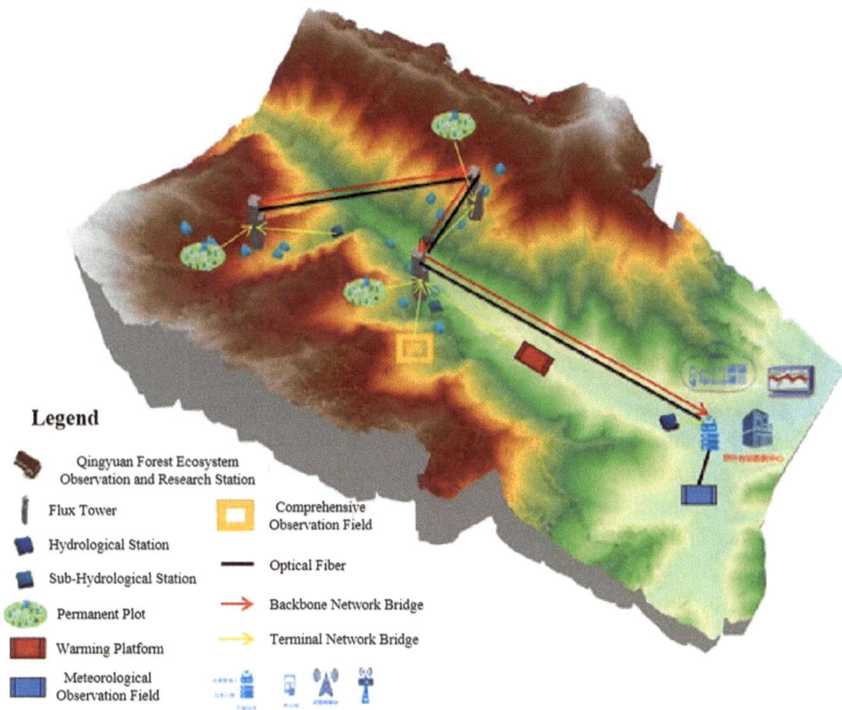

Fig. 3 IoT system deployment and outcomes at the Qingyuan Forest CERN [17]

4.1.2 IoT System Architecture at the Qingyuan Forest CERN

Focusing on the core requirements of IoT applications for ecological monitoring at the Qingyuan Forest CERN, the research team has developed key technological solutions resolving challenges in the system architecture, data collection, signal transmission, and sharing service. A standardized ecological monitoring IoT system was established to ensure seamless integration and interoperability. The team conducted research on multi-source data integration, stable data transmission, and remote networked data logging and transmission technologies. Additionally, we developed an ecological monitoring information system that enables fully automated data acquisition, real-time data sharing, and cloud-based analytical functionality, ultimately forming an intelligent ecological monitoring IoT system.

(1) All-Element Access Subsystem (Sensing Layer)

The comprehensive access subsystem primarily ensures the seamless adaptation of different types of equipment interfaces, IoT modules, and data transmission strategies, considering factors such as the characteristics of field sampling sites, observation elements, and the collection frequency and data volume of field observation equipment. It consists of three main components: First, it integrates existing equipment by converting multi-source data interfaces into standardized hardware and software interfaces, enabling uniform digital management. Second, it provides local network coverage for field sites by incorporating uplink 4G/5G communication modules, fiber optics, and Ethernet interfaces, addressing equipment networking challenges and ensuring reliable internet access for field researchers. Third, for sensors located in remote areas beyond public network coverage, it deploys low-power, integrated collection, storage, and transmission devices to facilitate data acquisition.

(2) Intelligent Gateway Subsystem (Network Layer)

This subsystem is designed to connect field ecological observation instruments to the network, ensuring smooth data transmission and generating alerts in the event of sensor malfunctions. It features local data storage capabilities, allowing data to be temporarily stored during network outages and automatically uploaded to the platform once connectivity is restored. Additionally, in scenarios requiring high real-time processing, large-volume data transmission, or in cases of platform network disruptions, the subsystem can establish localized wireless configurations to maintain complete data transmission, efficient processing, and alert dissemination. This design employs a dual-redundancy transmission protocol that integrates optical fibers and wireless network bridges to enhance the stability and reliability of data transmission. If one transmission path fails or is interrupted, the other immediately takes over the data transmission task, ensuring data continuity and integrity while significantly reducing the risk of service interruptions.

(3) Data Integration Subsystem (Application Layer)

The IoT platform facilitates the integration of a vast number of multi-source heterogeneous sensors for data aggregation while supporting the development of application service platforms, adaptive monitoring interfaces, and virtualized operation and maintenance for customized application services. Devices and sensors in the sensing layer establish secure, authenticated connections with the information center and maintain bidirectional communication with the IoT cloud. The monitoring and maintenance module enables real-time monitoring of connected devices, log processing, firmware upgrades, alert handling, and other essential functions. Based on forest carbon sink observation indicators and the specifications of automated observation equipment, the research team has integrated high-precision micro data acquisition and storage devices, GPS positioning systems, and adjustable multi-type wireless transmission and control modules. Additionally, we have developed software systems for data acquisition, system control, and data verification and management.

4.1.3 Functional Capabilities of the Qingyuan Forest CERN IoT

The research team developed the Qingyuan Forest CERN IoT platform based on multi-source ecological data monitoring, reducing the complexity of field data monitoring and collection, optimizing labor efficiency, and enabling key functions such as intelligent data sensing, automated recognition, and cloud-edge computing.

Through intelligent data collection terminals, automated collection boxes, and other methods, the research team acquires key monitoring data, including carbon flux, micro-meteorology, hydrology, vegetation phenology, and animal activity. To enable IoT connectivity in remote base-station-deficient environments, we developed low-power, long-distance wireless networking transmission devices and relay equipment, enabling unified management of field-deployed monitoring equipment and visualization of observation and monitoring content. Supported by the IoT platform, field monitoring data can be transmitted in real time, and forest carbon flux and hydrological station indicators can be dynamically displayed. When combined with forest structure data, this facilitates research on the carbon-water cycle, hydrological processes, and regulating mechanisms of forest ecosystems. The platform also integrates atmospheric, biological, and soil monitoring data, supporting the intelligent management of primary plots, equipment and datasets. Additionally, the research team deployed infrared sensors, video surveillance, wireless communication, and mobile internet infrastructure, leveraging cloud-edge computing for image recognition and big data analysis. This system enables automated identification of animal species and population counts, forming an intelligent management framework for wildlife resources. Ultimately, through the IoT platform, data such as carbon flux, soil temperature and humidity, observation plot phenology, and animal activity are seamlessly integrated into the data center, achieving digitalization, networking, and intelligent automation. This significantly enhances the real-time capabilities, comprehensiveness, accuracy, and reliability of ecological monitoring.

4.1.4 Database Integration and Data Security

To efficiently manage the data collected through the IoT, the research team performed standardized processing and structured its storage within the database based on predefined rules and relationships. Utilizing multi-source heterogeneous fusion technology, we established the Qingyuan Forest CERN database, which consolidates multiple sub-databases to fulfill data management and analytical demands for multidisciplinary research and observation domains. The platform incorporates a suite of database technologies, enabling centralized data storage and seamless access. It features a comprehensive set of modules for data uploading, parsing, visualization, statistical analysis, downloading, and access control, significantly improving data utilization efficiency and operational oversight.

The research team designed and implemented a high-performance data storage model and a fast-access architecture, distributing read operations across distributed replica servers to alleviate the load on the main server, enhance the system's concurrent processing capability, and improve disaster recovery resilience. A data sharding protocol specific to ecological datasets was implemented through horizontal partitioning of oversized tables, preventing performance degradation from monolithic table architectures. To accommodate high-concurrency user access patterns, the team optimized indexing techniques, enhancing query efficiency under simultaneous multi-user access. Additionally, we achieved the efficient development of the Qingyuan Forest CERN database platform's visual interface, employing a reactive programming framework and implementing a well-structured modular design. This approach facilitates ease of maintenance and scalability while ensuring a seamless and consistent user experience.

Based on the database platform, the system enables dynamic visualization of observation data (Fig. 4) and facilitates its upload to the CERN data center via the internet [15]. The data center is interconnected with CERN and the ChinaFlux ensuring continuous synchronization of the Qingyuan Forest CERN's data status.

In terms of data security, the research team provides technical support throughout the entire data lifecycle, encompassing key aspects such as data protection, operational security and governed sharing. At the infrastructure level, physical resources and logical operations are fully separated to ensure the security and independence of different modules, facilitating the rapid identification and resolution of security issues while enhancing data isolation during operation. For storage security, cryptographic isolation renders data irretrievable even under physical media breaches. In the data transmission phase, the platform implements end-to-end encryption technology to prevent data interception or tampering during network transmission. Additionally, a defense mechanism against man-in-the-middle attacks has been integrated to further enhance transmission security. In the data sharing and open-access phase, the platform adopts a "data provenance framework" to regulate the entire process of data ownership, authorization and distribution. Through data usage licensing and circulation control, the platform dynamically adjusts access permissions during data

Fig. 4 Real-time observation data (**a**) and trend analysis chart (**b**) of the Ker Towers displayed via the internet of things

sharing, ensuring that data is only utilized within the authorized scope. Furthermore, to enable the traceability of security incidents, the platform integrates a log auditing function that records all operational logs in real time, storing them in an encrypted format to provide verifiable evidence for data security management.

4.2 Digital Twin and Analysis

Digital twin technology synthesizes physical-based models, real-time sensor updates, operational historical logs, and other source data to integrate multidisciplinary, multi-physics, multi-scale, and probabilistic simulations. It constructs a cyber-physical mirror that digitally replicates entity lifecycles. This technology enables bidirectional data feedback, virtual-real interaction, and prescriptive analytics. Building upon high-precision forest and grassland survey data and using digital forest environments as a foundation, the research team integrates forest growth, management, and ecosystem process models. By incorporating IoT, cloud computing, virtual reality, and AI technologies, we have developed a forest digital twin system that enables coordinated simulations of forest structure, ecosystem functions, tree dynamics, and management processes. This case study documents the digital twin implementation at the Saihanba Forestry Center in Hebei Province, focusing on four pillars: (1) the digital twin data infrastructure, (2) digital twin model library development, (3) forest structure and function simulations, and (4) management decision-making (Fig. 5).

4.2.1 Construction of the Digital Twin Data Foundation

The data foundation of the Digital Twin integrates multi-source datasets through satellite-airborne-ground campaigns, capturing forestry multispectral and hyperspectral imagery, aerial surveys, ground surveys, IoT sensing data, and other types of datasets. Raw data are cleaned, transformed, and managed, ultimately enabling their

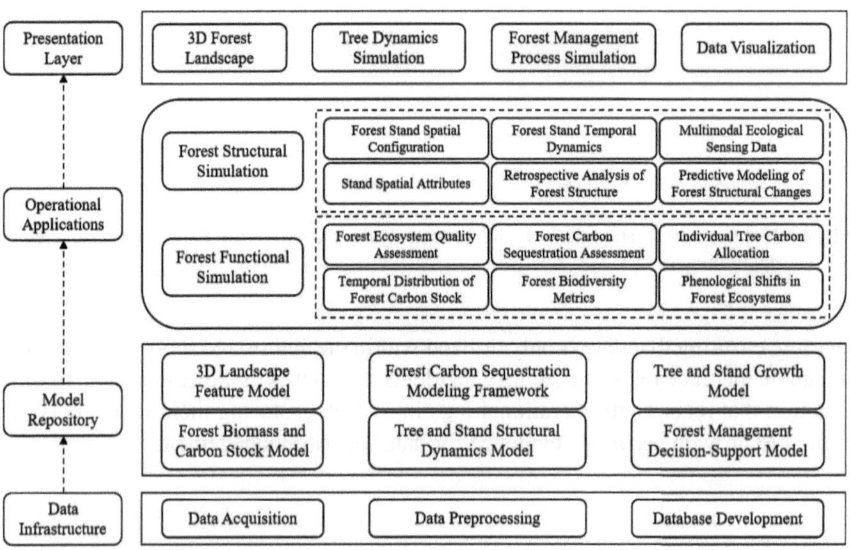

Fig. 5 Framework of the digital twin for Saihanba Forestry Center

processing and integration into the database. In the Saihanba Forestry Center, the research team installed 52 forest growth monitoring devices, 1 forest habitat monitoring device, 5 infrared cameras, 1 phenology observation device, 5 fire detectors and 1 network transmission system. Through ground surveys and unmanned aerial vehicles (UAVs) aerial imaging, data were collected for 41,610 individual trees in the Shanghai Forest of the Saihanba Forestry Center, including attribute data, digital elevation models, orthophotos, oblique imagery and point cloud data representing the structure of 1,255 individual trees. The team provides services for multi-dimensional and multiscale spatio-temporal data aggregation, cleaning, transformation, sharing, visualization, computation and updating for various types of collected data. The data are logically categorized using the "database → sub-database → dataset → layer (table/model)" structure. Database construction is customized according to data type, enabling database parameter settings, data quality checks, data entry, organization and security configurations. This data foundation supports subsequent algorithmic models and functional applications for the Digital Twin.

4.2.2 Construction of the Digital Twin Model Library

For the virtual model entities in the digital twin scenario and the various ecosystem process models that drive scene operations, the research team developed a standardized and parameterized model repository to facilitate model maintenance, management, performance optimization, integration and parameter adjustment. The team constructed three-dimensional models of common vegetation (trees, shrubs and grasses), landscape elements and terrain features within the region. These models are mapped to real-world geographic coordinates using geospatial data. Environmental lighting and texture mapping are optimized to create a simulated regional digital twin. By integrating UAV- and ground-based LiDAR point cloud data, individual trees were segmented, tree skeleton structures were extracted, and mesh models of tree trunks and branches were generated. Additionally, leaf attachment points were computationally derived, and the rendering of trunks, branches and foliage was achieved by incorporating texture atlases from the centralized data repository. Furthermore, by integrating three-dimensional tree models, terrain models, herbaceous vegetation distribution data and tree attribute data including coordinates, tree height, diameter at breast height (DBH), crown width and branch height, the team constructs a detailed stand-level forest structure scene incorporating trees, shrubs, and grasses. In addition, using techniques such as level-of-detail models, occlusion culling and ray tracing, we developed a digital twin model at the stand level.

The research team aggregated carbon sink estimation models for local living trees, dead wood and litter. Based on the 2020 forest inventory survey data, stem analysis data of tree species and management plan data from the Saihanba Forestry Center, we developed ecosystem models for species such as *Larix principis-rupprechtii*, *Pinus sylvestris* var. *mongolica*, and *Betula platyphylla*. These models include growth models, volume models, stand structure models and management decision models, which serve as the foundational drivers of the digital twin system.

4.2.3 Simulation of Forest Structure

Through integration of multi-source survey data and AI analytical algorithms, a comprehensive analytical framework is established for afforestation process at Saihanba Forestry Center, encompassing forest structure simulation, IoT-enabled environmental monitoring telemetry and dynamic stand structural analysis. The forest structure simulation applies AI algorithms to analyze historical remote sensing imagery and operational data, quantifying the spatial distribution patterns of attributes such as stand layout, tree density, DBH and tree height for *Larix principis-rupprechtii* at key time points, including 1962, 1964, 1987, 2000, 2014 and 2024. This enables the reconstruction of the ecological system structure of the larch plantation at critical time nodes. For the spatial and temporal distribution of DBH, micro-meteorological factors and animal diversity within digital plots, the system integrates real-time data from sensors measuring tree DBH, micro-meteorological conditions and infrared camera outputs. By performing AI analysis of multi-modal sensing data, the system facilitates the coordinated analysis of forest environmental factors, tree growth dynamics and animal distribution patterns. Based on survey and retrospective data, parameters including stand density, timber volume, tree height, DBH, species mixing degree and angular scale are computed and compared, allowing for the dynamic simulation and evaluation of forest structural changes within the study area (Fig. 6).

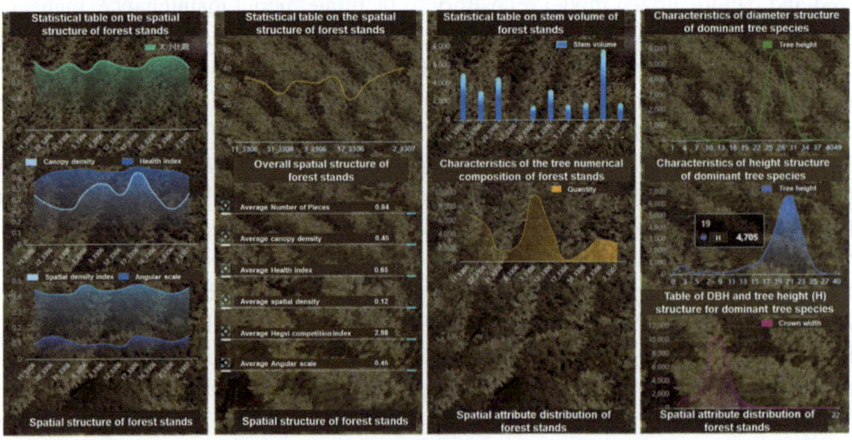

Fig. 6 Dynamic analysis of forest structure using a digital twin

4.3 Supporting Forest Management Decision-Making

4.3.1 Competition-Based Thinning Management for Plantation Forests

A primary objective of competition studies in thinning-mediated stand structure optimization is to support forest management. Conventional forestry operations frequently depend on empirical judgment, which makes it challenging to establish quantitative silvicultural protocols. This study proposes KHegyi, a distance- and tree species-specific shade tolerance-informed competition index, and employ LiDAR point cloud data to simulate spatial structure changes in forest stands before and after thinning. This methodology enables data-driven identification of scientifical thinning strategies.

The research team focused on pure larch plantations and typical larch-broadleaf mixed forests in northeastern China, including sites in Qingyuan (Liaoning province), Changbaishan (Jilin province) and Maoershan (Heilongjiang province). The primary research objective was to analyze individual tree competition. The team applied voxel-based point cloud analytics to quantify individual tree architectural complexity. The methodological innovation lies in synthesizing the distance-dependent Hegyi competition index with a novel shade tolerance-modified KHegyi, subsequently integrating these with forest density metrics to establish a new stand-level competition index. Digital methods are employed to calculate and compare the comprehensive structural index of forest stands under different thinning schemes to determine the optimal thinning strategy. Key findings demonstrate that thinning intensities of 5, 10, 20 and 30% improved the overall spatial structure of forest stands to varying degrees, evidenced by increases in the spatial structure index. During the thinning simulation, the individual tree size parameters exhibited minimal variation, with a negligible effect on the overall model. This suggests that individual tree size should not be considered a critical criterion for selecting target trees for thinning. This case study simulated the effects of thinning on forest stand dynamics, providing practical insights for improving stand management practices and enhancing thinning efficiency.

4.3.2 Digital Twin-Based Management Decision-Making

To address the multifunctional management requirements of the Saihanba Forestry Center, a digital twin-based management decision-making framework comprehensively integrates local operational conditions, policy requirements and regulatory constraints to establish multi-objective expert decision support analysis. This innovative approach enables multi-objective expert decision analysis aimed at enhancing plantation productivity, mitigating pest and disease risks, and reducing wildfire potential. Through parametric simulations of management processes including harvesting methods, harvesting intensity, underplanting species selection and planting density, the digital twin system incorporates specific silvicultural strategies within forest management pilot areas, such as density regulation, underplanting in natural secondary forests, and thinning-with-underplanting in *Larix-Betula* mixed forests. By employing data comparison, multimedia visualization and scenario analysis, the system can simulate the characteristics and variations of different management plans. Particularly focusing on carbon sequestration enhancement objectives, it enables systematic comparison of operational outcomes across different management scenarios, facilitating expert decision-making.

The current digital twin system of the Saihanba Forestry Center integrates unified management of fundamental data and twin models, simulating forest ecosystem structure and function, as well as the dynamic changes in forests under different management decision scenarios. Through continuous advancements in the accuracy and efficiency of three-dimensional modeling, this technology demonstrates dual optimization capacities: (1) reducing labor and time costs while providing forest managers with high-precision tree structural characteristics and growth information, thereby enhancing the ability to accurately simulate complex forest ecosystems; (2) strengthening decision-making capabilities for forest management planning, thus increasing the efficiency of forest resource conservation and sustainable management. This research methodological framework has applications in protected area management and decision-making processes, particularly for national parks, nature reserves, state-owned forests and other forest management entities, offering technical support for precision-based and intelligent forest management decision-making, networking systematization and intelligent automation.

5 Prospect

CSF represents an emerging paradigm for sustainable forest ecosystem management. This paradigm leverages digital informatization, networking systematization, and intelligent automation to develop innovative and forward-looking forest management approaches and decision-making frameworks that address the challenges of climate change [15, 16]. The author emphasizes that advancing CSF requires the development and application of emerging information technologies, including the IoT, AI and digital twin technology. Although substantial progress has been achieved in CSF research, current advancements demonstrate significant disparities across different research directions and domains. Further research should focus on the following key aspects.

In the data acquisition infrastructure, forest-specific IoT devices and sensor networks are currently being deployed across representative field research stations in China. However, challenges remain regarding their scalability and integration with sensors from different manufacturers. Additionally, the operational reliability and long-term performance of these systems under harsh environmental conditions, particularly in tropical forests characterized by persistent high temperature and humidity, require fully validated. UAVs have emerged as a core tool for near-surface remote sensing, demonstrating extensive utility in monitoring forest structure and health. Nevertheless, current UAVs implementations face threefold constraints: (1) substantial cost barriers, (2) operational and technological complexity, and (3) technical limitations in obtaining key parameters (e.g. DBH and canopy height). Further research is essential to ensure operational safety and reliability, particularly for under-canopy flight operations in mountainous natural forests.

In sensor technology development, the current sensors employed for forest attribute acquisition remains limited diversity. Advanced sensor development is essential to capture a broader spectrum of forest attributes. Additionally, enhancing the robustness of deep learning models for forest applications requires training on large volumes of labeled data [22]. For larger spatial scales, the integration of technologies such as synthetic aperture radar (SAR), LiDAR, optical remote sensing and the IoT can facilitate more comprehensive forest monitoring across broader spatial extents.

In the field of analytical methodologies, advancements in big data technologies have empowered machine learning and big data analytics to process and predict vast amounts of data generated from multiple sources, including sensors, UAVs and satellites. These advancements can significantly accelerate the development of CSF and enhance the efficiency of forest information management. High-performance big data architectures can greatly improve the efficiency of large-scale forestry research, while AI facilitates the extraction of vegetation features and ecological parameters from remote sensing data [27]. Consequently, the development, integration and application of big data and AI technologies have become key priorities in CSF research. Parallel attention is directed toward data-driven and process model-based approaches in forest plant studies.

In modeling forest ecosystems, the inherent complexity and long-term dynamics of these ecosystems pose significant challenges in developing models capable that integrate multiple ecological processes while managing large volumes and complex data. Addressing these challenges requires a mechanistic understanding of critical forest ecosystem processes, including biological community succession, biogeochemical cycling and biogeophysical processes. Additionally, the establishment of standardized data interoperability protocols is essential to facilitate cross-disciplinary collaboration [28]. Currently, the research team is advancing AI-driven models for predicting tree growth and assessing forest health.

Current decision-making systems for CSF remain in the early stages of operationally development. Future efforts should focus on developing a sustainable management decision-making system for forest ecosystems and biodiversity conservation through "human-AI" collaborative interaction. This system should optimize the allocation of forest resources, preserve and quantitatively augment ecosystem services, protect and restore forest ecosystems, and support sustainable management strategies in response to climate change. Given the high costs associated with advanced technological applications, CSF should prioritize cost-effective alternatives, seek government subsidies, and promote public–private partnerships to reduce costs and improve accessibility. Additionally, strengthening specialized training for forestry personnel is essential to enhance the effectiveness of decision implementation.

Acknowledgements This research was supported by the National Natural Science Foundation of China (Grant No. 32192435), the Special Application Demonstration Project for Cybersecurity and Informatization of the Chinese Academy of Sciences (CAS-WX2022SF-0101), and the Xingliao Talent Program (Grant Nos. XLYC2201002, YS2023006). The authors would also like to thank Dr. Ning Zheng and Engineer Linsen Jin from Beijing LiGA United Technology Co., Ltd. for their assistance with the illustrations.

References

1. Ding ZL, The Key To Achieving Carbon Neutrality Is To Build a "Three Terminal Force" System [J]. China Petroleum Enterprise, 2021(6): 10–11+111. (in Chinese)
2. Yu GR, Zhu JX, Xu L, et al. Technological approaches to enhance ecosystem carbon sink in China: Nature-based solutions [J]. Bulletin of Chinese Academy of Sciences, 2022, 37(4): 490–501. (in Chinese)
3. Benyus JM. Biomimicry: Innovation inspired by nature [C]. New York: William Morrow, 1997.
4. Cohen-Shacham E, Janzen C, Maginnis S, et al. Nature-based solutions to address global societal challenge s[C]. Gland: IUCN, 2016.
5. European C. Horizon 2020 work programme 2016–2017 [S]. Cross-cutting Activities (Focus Areas), 2016.
6. Baumgartner RJ. Sustainable development goals and the forest sector—a complex relationship [J]. Forests, 2019, 10(2): 152.
7. Zhang J, Fu BJ, Stafford-Smith M, et al. Improve forest restoration initiatives to meet sustainable development goal 15 [J]. Nature Ecology & Evolution, 2020, 5(1): 10–13.
8. Friedlingstein P, Jones MW, O'Sullivan M, et al. Global carbon budget 2021 [J]. Earth System Science Data, 2022, 14(4): 1917–2005.
9. Yang Y, Shi Y, Sun W, et al. Terrestrial carbon sinks in China and around the world and their contribution to carbon neutrality [J]. Science China Life Sciences, 2022, 65(5): 861–895.
10. Chapin FS, Matson PA, Vitousek PM. Principles of terrestrial ecosystem ecology [M]. New York, NY: Springer New York, 2011.

11. Bonan GB. Forests and climate change: Forcings, feedbacks, and the climate benefits of forests [J]. Science, 2008, 320(5882): 1444.
12. Curtis PS, Gough CM. Forest aging, disturbance and the carbon cycle [J]. New Phytol, 2018, 219(4): 1188–1193.
13. Zhu JJ, Gao T, Yu LZ, et al. Carbon sink of forest ecosystems: Concept, time effect and improvement approaches [J]. Chinese Journal of Applied Ecology, 2024, 35(9): 2313–2321. (in Chinese)
14. Harris NL, Gibbs DA, Baccini A, et al. Global maps of twenty-first century forest carbon fluxes [J]. Nature Climate Change, 2021, 11(3): 234–240.
15. Zhu JJ, Wang GG, Zhang HQ, et al. On the Research of Climate-Smart Forestry [J]. Scientia Silvae Siniae, 2024, 60(7): 1–7. (in Chinese)
16. Wang GG, Lu DL, Gao T, et al. Climate-smart Forestry: An AI-enabled sustainable forest management solution for climate change adaptation and mitigation [J]. Journal of Forestry Research, 36(1): 7.
17. Gao T, Zhu JJ, Zhang JX, et al. Estimation of Carbon Flux of a Temperate Forest Ecosystem Based on Next-Generation Information Technologies [J]. Frontiers of Data & Computing, 2023, 5(2): 60–72. (in Chinese)
18. Buonocore L, Yates J, Valentini R. A proposal for a forest digital twin framework and its perspectives [J]. Forests, 2022, 13(4): 498.
19. Qiu H, Zhang H, Lei K, et al. Forest digital twin: A new tool for forest management practices based on spatio-temporal data, 3d simulation engine, and intelligent interactive environment [J]. Computers and Electronics in Agriculture, 2023, 215: 108416.
20. Lederman NG. Students and teachers conceptions of the nature of science: A review of the research [J]. Journal of Research in Science Teaching, 1992, 29(4): 331–359.
21. Tan JW, Zhang HQ, Liu Y, et al. Question-answering Forestry Pre-trained Language Model: ForestBERT [J]. Scientia Silvae Siniae, 024, 60(9): 99–110. (in Chinese)
22. Chen QD, Gao T, Zhu JJ, et al. Individual tree segmentation and tree height estimation using leaf-off and leaf-on uav-lidar data in dense deciduous forests [J]. Remote Sensing, 2022, 14(12): 2787.
23. Zhao Q, Yu S, Zhao F, et al. Comparison of machine learning algorithms for forest parameter estimations and application for forest quality assessments [J]. Forest Ecology and Management, 2019, 434: 224–234.
24. Wang J, Zhang H, Liu Y, et al. Tree-level Chinese fir detection using UAV RGB imagery and yolo-dcam [J]. Remote Sensing, 2024, 16(2): 335.
25. Zhu NF, Zhang HQ, Cui ZY, et al. Simulation of operation for forest management based on body action interaction [J]. Forest Research, 2021, 34(5): 95–103.
26. Li YL, Shen K, Zhang YQ, et al. Forest Virtual Simulation System Based on CAVE2 [J]. Forest Resources Management, 2019, 2: 123–131+136. (in Chinese)
27. Jing W, Kuang Z, Scherer R, et al. Editorial: Big data and artificial intelligence technologies for smart forestry [J]. Frontiers in Plant Science, 2023, 14: 1149740.
28. Yun T, Li J, Ma L, et al. Status, advancements and prospects of deep learning methods applied in forest studies [J]. International Journal of Applied Earth Observation and Geoinformation, 2024, 131: 103938.

Dr. Jiaojun Zhu is an Academician of Chinese Academy of Engineering, a professor researcher at Institute of Applied Ecology (IAE), Chinese Academy of Sciences (CAS). He currently serves as the Director General of IAE, CAS, and the Director of Qingyuan Forest CERN, National Observation and Research Station. He is the President of Liaoning Ecological Society, and the Co-Editor-in-Chief of "Ecological Processes".

Dr. Zhu has been major in Ecology of Protective Forests and Forestry eco-engineering. He has led over 20 major projects, including those funded by National Natural Science Foundation of China (Distinguished Young Scholars, Major Program, and Key Program), and Ministry of Science and Technology of China (National Program on Key Basic Research Project (973 Program), and National Key Research and Development Program of China). Aiming at the needs of national major ecological engineering construction and the scientific and technological frontiers, he has centered his work on the world's largest ecological engineering, the Three-North Program, and innovated a theoretical and technical framework of "Establishment, Management, and Assessment" for the Three-North Protective Forest Program. He pioneered a new silviculture theory and directed a management technology for different growth stages, filling the theoretical gap in the ecological function-oriented protective forests management. He was the first to propose a construction concept of protective forests that is based on "total water resources and system coordination". He established an integrated evaluation system of Space-Sky-Tower-Field Assessment Techniques. Through his commitment to scientific theoretical and technological innovation, Dr. Zhu has steered the construction practices of forestry ecological programs, promoted the development of Applied Ecology, and provided the scientific support for the construction of national ecological security.

As the principal investigator, Dr. Zhu has been awarded two National Science and Technology Progress Second Prizes and four Provincial and Ministerial First Prizes. His contributions have also earned him over 20 personal awards, including the IUFRO (International Union of Forest Research Organizations) Scientific Achievement Award. He has authored six monographs, such as "Ecology of Protective Forests", which has been chosen as a graduate textbook. He has published more than 300 peer-reviewed papers as first or corresponding author, including the article in "Nature Climate Change". He has written more than 10 Consulting Reports adopted by the National Government, and assisted in the formation of national laws, regulations, industry standards, and plans.

Tian Gao Ph.D., is a Research Professor and Ph.D. advisor at the Institute of Applied Ecology, Chinese Academy of Sciences. He currently serves as Deputy Director of the Liaoning Terrestrial Ecosystem Carbon Neutrality Key Laboratory, Executive Director of the National Innovation Alliance for 3D Visualization Technology Applications in Forestry and Grassland, and a member of the Ecological Remote Sensing Professional Committee of the Ecological Society of

China. His research focuses on forest remote sensing, carbon-water flux observations in ecosystems, and ecosystem service assessments. He has published over 40 papers and has led research projects, including National Key Research and Development Program projects and National Natural Science Foundation of China (NSFC) projects. Dr. Gao has been recognized with several prestigious awards, including the Second-Class National Science and Technology Progress Award in 2019 and the Chinese Academy of Sciences Science and Technology Advancement Award in 2016.

Huaiqing Zhang Ph.D., is the Chief Scientist at the Chinese Academy of Forestry and Deputy Director of the Institute of Forest Resource Information Techniques, Chinese Academy of Forestry. He is a Research Professor and has been selected for China's National Ten Thousand Talents Program. He has also been recognized as a National Young and Middle-aged Expert with Outstanding Contributions and a Leading Talent in Forestry and Grassland Science and Technology Innovation. Dr. Zhang has long been engaged in research on smart forestry, artificial intelligence, and visualization technology. He has led over 50 national research projects, authored eight academic monographs, and published more than 220 research papers. He holds 22 international and national invention patents, has formulated eight industry standards, and has registered 52 software copyrights. His contributions have been recognized with several prestigious awards, including one Second-Class National Science and Technology Progress Award, three First-Class Provincial and Ministerial Science and Technology Progress Awards, and one Second-Class Provincial and Ministerial Science and Technology Progress Award.

Yangang Wang Ph.D., is a Research Professor and Director of the Artificial Intelligence Technology and Application Development Department at the Computer Network Information Center, Chinese Academy of Sciences. His research focuses on AI computing and data service platform development, as well as AI-driven scientific discovery applications and parallel computing software. Dr. Wang has published over 80 research papers in leading international conferences and journals, including *PPoPP* and *SC*. He holds more than 30 patents and has authored three academic monographs. He has also led several national research projects, including the National Key Research and Development Program, the Chinese Academy of Sciences Strategic Priority Research Program (Category B), and the Chinese Academy of Sciences Informatization Special Project.

Zongguo Wang Ph.D., is an Associate Researcher and Deputy Director of the Applied Software Development Laboratory in the Artificial Intelligence Technology and Application Development Department at the Computer Network Information Center, Chinese Academy of Sciences. She is also a member of the Chinese Academy of Sciences Youth Innovation Promotion Association and a representative of the Chinese delegation at the 8th BRICS Young Scientist Forum. Her research focuses on the application of artificial intelligence in various scientific disciplines. Dr. Wang has led more than ten research projects at the national, academy, and regional levels. In the field of materials science, she developed *MatChat*, a large language model for inorganic material synthesis. Additionally, her research team designed *ANT-MOC*, a large-scale neutron transport equation solver, which was nominated for the Best Paper Award at the 35th International Conference for High-Performance Computing, Networking, Storage, and Analysis (SC23), a prestigious CCF Class-A conference.

Yirong Sun Ph.D., is a Senior Engineer at the Institute of Applied Ecology, Chinese Academy of Sciences. His research focuses on forest ecology and tree physiology. Dr. Sun has led multiple research projects, including a sub-project of the Chinese Academy of Sciences Strategic Priority Research Program, a specialized project under the National Key Research and Development Program, and a key project of the Chinese Academy of Sciences Science and Technology Service Network Initiative. He has published 14 papers in SCI-indexed journals such as

Ecological Processes, Forest Ecology and Management, and *Environmental and Experimental Botany*, including two as the first author. His contributions have been recognized with the Chinese Academy of Sciences Science and Technology Advancement Award in 2016 (ranked 7th/10) and the First-Class Liaoning Provincial Science and Technology Progress Award in 2020.

Jinxin Zhang is a Senior Engineer at the Institute of Applied Ecology, Chinese Academy of Sciences. Her research focuses on long-term ecological monitoring, data quality control and management, platform development, and the application of next-generation information technology in forest management and restoration. Dr. Zhang has led a sub-project of the National Key Research and Development Program and has been a key contributor to major and key projects under the National Natural Science Foundation of China (NSFC) and the Chinese Academy of Sciences Informatization Demonstration Project. She was a principal contributor in the design and construction of the Qingyuan Forest Station Informatization Demonstration Platform. Her work has been recognized with two First-Class Liaoning Provincial Science and Technology Progress Awards (2021) and the Chinese Academy of Sciences Science and Technology Advancement Award (2016).

Deliang Lu Ph.D., is an Associate Researcher at the Institute of Applied Ecology, Chinese Academy of Sciences. His research interests focus on silviculture and forest ecology. He has led multiple research projects funded by the National Natural Science Foundation of China (NSFC) and the National Key Research and Development Program. Dr. Lu has published 17 research articles as the first or corresponding author in reputable journals, including *Agricultural and Forest Meteorology* and *Forest Ecology and Management*. He also holds two authorized patents. He is a member of the Youth Innovation Promotion Association of the Chinese Academy of Sciences and has been honored as a Young and Middle-Aged Science and Technology Innovation Talent in Shenyang.

Tingdong Yang is an Assistant Researcher at the Institute of Forest Resource Information Techniques, Chinese Academy of Forestry. His research focuses on digital twin technology for forestry and grassland ecosystems, as well as 3D forest simulation. He has led a sub-project under the National Key Research and Development Program and an institutional research fund project. As a key researcher, he has participated in six major projects, including the National High Technology Research and Development Program ("863 Program"), the National Key Research and Development Program of China, the Science and Technology Basic Resources Investigation Program, and projects funded by the National Natural Science Foundation of China (NSFC). He has received recognition as the second contributor for one forestry society research achievement and one industry standard. He has published five academic papers and holds three software copyrights and eight authorized invention patents.

Dexiong Teng Ph.D., is an Assistant Researcher at the Institute of Applied Ecology, Chinese Academy of Sciences. His research focuses on machine learning-based forest carbon sequestration measurement and the carbon-water cycle of forest ecosystems. Dr. Teng has led a Young Scientist Fund Project under the National Natural Science Foundation of China (NSFC). In recent years, he has published four SCI-indexed papers as the first author and has obtained two software copyrights.

Shuailing Hao is an Engineer at the Shenyang Artificial Intelligence Computing Center. His research focuses on image understanding, image-text retrieval, and image generation. He has published research on image description in *AAAI*, a top-tier international conference in artificial intelligence.

Fengyuan Yu is an Engineer at the Institute of Applied Ecology, Chinese Academy of Sciences. His research focuses on forest ecology and forest flux dynamics. He has led two tasks under the National Key Research and Development Program of the Ministry of Science and Technology. Additionally, he has contributed to projects under the Chinese Academy of Sciences Strategic Priority Research Program, the National Natural Science Foundation of China (NSFC), and the Cybersecurity and Informatization Special Application Demonstration Project. Dr. Yu has published five papers in SCI-indexed journals, including *Agricultural and Forest Meteorology*, *Journal of Hydrology*, and *Remote Sensing*.

A Large-Model-Driven Robotic AI Chemist with the Fusion of Theory and Experiment for Research

Jun Jiang, Shuo Feng, and Chuxuan Ding

Abstract The exhaustive trial-and-error approach struggles to perform a global optimal solution in increasingly complex and high-dimensional chemical research. Artificial intelligence (AI), with its ability to decipher the high-dimensional correlations in complex data, offers a transformative opportunity to address this challenge. The automated experimental platform can produce sufficiently standardized data for intelligent models by executing high-throughput experimentation, and equip these models with practical research capabilities. By combining AI with automated experiments, the AI-Chem team at University of Science and Technology of China has developed the first all-around and data-intelligence-driven robotic AI chemist in the world. Using predictive models based on theoretical computations and experimental feedback, the robotic AI chemist has achieved the global optimal synthetic formula, along with a series of benchmark examples for the new data-driven research paradigm. To further promote this paradigm, a cloud infrastructure for AI chemist is proposed, which includes multi-domain, multi-modal, and standardized intelligent databases; large models based on mathematical logic; distributed and fully automated intelligent experimental facilities; and an intelligent chemistry cloud platform for resource allocation and sharing. This cloud infrastructure can guide a new research paradigm that combines intelligence and materials science, thereby propelling China's chemistry discipline to the forefront of the international stage.

Keywords Robotic AI chemist · Fusion of theory and experiment · Large model · Automated experimental platform · Cloud infrastructure for AI chemist

J. Jiang (✉) · S. Feng · C. Ding
State Key Laboratory of Precision and Intelligent Chemistry, University of Science and Technology of China, Hefei, P.R. China
e-mail: jiangj1@ustc.edu.cn

China's e-Science Blue Book 2025, https://doi.org/10.1007/978-981-96-8252-2_12

1 Introduction

1.1 Urgent Need for Innovation in Chemical Research Methodology

Chemistry research in the present era is beset by numerous challenges, including low experimental reproducibility, inconsistent data quality, and the increasing complexity of research subjects. Conventional research paradigms primarily rely on exhaustive and trial-and-error approaches, which are time-consuming and inefficient in uncovering the intricate mechanisms of materials within the vast chemical space. Moreover, searching for optimal formulations and processes often becomes trapped in local optima, thereby limiting the potential for global exploration. For instance, drug discovery is a complex and resource-intensive process, where conventional methods struggle to efficiently navigate the extensive chemical space. Similarly, chemical synthesis remains heavily dependent on expert intuition, falling short of intelligent optimization. Furthermore, incomplete materials data and unclear structure-property relationships hinder inverse design strategies. To address these challenges, a paradigm shift in chemical research is necessary to enhance efficiency and reduce costs in discovering and creating new substances.

1.2 Artificial Intelligence Brings Opportunities for Paradigm Shifts

Following its introduction by John McCarthy in 1956, artificial intelligence (AI) has become a key driving force behind transformative advancements in chemical science. AI has demonstrated proficiency in uncovering relationships between variables within high-dimensional and highly complex datasets, thereby providing new tools for addressing scientific challenges in complex systems. For instance, in drug discovery, Barzilay et al. at the Massachusetts Institute of Technology trained a deep neural network that identified a novel and effective antibiotic molecule, halicin, from a library of over 107 million compounds [1]. In materials structure prediction, DeepMind developed the GNoME intelligent model in 2023, which successfully predicted stable crystal structures of 380,000 new materials, thereby increasing the number of known stable materials nearly tenfold to 421,000 [2]. In molecular synthesis, the Polish Academy of Sciences and the Ulsan National Institute of Science and Technology in South Korea jointly developed the retrosynthesis software Chematica (now known as SYNTHIATM), which assists chemists in predicting and designing synthetic routes for complex molecules [3]. Utilizing the SYNTHIA™ software, Tim Cernak et al. at the University of Michigan conducted retrosynthetic analyses on 12 anti-COVID-19 drugs, successfully identifying more efficient synthetic pathways for 11 of them [4]. In summary, AI, as an emerging scientific tool, is reshaping the paradigms of discovery, transformation, and functional research in chemical science, aiding scientists in making discoveries and driving technological innovation.

1.3 The Developments in Automated Experiments

The practical training of AI algorithms relies on large volumes of high-quality and structured data. However, traditional chemical experiments often suffer from inconsistent data formats, fragmented results, and issues with reproducibility and precision, which impede AI's capacity to leverage data analysis and processing. By conducting high-throughput experiments, automated experimental platforms can rapidly generate precise and consistent data, thereby unlocking AI's full potential in chemical research. As a result, such platforms have garnered significant attention from researchers worldwide (Fig. 1).

As early as 2009, King et al. at Chalmers University of Technology in Sweden designed and built an automated robot named "Adam" [5]. Using high-throughput experiments, Adam conducted microbial batch growth experiments in micro-titer plates, measuring growth curves across different culture media. It successfully validated 20 gene hypotheses, with 12 of them yielding statistically significant results ($P < 0.05$) for the null hypothesis, demonstrating its vast potential in automated experimentation and knowledge discovery. Furthermore, in 2015, King et al. developed a second-generation robot named "Eve", which was primarily designed for drug synthesis [6]. Through high-throughput and cost-effective intelligent screening, Eve facilitated the docking of compounds with malaria target enzymes, ultimately identifying triclosan as a potential anti-malarial agent.

In addition to the pursuit of robotic experimentation, researchers have also sought to automate chemical laboratory equipment. In this regard, Cronin et al. at the University of Glasgow have made a series of groundbreaking advancements. In 2014, they

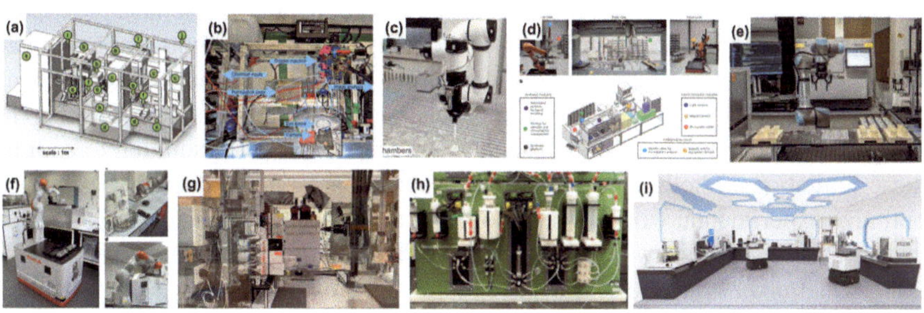

Fig. 1 Automated and intelligent robotic experimental platforms developed by research teams worldwide: **a** Chalmers University of Technology, Sweden; **b** University of Glasgow, UK; **c** Peking University, China; **d** Shenzhen Institute of Advanced Technology, China; **e** DeepMind in collaboration with Berkeley Lab, USA; **f** University of Liverpool, UK; **g** Massachusetts Institute of Technology, USA; **h** University of Illinois, USA; **i** University of Science and Technology of China

developed a droplet-based robot built on the RepRap 3D printer architecture to investigate how mixing ratios of four compounds influence the composition and properties of oil droplets [7]. In 2019, they developed the "Chemputer" system, an integrated platform capable of literature mining, experimental design, synthesis, and characterization [8]. This system translates chemical reactions into χDL (chemical description language) code, making complex molecular synthesis fully automatable. Its database contains χDL codes for 103 reactions, with half of the compounds achieving yields and purities comparable to manual synthesis. In 2022, Cronin et al. further introduced an autonomous portable platform for universal chemical synthesis based on the "Chemputer" [9]. In 2024, they expanded its capabilities to investigate amino acid reactions in mineral environments. This system analyzed nearly 1.4 million sequences and identified over 550 specific sequences [10].

In 2022, Mo et al. at Peking University developed an automated robotic platform for thin-layer chromatography (TLC) analysis [11]. TLC is widely used in synthetic laboratories to estimate molecular polarity. However, traditional techniques often suffer from a lack of standardization and reproducibility. The automated platform developed by Mo et al. efficiently collects standardized TLC data, generating a large dataset represented in SMILES format and retardation factor (R_f). Utilizing this comprehensive dataset, they have developed machine learning (ML) models capable of predicting the R_f values of organic compounds in diverse solvent combinations with a high degree of accuracy. This development has not only accelerated the generation of high-quality TLC data but also provided comprehensive guidelines for the selection of purification conditions. However, Mo et al. also noted that factors affecting TLC analysis could reduce prediction accuracy in different laboratories, limiting the system's broader applicability.

1.4 Intelligent-Driven Robotic Experiment Systems

Although automated chemical experiments have significantly accelerated the experimental process and improved result reproducibility, traditional robots remain limited to executing preset programs, lacking flexibility and autonomous decision-making abilities, making them ill-suited for complex conditions or real-time adjustments. The integration of intelligent technologies has emerged as a pivotal solution to address these limitations. Intelligent-driven robotic systems can autonomously design and optimize experimental workflows, uncover patterns through data analysis, and even generate and validate new experimental hypotheses. This paradigm shift signifies a transformation in which experimental robots evolve from mere task executors to intelligent assistants, thereby enabling scientific exploration and significantly enhancing research efficiency and innovation potential.

In 2020, Cooper et al. at the University of Liverpool developed a mobile robot capable of searching for improved photocatalysts for hydrogen production from water [12]. This robot, equipped with a Bayesian optimization algorithm, can independently perform tasks including liquid addition, weighing, and catalytic reactions while moving freely in the laboratory. It completed 688 experiments and investigated 1,000 catalytic formulations, achieving 17 times the efficiency of manual work. However, during Bayesian optimization, the system lacked chemical knowledge and computational capabilities, preventing it from independently generating and validating scientific hypotheses. In 2024, the research team proposed a fully autonomous solid-state workflow for powder X-ray diffraction (PXRD) experiments [13]. The workflow involves 12 steps performed by a team of three multipurpose robots, and can match or even surpass manual data quality, illustrating the power of flexible, modular automation to integrate complex, multitask laboratories. Compared to the photocatalyst research in 2020, PXRD experiments involve more complex sample handling and experimental steps. This autonomous system is among the most complicated in chemistry, relying only on off-the-shelf robots and standard laboratory hardware.

In 2022, Jiang et al. at the University of Science and Technology of China (USTC) developed the first data-driven AI-Chemist, which fully automates the process from synthesis and characterization to testing and data processing [14]. This system can independently perform almost all key steps in chemical research, with continuous optimization and iteration. The AI-Chemist breaks the traditional "trial-and-error" approach, showcasing the advantages of an intelligent research paradigm led by a "chemical brain." Subsequent studies have demonstrated that the AI-Chemist, driven by large models, enhances research efficiency in high-entropy electrocatalysts, functional molecules, and energy materials, accelerating innovation in cutting-edge domains.

In 2018, Jensen et al. at the Massachusetts Institute of Technology (MIT) designed a plug-and-play automated synthesis platform for unit operations [15]. In 2022, they further integrated an automated electrochemical experimental platform with intelligent models, screening 38,865 potential electrochemical reactions, 80% of which are competent [16]. In the same year, Burke et al. at the University of Illinois developed an automated chemical synthesis machine to synthesize complex three-dimensional molecular structures, automating the synthesis process and closed-loop optimization of reaction conditions [17]. In 2023, Zhao et al. at the Shenzhen Institutes of Advanced Technology used a robotic platform to improve traditional trial-and-error synthesis and characterization, enabling the morphology-oriented inverse design of colloidal nanocrystals with unique properties [18]. In the same year, DeepMind developed the A-Lab in collaboration with the Berkeley Lab. Guided by intelligent models, the A-Lab operated continuously for over 17 days, completing 355 experiments and synthesizing 41 new materials from a set of 58 targets, with a success rate of 71% [19].

Overall, intelligent-driven robotic systems are propelling a paradigm shift in chemical research globally. Figure 2 shows the experimental capabilities, types of workstations, coverage of experimental processes, and applications of several groups worldwide. In general, significant progress has been made in experimental process coverage, data processing, and intelligent optimization. However, most groups are still in the single-function integrated workstation stage and have not achieved full process integration. In contrast, the AI-Chemist developed by Jiang et al. at USTC

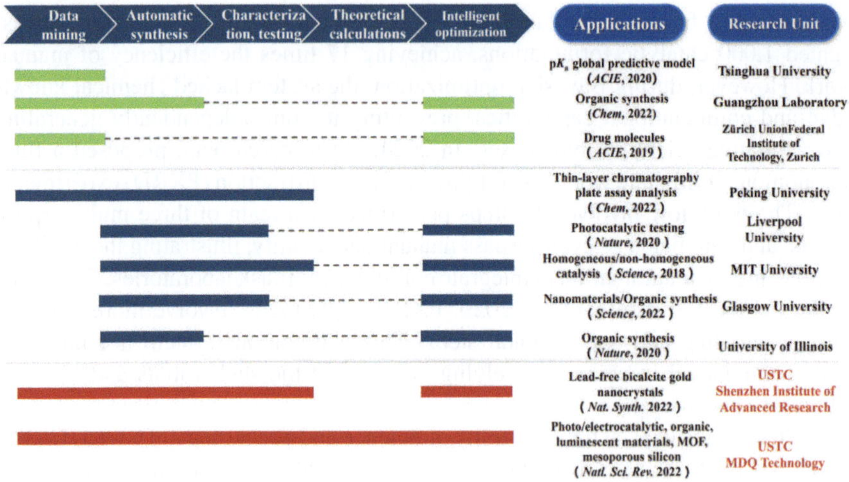

Fig. 2 Research progress of domestic and international teams in the field of intelligent robotic experimentation

has achieved full-process automation from synthesis to data processing, indicating the potential of China to take a leading role in this field.

2 AI-Chemist Platform that Fuses Theory and Experiment

In chemistry research, experimental studies of complex systems often provide limited small datasets, while theoretical calculations, though capable of generating large amounts of data, introduce systematic errors due to idealized approximations. Relying solely on theoretical or experimental data is insufficient to harness the potential of data intelligence fully. Addressing this limitation, Jiang et al. at USTC have proposed a new research paradigm of "the fusion of theory and experiment".

This approach fuses theory and experiment via intelligent models, using large datasets for pre-trained models and transfer learning for small experimental data. The goal is to incorporate rules from theoretical data and reflect the real-world complexity through experimental data, leading to the development of "the fusion of theory and experiment" for complex systems, paving the way for a new research paradigm (Fig. 3). Driven by this model, the AI-Chemist platform achieved full automation of synthesis, characterization, and testing, significantly enhancing the research efficiency of photocatalysts, functional molecules, and energy materials.

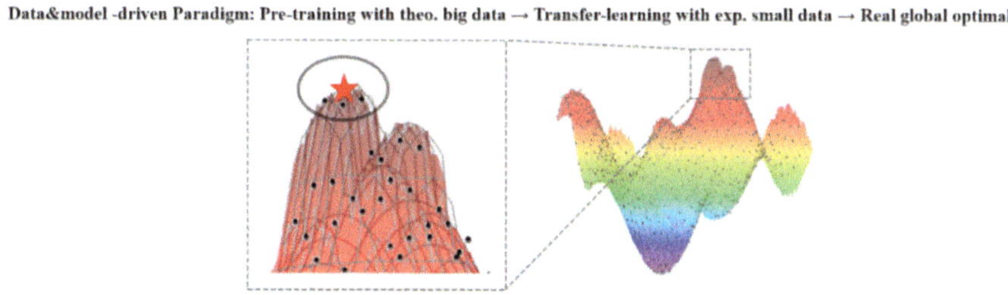

Fig. 3 A new research paradigm fusing theory and experiment

2.1 A Brief Introduction of the AI-Chemist Platform

The AI-Chemist platform at USTC is the world's first intelligent system integrating literature review, experimental design, and autonomous optimization, covering the entire chemical development process. It comprises three main modules: a machine reading (experience) system, a machine thinking (calculate) system, and a machine doing (practice) system (Fig. 4). The machine reading system automatically processes vast amounts of chemical literature to extract existing knowledge, formulate scientific hypotheses, and design experiments. The machine thinking system conducts high-throughput theoretical simulations, builds pre-trained models based on theoretical data, and integrates experimental data to establish a theory-experiment fusion model. The machine doing system, consisting of mobile robots and 15 workstations, autonomously executes synthesis, characterization, and performance testing, encompassing a wide range of chemical tasks. The AI-Chemist has improved research efficiency by two to five orders of magnitude in fields such as photocatalysis, electrocatalysis, luminescent molecules, optical thin-film materials, and energy materials. As the platform undergoes further development, its scope is expected to expand, supporting paradigm-shifting innovations across multiple research domains.

2.2 Typical Cases of AI Chemist Platforms Implementing New Paradigms

2.2.1 Efficient Design of Non-precious Metal High-Entropy Oxygen Evolution Catalysts

Leveraging its reading, thinking, and doing systems, the AI-Chemist first explored its potential for designing high-entropy, non-precious metal oxygen evolution catalysts. High-entropy catalysts, characterized by complex multi-metallic compositions, offer diverse active sites, enhancing catalytic activity in electrocatalysis and photocatalysis. Additionally, multi-element synergy significantly improves structural stability. However, the highly disordered mixing of multiple elements poses a considerable

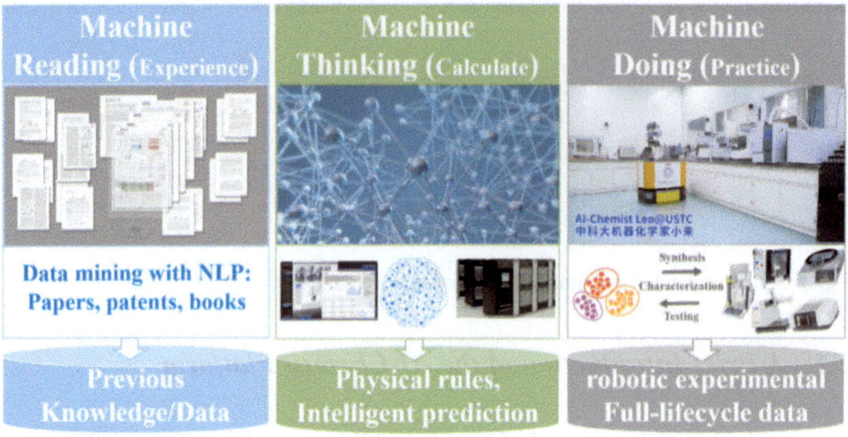

Fig. 4 A data-driven all-round robotic AI-Chemist

challenge for manually identifying the optimal composition, as it requires navigating an immense chemical space, with conventional research paradigms often converging on local optima. To address this challenge, the AI-Chemist, driven by big data and intelligent models, established a global search framework for high-entropy catalyst systems, successfully identifying globally optimal solutions [14].

The AI-Chemist autonomously analyzed ~16,000 scientific papers, selecting five non-precious metal elements, and integrated 20,000 theoretical calculations with 207 full-process machine experiment datasets to construct a theory-experiment fusion intelligent model. The analysis of the 207 experimental datasets revealed that relying solely on theoretical models led to underfitting while using only experimental data risked overfitting, potentially misinterpreting noise and errors as valid patterns. Only a model integrating both theoretical and experimental data achieved an optimal fit. During the subsequent Bayesian optimization process, the fusion model was iteratively utilized, ultimately guiding the optimization algorithm to identify the most effective high-entropy catalyst from 550,000 possible metal compositions—reducing the exhaustive "trial-and-error" search time from 1,400 years to just five weeks. This work represents an initial exploration of a data-driven paradigm in chemical research, significantly accelerating material discovery by rapidly identifying optimal compositions from millions of possibilities, and advancing chemistry toward the digitalization of knowledge, automation of operations, and intelligent material design (Fig. 5).

2.2.2 Digitized Synthesis of Colloidal Nanocrystals

In the previous study, the AI-Chemist leveraged literature data mining to construct a database, enabling full-process synthesis-characterization-testing development under the dual drive of big data and intelligent models. A global search model was established for high-entropy catalyst systems, successfully identifying the global

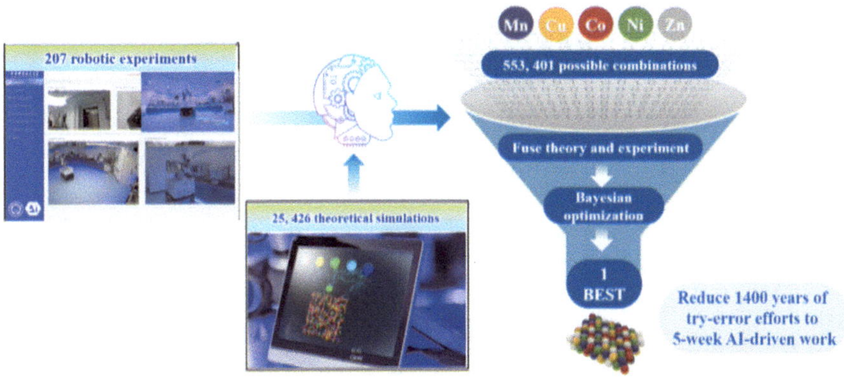

Fig. 5 AI-Chemist enable the efficient development of high-entropy non-precious metal oxygen evolution catalysts

optimum. This research paradigm is both generalizable and broadly applicable, capable of facilitating the design of various materials and extending to the digitalized synthesis of colloidal nanocrystals.

Colloidal nanocrystals hold great potential in optical and electrochemical applications. A key objective in their synthesis is precise morphological control to achieve desired physicochemical properties. However, traditional methods are inefficient, hindering the development of nanocrystals. To address this challenge, Zhao et al. from the Shenzhen Institutes of Advanced Technology, in collaboration with Jiang et al., established a robotic-assisted digital manufacturing platform for colloidal nanocrystals. This platform integrates data-driven automated synthesis, robot-assisted controlled synthesis, and morphology-oriented inverse design [18].

Gold nanocrystal and bis-calcite were used as model systems in this study. For the gold nanocrystal, the AI-Chemist employed data mining techniques to extract key synthetic formulation information from 1,300 published articles. Based on previous research, it identified a recommended range for the formulations, enabling exploration within a smaller but reasonable formulation space. In the case of bis-calcite, the AI-Chemist used data mining to select 48 potential solvents and 61 surfactants that could regulate the size and morphology of bis-calcite. By combining high-throughput in situ synthesis and characterization techniques, the optimization and screening of solvents and surfactants were rapidly completed. Furthermore, based on the large amount of data obtained from high-throughput experiments and in situ characterization, ML techniques were employed to construct relationship models between key synthesis parameters and absorption spectra, nanocrystal size, as well as the correlation between color information and nanocrystal size. Using these models, the platform can provide corresponding synthesis parameters based on the target product size, thus enabling the reverse design and synthesis of colloidal nanocrystals and successfully achieving their digital fabrication (Fig. 6).

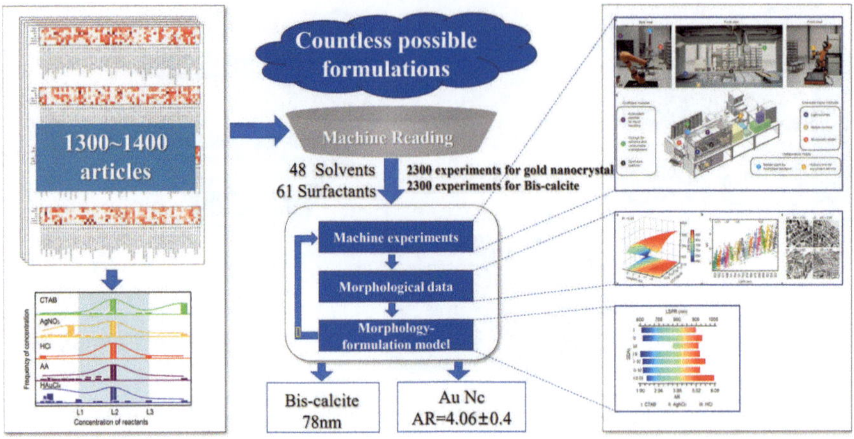

Fig. 6 AI-Chemist enable the digital synthesis of colloidal nanocrystals

2.2.3 Precision Fabrication of Chiral Optical Films

For complex systems that are difficult to describe with simple formulations, for example, chiral optical films, the AI-Chemist can still autonomously calibrate models through experimental validation and iterative optimization. This approach allows for the adjustment of process formats, screening of process conditions, and effective prediction of material synthesis pathways, enabling on-demand design by controlling process parameters.

Chirality, a fundamental property in nature, has emerged as a significant area of research in polymer science, particularly in the context of fabricating chiral polymers and studying their properties. In the development of chiral optical films, the core challenge is to explore the structure-function relationship between material structure and chiral optical properties, and to achieve precise construction of chiral structures through reverse design, enabling the manipulation of chiral optical functions at will. However, the complex process parameters and vast design space of organic polymer materials make it difficult to conduct a global exploration using traditional research paradigms. To address these challenges, Zou et al. and Jiang et al. have collaborated to use the AI-Chemist platform to achieve reverse design-driven material creation based on generative adversarial networks (GAN) [20].

They designed a chiral functional film construction scheme using a hetero-twisted-stacking strategy, and established a film preparation and characterization experimental platform. The AI-Chemist performed precise experimental operations on this platform, starting from the material preparation process parameters and spectroscopic properties. By controlling the film preparation parameters and measuring its optical response, approximately 1,500 experimental data were collected to construct a film spectroscopy-performance database as training samples. Subsequently, combining theoretical and experimental data, they trained a neural network to achieve end-to-end accurate prediction of material optical properties. Furthermore, they used a GAN

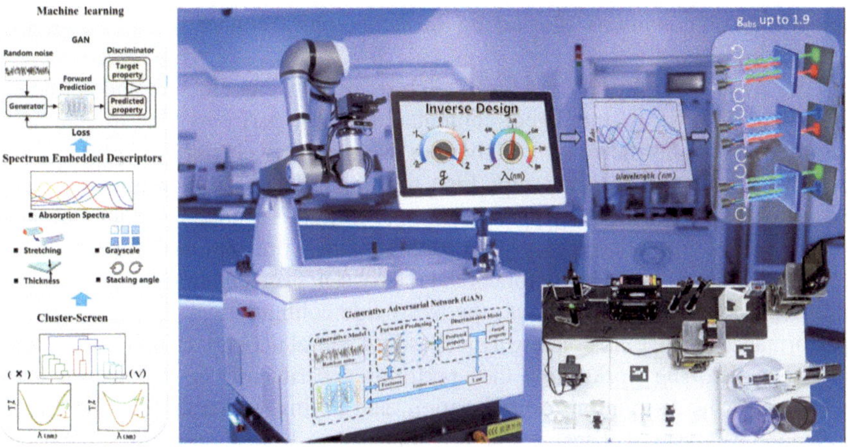

Fig. 7 The AI-Chemist enables the forward precise prediction and inverse design of chiral polymer functional films

to personalize the film preparation process based on actual needs, and perovskite quantum dot composite films with high chiral conversion efficiency were reverse designed and synthesized. The films exhibited a dissymmetry factor as high as 1.9, approaching the theoretical limit, and have been tested for anti-counterfeiting effectiveness at the China Banknote Printing and Minting Corporation (Fig. 7).

2.2.4 Intelligent Automated Synthesis of Oxygen-Producing Catalysts from Mars Meteorites

In addition to efficiently predicting material properties and performing reverse design based on existing theoretical frameworks, when faced with unknown material formulations, complex material systems, and unclear internal structures, the AI-Chemist can still intelligently create new materials by autonomously adjusting the proportions of unknown substances. For example, in the in-situ resource utilization task for deep space exploration, the robotic chemist achieved the unmanned intelligent creation of Martian meteorite-based oxygen evolution catalysts. This achievement not only broadens the scope of material design but also provides new solutions for studying complex material systems.

In the journey to human settlement on Mars, creating a habitable environment for humans in the Martian atmosphere presents a significant challenge. One of the primary tasks is to produce oxygen locally from Martian resources. Given the recent discovery of water resources on Mars, it is thus possible to use electrochemical water oxidation processes to generate oxygen through catalytic reactions. However, transporting catalysts or catalyst materials from Earth to Mars is not feasible, so developing efficient oxygen evolution reaction (OER) catalysts locally on Mars is essential. The AI-Chemist can autonomously and intelligently create OER catalysts under Martian conditions, thus laying the technological foundation for future Mars exploration [21].

The AI-chemist for creating OER catalysts relies on an efficient and automated workflow (Fig. 8). Its intelligent computational core performs nine digital calculation steps, while the robot collaborates with various chemical workstations to complete twelve experimental steps. First, the robotic chemist performs laser-induced breakdown spectroscopy (LIBS) analysis on Martian ores to obtain their compositional information. Subsequently, the computational core uses molecular dynamics (MD) simulations and density functional theory (DFT) calculations to select 29,902 different metal ratios of high-entropy hydroxide configurations for OER catalytic performance evaluation, creating a database for neural network training of the theoretical model. The workstation then performs experimental operations, including weighing, acid dissolution, centrifugation, and drying, to synthesize and characterize the catalyst. OER test results are uploaded to a cloud server for analysis by the computational core. After further optimization of experimental data, a theory-experiment fusion ML model is developed to predict catalyst performance accurately. Finally, the robot selects the optimal Martian ore formula using a Bayesian algorithm and conducts experimental validation.

Faced with the vast chemical space of up to 3.76 million potential formulations, the traditional trial-and-error approach would require approximately 2,000 years to identify the optimal solution. In contrast, the AI-chemist, leveraging intelligent computational frameworks and high-throughput experimental operations, integrate theoretical big data with experimental small data through machine learning models. By employing Bayesian optimization algorithms, they successfully predicted and validated the globally optimal catalyst formulation, ultimately achieving the fully autonomous intelligent design of an oxygen evolution catalyst on Mars. This work demonstrates that the AI-chemist can autonomously develop tailored chemical products on extraterrestrial bodies, providing a novel technological pathway for future extraterrestrial civilization exploration and establishing a unique strategy for the in situ utilization of interstellar resources.

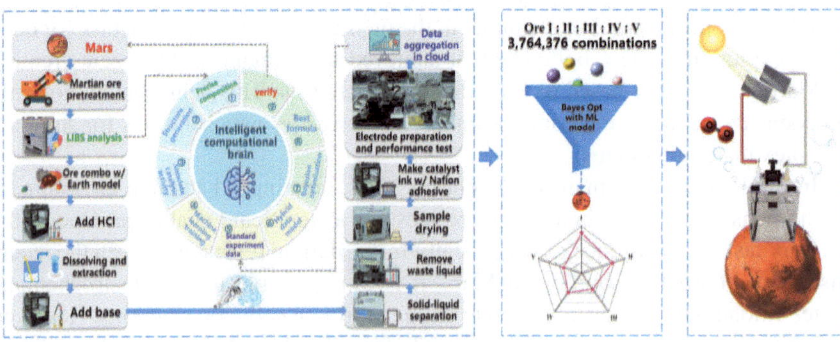

Fig. 8 The AI-chemist enables the fully autonomous intelligent design of an oxygen evolution catalyst derived from Martian meteorites

2.2.5 Material Science Studies Assisted by Large Language Models

Jiang et al. have developed an advanced automation system based on Large Language Models (LLMs), building upon their AI-chemist system. This new system includes two robots, twenty automated stations, and a backend control system. LLMs exhibit strong natural language processing (NLP) capabilities, and their integration into laboratory workflows has the potential to accelerate the discovery of new materials. However, using a single LLM to drive machine experiments makes executing complex chemical tasks on demand challenging. To overcome this limitation, the Jiang group has created a multi-agent system based on LLMs. This system enables collaboration among intelligent agents with different roles, allowing them to respond to more complex scientific research tasks as needed flexibly (Fig. 9).

After building the intelligent agents, the Task Manager coordinates with the Literature Reader, Experiment Designer, Robot Operator, and Computation Performer to work with human researchers. All agents in the system are powered by the Llama-3.1-70B large language model, with decision-making processes built on the LLM's ability to understand context, plan actions, execute functions, and utilize tools. In this setup, the Task Manager operates at the top level, while each specialized agent acts as a callable, predefined tool. Thanks to the inherent tool-calling capabilities of the LLM, the Task Manager can determine which tools to utilize and effectively activate them. When researchers submit a request, the Task Manager dynamically plans and decides how and when to engage these agents based on the requirements of the task. The system can perform a wide range of tasks by leveraging four core resources—literature databases, experimental protocol libraries, automated laboratories, and model

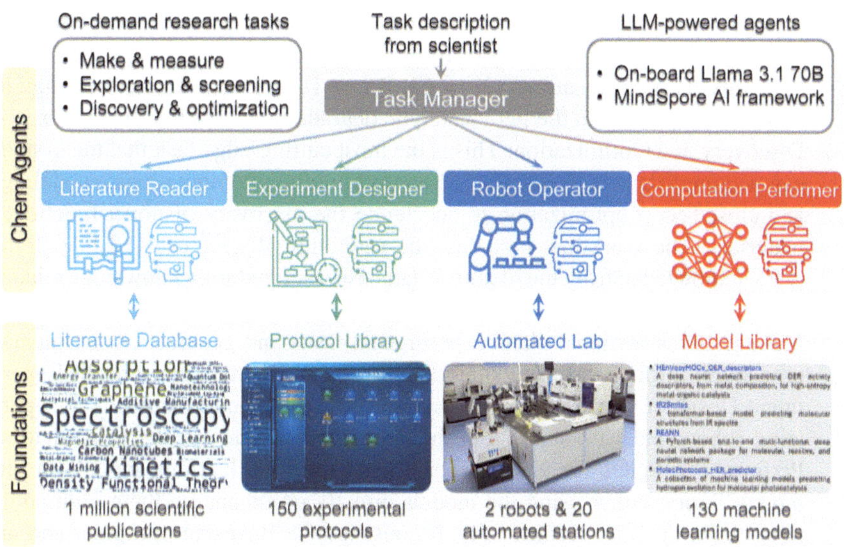

Fig. 9 The multi-agent-driven robotic AI chemist's architecture

libraries. These tasks vary from simple synthesis and characterization to intricate experimental parameter exploration and screening, all with minimal human intervention. Ultimately, this system aims to facilitate the discovery and optimization of functional materials. The main types of tasks completed by the system include "Preparation and Measurement," "Exploration and Screening," and "Discovery and Optimization."

(1) Preparation and Measurement: This is the most fundamental task executed by the system, mainly involving the synthesis or characterization of target compounds. Upon receiving the task description, the Task Manager parses and understands the instructions, then relays the necessary information to the Experiment Designer. The Experiment Designer determines the appropriate experimental protocols and generates the corresponding experimental steps based on this information. The Robot Operator then translates these steps into coded instructions and executes the corresponding synthesis and testing operations in the automated laboratory, ultimately returning the experimental results to the Task Manager. Examples of tasks include characterizing three azobenzene molecules using Fourier-transform infrared spectroscopy (FT-IR), synthesizing specified metal oxides and characterizing them through powder X-ray diffraction (PXRD), or obtaining color information from normalized fluorescence emission spectra of perovskite quantum dot films.
(2) Exploration and Screening: This task is more demanding, as it involves exploring and screening various experimental parameters to identify desirable functional materials. The Task Manager must understand complex domain knowledge and recognize the importance of systematically varying multiple variables to assess their impact. For instance, the system can synthesize graphitic carbon nitride (g-C_3N_4) using factorial experiments to evaluate how different heating temperatures and durations affect its performance. Additionally, the system can examine how various halogen atoms (X = Cl, Br, or I) influence bismuth oxyhalide (BiOX) compounds in the photocatalytic degradation of tetracycline (TC).
(3) Discovery and Optimization: This is the most cutting-edge task that the system can accomplish, which integrates literature mining, computational modeling, and closed-loop optimization to accelerate the discovery of novel functional materials. The workflow includes: the Literature Reader uses NLP tools to extract knowledge from the literature database, accumulating knowledge related to the experimental objectives; the Experiment Designer, following the Task Manager's instructions, plans experimental tasks and generates experimental procedures through the protocol library; the Robot Operator converts the Experiment Designer's experimental procedures into code and commands for operating robots and workstations in the automated laboratory; the Computation Performer searches for suitable pre-trained machine learning models in the model library, either using the models directly or enhancing them with additional training data. Additionally, it consists of a Bayesian optimizer and an LLM managed by the Computation Performer that can fit historically observed experimental data and predict future outcomes. This optimizer selects the next

experimental conditions by balancing the predictive performance and uncertainty of the future data. This algorithm allows the system to utilize real-time data interactions to iteratively optimize experimental conditions, thereby accelerating material screening. A "Reviewer" agent has been added to the system to verify and refine the LLM's output using predefined expert rules. This additional layer ensures that suboptimal decisions are corrected before execution. Based on this workflow, the system can discover metal-organic high-entropy catalysts for oxygen evolution reactions, achieving efficient material exploration and optimization.

This multi-agent intelligent system has been demonstrated to exhibit adaptability and precision in generating experimental procedures, conducting robotic experiments, and applying computational models for discovery and optimization tasks by effectively completing these three activities. As automated laboratories transition from isolated systems to coordinated cloud platforms, potentially evolving into advanced intelligent scientist cloud platforms, the AI chemist driven by multi-agent intelligent systems aligns closely with this developmental trajectory and will significantly accelerate this process. Furthermore, this multi-agent intelligent system can be continually updated and improved. For instance, it lacks an anomaly feedback mechanism, which may lead to suboptimal performance or errors during broader applications. Therefore, future designs for feedback mechanisms in automated laboratories should consider this gap and aim for comprehensive improvements. In terms of task scheduling, prospective enhancements should concentrate on the refinement of scheduling algorithms to accommodate a more extensive range of experimental conditions and constraints. These enhancements should also involve the incorporation of advanced robotics technology and sensor networks in the system to dynamically adjust subsequent tasks based on real-time feedback of reaction states, and to cover the scenarios of long-term chemical reactions or multi-step experiments. Additionally, the integration of reinforcement learning and optimization algorithms can balance the workload among robots, improving resource utilization and operational efficiency through better task allocation methods.

3 Suggestions for Building a Large Model-Driven AI Chemist Cloud Infrastructure

To advance the application of the AI for Science paradigm in chemical research, it is essential to establish intelligent laboratories covering diverse application scenarios, develop large models that integrate multidisciplinary knowledge principles to accurately predict material evolution, and enable the sharing of data and knowledge resources. To this end, this article proposes promoting the standardization of intelligent experimental systems and expanding the construction of distributed facility clusters. These efforts aim to aggregate vast standardized datasets, develop large novel models capable of accurately predicting the evolution of complex systems, and

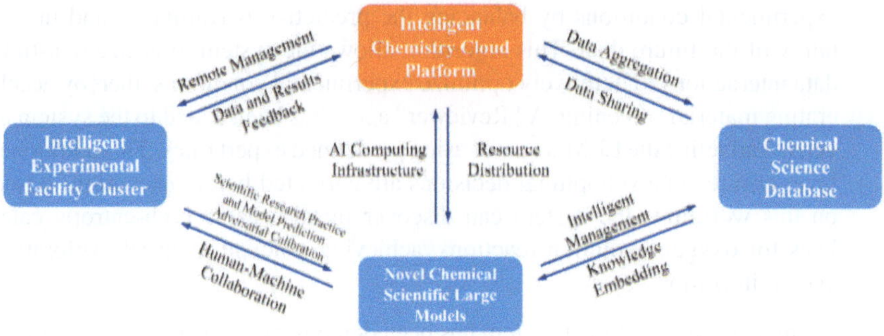

Fig. 10 Layered architecture of AI chemist cloud infrastructure

leverage cloud platforms for human-machine interaction and resource sharing, ultimately establishing a cloud-based robotic chemist infrastructure with a novel research model. This infrastructure includes a chemical science database, novel chemical scientific large models, an intelligent experimental facility cluster, and an intelligent chemistry cloud platform. Theoretical and experimental data are aggregated into the chemical science database via the cloud platform, where they are systematically managed by intelligent models and embedded into scientific large models as knowledge and governing principles. Supported by these novel large models, human-machine collaborative workflows iteratively optimize experimental designs, driving the intelligent experimental facility cluster to conduct scientific investigations while refining and validating model predictions through adversarial calibration and iterative evolution. Serving as the core of the cloud infrastructure, the intelligent chemistry cloud platform enables data sharing within the chemical science database, remote management of the intelligent experimental facility cluster, and centralized allocation of computational resources (Fig. 10).

In this framework, the chemical science database and the new-generation chemical science model will augment researchers' cognitive capabilities. At the same time, the intelligent experimental facility cluster will alleviate the need for manual labor. The intelligent chemistry cloud platform interconnects individual researchers, continuously optimizing through iterative calibration between data and the scientific model to drive scientific breakthroughs. This infrastructure will transform China's research organization, enabling cross-disciplinary integration and expanding the depth and breadth of chemical research.

3.1 Chemical Science Database

In the data-driven research paradigm, the core of innovation lies in the efficient integration and utilization of scientific data. However, a major challenge is the lack of unified standards, the inconsistency of data quality, and the isolation of data from different sources, resulting in "data silos." These issues limit the ability of data to drive chemical science research. Therefore, it is imperative to develop a standardized, multimodal, and cross-disciplinary AI-driven chemical science database. The establishment of such a database involves three key aspects:

(1) Multimodal Data Integration: Chemical data encompass a wide range of resources, including scientific literature, theoretical calculations, and experimental results, which are scattered across various databases and laboratories. To enable comprehensive data aggregation in chemical sciences, the development of machine-readable tools is imperative. These tools should leverage data mining techniques, such as natural language processing and image recognition, to automatically extract multimodal data—including text, images, and tables—from diverse sources, including textbooks, scientific journals, and research papers. Additionally, computational chemistry simulations can be employed to generate physicochemical databases for molecules and materials. For experimental data, standardized data acquisition channels should be established, and electronic laboratory notebooks should be utilized to capture full-lifecycle data, enabling intelligent data collection.
(2) Intelligent Data Management: Embedding intelligent models within the database enables automated annotation and structured organization of both experimental and computational data. Data classification and structuring could also enhance accessibility and usability. To ensure high-quality databases, interpretable model-based techniques should be developed for data validation and quality assessment, facilitating intelligent data cleaning and filtering. These techniques help eliminate erroneous data, supplement missing information, and infer uncharted data points, thereby constructing a high-quality database for materials science.
(3) Construction and Application of Knowledge Graphs: Systematic analysis of interrelations among diverse datasets enables the construction of structure–property relationships and evolutionary correlation models, forming the basis for knowledge graph development. Integrating knowledge graphs facilitates the fusion of multimodal data, unifying heterogeneous datasets into a structured, scalable, and efficient storage format. This approach enhances data organization and usability, simplifying subsequent data retrieval and analysis. Moreover, embedding knowledge graphs into scientific foundation models using advanced tools such as pre-trained models can significantly improve knowledge acquisition and utilization efficiency.

3.2 *Novel Chemical Scientific Large Models*

General-purpose language models often face challenges in handling complex chemical problems due to their limited interpretability, lack of tunability, and insufficient depth in semantic understanding. These limitations result in suboptimal predictive accuracy and reasoning capabilities. Therefore, it is essential to develop scientific foundation models based on mathematical and logical principles, integrating data-driven neural network models with knowledge-driven symbolic reasoning engines. This deep fusion approach will enable advancements in intelligent scientific research across disciplines such as chemistry, physics, and mathematics.

Traditional information systems are centered around structured databases, which primarily support data querying through application programs built on top of them. In contrast, unstructured databases encompass data sources such as text, images, and videos. Large models can extract information from these sources, empowering intelligent systems with reasoning and decision-making capabilities. The framework of the next-generation chemical science foundation model integrates structured databases, unstructured databases, knowledge bases, and large models while incorporating a knowledge-driven reasoning engine. This engine, built upon ontologies and domain-specific knowledge bases in materials science, enables logical reasoning by leveraging Prolog, SOAR, first-order logic, emerging reasoning paradigms such as methodological reasoning, reinforcement learning-based reasoning, and reactive reasoning. It ensures the interpretability of inference results, when interfaced with databases, it can simulate human cognition, reasoning, and decision-making processes. By fusing the reasoning engine with large models, this novel framework achieves a dual knowledge- and data-driven approach, providing more intelligent and efficient solutions to complex scientific challenges and effectively overcoming the limitations of conventional large models.

In the context of chemical science research, this framework can be further enhanced by integrating knowledge graphs and chemical descriptors into next-generation scientific foundation models. The application of knowledge graphs enables the systematic organization and representation of chemical information, while chemistry-aware knowledge augmentation algorithms enhance the model's comprehension of chemical phenomena. Additionally, the incorporation of specialized chemical descriptors optimizes the model's performance by capturing critical features in chemical data. By combining knowledge graphs, knowledge-augmented algorithms, and expert-driven chemical insights, AI algorithms grounded in chemical principles can be developed, ultimately creating an intelligent brain for machine chemists.

3.3 *Intelligent Experimental Facility Cluster*

The intelligent experimental facility network consists of a distributed full-process intelligent laboratory system, where each laboratory integrates fully autonomous experimental mobile robots, fully automated experimental workstations, and an intelligent management system. This infrastructure aims to provide efficient and precise experimental and data processing solutions, requiring the development of the following three key components:

(1) Fully Autonomous Experimental Mobile Robots: These robots should be equipped with six-degree-of-freedom robotic arms and dexterous control mechanisms with real-time force feedback to perform experimental operations with high precision. Integrating deep reinforcement learning and adaptive control enhances the robotic arms' ability to learn behavioral strategies in complex environments. Robots must incorporate multimodal perception capabilities for positioning, fusing data from LiDAR, depth cameras, and infrared spectrometers to achieve precise localization and tactile perception in unstructured environments. Additionally, robots should integrate omnidirectional mobile platforms, advanced visual perception algorithms, and dynamic obstacle avoidance techniques for autonomous mobility, enabling them to construct high-precision multimodal maps of laboratory environments and ensure accurate vision-guided operations. A task management system is also required to facilitate autonomous mobility and experimental execution in complex chemical settings.
(2) Fully Automated Experimental Workstations: Robotic experimental operations must be coordinated with various automated workstations. Laboratories should be equipped with proprietary fully automated chemical workstations, including liquid handling stations, automated sealing machines, automated spectroscopic analysis stations, microchannel continuous-flow reaction workstations, and electrochemical performance testing stations. These workstations facilitate automated sampling, product collection and post-processing, spectral characterization, and real-time online detection, ensuring seamless execution of chemical processes.
(3) Fully Automated Experimental Workstations: Long-term sequential experiments require scheduling by an intelligent management system. This system integrates multiple functional subsystems, including a central control and human-machine interaction system, reaction temperature regulation, and automated multi-channel reactant switching. Additionally, each subsystem is equipped with redundant sensors and employs vision recognition technology to automatically filter anomalous data and provide real-time process feedback, ensuring the accuracy and reliability of experimental results. Through the intelligent management system, the collaborative control of laboratory instruments and task scheduling of experimental workflows can be achieved. By integrating fully autonomous experimental mobile robots and fully automated chemical workstations, the system establishes an intelligent chemical laboratory that comprehensively covers chemical synthesis, sample characterization, and performance testing. Ultimately, the nationwide deployment of distributed full-process intelligent laboratories will collectively form an intelligent experimental facility network, driving advancements in automated and intelligent chemical research.

3.4 *Intelligent Chemistry Cloud Platform*

The intelligent chemistry cloud platform integrates the robotic chemist intelligent operating system and a federated learning algorithm framework, collectively facilitating knowledge and data sharing across different experimental systems and research tasks. This enables robotic chemists to transfer and learn across diverse laboratories and scientific endeavors while leveraging the centralized intelligence of large models to drive scientific innovation in distributed laboratory environments.

To standardize intelligent laboratories, the nation's leading intelligent chemistry laboratories should serve as templates for establishing standardized instruction sets, interface functions, experimental templates, and specifications for intelligent equipment. These standards should be widely disseminated to promote the large-scale construction of intelligent experimental facilities. On this foundation, a cloud platform should be developed to interconnect laboratories across different locations, facilitating resource scheduling through a unified intelligent operating system while providing user services. This operating system should feature intuitive data visualization and remote experiment management capabilities, enabling researchers to conduct experiments beyond physical constraints.

Ensuring data privacy is paramount in this framework. Federated learning, a distributed machine learning approach, allows individual nodes to perform computations and model updates locally, with only the updates being aggregated by a central server to refine the global model. This method eliminates the need for centralized data storage or transmission, significantly enhancing data privacy protection. By embedding federated learning into the intelligent chemistry cloud platform, researchers can ensure data usability while maintaining confidentiality, collaboratively building models, and enabling knowledge sharing. This mechanism effectively mitigates the privacy risks and costs associated with traditional centralized machine learning and data science methods, thereby strengthening data security and allowing seamless information exchange across different experimental systems while safeguarding sensitive data.

4 Perspective

The cloud-based robotic chemist infrastructure, which integrates centralized intelligence with distributed innovation, effectively drives the new research paradigm (Fig. 11). First, cloud users submit material innovation requests, after which the novel scientific large model intelligently recommends potential solutions and synthesis strategies. Subsequently, scientists refine and decide on the experimental plans

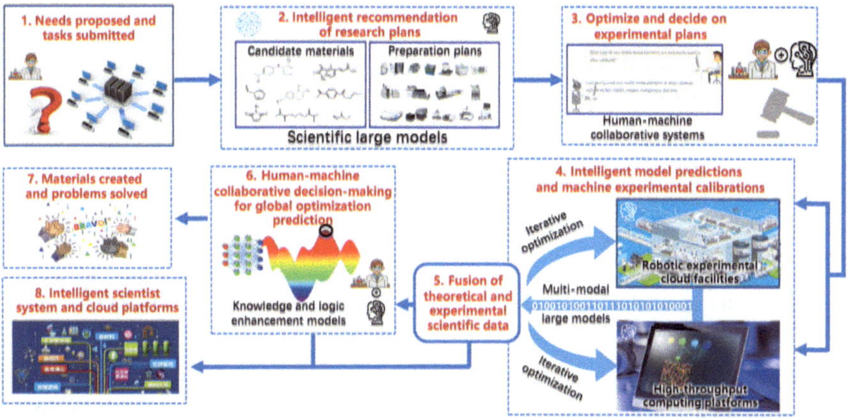

Fig. 11 The new paradigm of scientific research based on AI chemist cloud infrastructure

through a human-computer interaction system. Guided by these optimized experimental protocols, the robotic cloud infrastructure executes high-throughput experiments, while high-performance computing platforms perform theoretical simulations. Under the guidance of multimodal large models, iterative optimization is conducted, generating scientifically aligned big data that integrates theory and experiment.

In summary, the AI Chemist platform has introduced a new research paradigm driven by data intelligence, fusing theory with practice. This innovative approach involves generating interpretable, pre-trained models from extensive theoretical data, which are then calibrated and optimized using precise experimental data. This approach results in "theory-experiment fusion" models for complex systems, facilitating global optimization searches and enabling intelligent material creation. In the future, the development of standardized intelligent laboratories, multimodal intelligent databases, new scientific large models, and intelligent chemistry cloud platforms will lead to the establishment of an AI Chemist cloud infrastructure with entirely new forms of research organization. This infrastructure will promote the new data intelligence-driven research paradigm, significantly enhancing scientific research practices. It aims to create a research ecosystem characterized by centralized intelligence and distributed innovation, ultimately driving major scientific breakthroughs in chemistry, material science, and other fields across China.

References

1. Stokes J M, Yang K, Swanson K, et al. A deep learning approach to antibiotic discovery [J]. Cell, 2020, 180(4): 688–702.
2. Merchant A, Batzner S, Schoenholz S S, et al. Scaling deep learning for materials discovery [J]. Nature, 2023, 624(7990): 80–85.
3. Klucznik T, Mikulak-Klucznik B, McCormack M P, et al. Efficient syntheses of diverse, medicinally relevant targets planned by computer and executed in the laboratory [J]. Chem, 2018, 4(3): 522–532.
4. Lin Y, Zhang Z, Mahjour B, et al. Reinforcing the supply chain of umifenovir and other antiviral drugs with retrosynthetic software [J]. Nature Communications, 2021, 12(1): 7327.
5. King R D, Rowland J, Oliver S G, et al. The automation of science [J]. Science, 2009, 324(5923): 85–89.
6. Williams K, Bilsland E, Sparkes A, et al. Cheaper faster drug development validated by the repositioning of drugs against neglected tropical diseases [J]. Journal of the Royal Society Interface, 2015, 12(104): 20141289.
7. Gutierrez J M P, Hinkley T, Taylor J W, et al. Evolution of oil droplets in a chemorobotic platform [J]. Nature Communications, 2014, 5(1): 5571.
8. Steiner S, Wolf J, Glatzel S, et al. Organic synthesis in a modular robotic system driven by a chemical programming language [J]. Science, 2019, 363(6423): eaav2211.
9. Manzano J S, Hou W, Zalesskiy S S, et al. An autonomous portable platform for universal chemical synthesis [J]. Nature Chemistry, 2022, 14(11): 1311–1318.
10. Leonov A I, Hammer A J S, Lach S, et al. An integrated self-optimizing programmable chemical synthesis and reaction engine [J]. Nature Communications, 2024, 15(1): 1240.
11. Xu H, Lin J, Liu Q, et al. High-throughput discovery of chemical structure-polarity relationships combining automation and machine-learning techniques[J]. Chem, 2022, 8(12): 3202–3214.
12. Burger B, Maffettone P M, Gusev V V, et al. A mobile robotic chemist [J]. Nature, 2020, 583(7815): 237–241.
13. Lunt A M, Fakhruldeen H, Pizzuto G, et al. Modular, multi-robot integration of laboratories: an autonomous workflow for solid-state chemistry [J]. Chemical Science, 2024, 15(7): 2456–2463.
14. Zhu Q, Zhang F, Huang Y, et al. An all-round AI-Chemist with a scientific mind [J]. National Science Review, 2022, 9(10): nwac190.
15. Bédard A C, Adamo A, Aroh K C, et al. Reconfigurable system for automated optimization of diverse chemical reactions [J]. Science, 2018, 361(6408): 1220–1225.
16. Zahrt A F, Mo Y, Nandiwale K Y, et al. Machine-learning-guided discovery of electrochemical reactions [J]. Journal of the American Chemical Society, 2022, 144(49): 22599–22610.
17. Blair D J, Chitti S, Trobe M, et al. Automated iterative Csp^3-C bond formation [J]. Nature, 2022, 604(7904): 92–97.
18. Zhao H, Chen W, Huang H, et al. A robotic platform for the synthesis of colloidal nanocrystals [J]. Nature Synthesis, 2023, 2(6): 505–514.
19. Szymanski N J, Rendy B, Fei Y, et al. An autonomous laboratory for the accelerated synthesis of novel materials [J]. Nature, 2023, 624(7990): 86–91.
20. Xie Y, Feng S, Deng L, et al. Inverse design of chiral functional films by a robotic AI-guided system [J]. Nature Communications, 2023, 14(1): 6177.
21. Zhu Q, Huang Y, Zhou D, et al. Automated synthesis of oxygen-producing catalysts from Martian meteorites by a robotic AI chemist [J]. Nature Synthesis, 2024, 3(3): 319–328.

Jun Jiang doctoral supervisor, Chair Professor at the University of Science and Technology of China, and the Deputy Director of the State Key Laboratory of Precision and Intelligent Chemistry. He has long been engaged in theoretical chemistry research driven by artificial intelligence. He is a recipient of the National Science Fund for Distinguished Young Scholars of China, and the leader of the CAS Project for Young Scientists in Basic Research. He serves as the founding Editor-in-Chief of *Artificial Intelligence Chemistry*, which is Elsevier's flagship journal in the field of intelligence. He has received prestigious awards, including the Chinese Chemical Society Tang Ao-Chin Youth Award on Theoretical Chemistry, the Distinguished Lectureship Award from the Chemical Society of Japan, and the Outstanding Conclusion Award for the National Program on Key Basic Research Project. As a corresponding author, he has published over 100 high-impact papers in journals such as *Nature Energy*, *Nature Synthesis*, *Nature Computational Science*, *JACS*, *PNAS*, *National Science Review*, and *Nature Communications*.

Shuo Feng Associate Researcher at the University of Science and Technology of China. His primary research focuses on artificial intelligence-driven spectroscopy and the design of functional materials. He has presided over or participated in several research projects funded by the National Natural Science Foundation of China (NSFC), including the Youth and Major Programs. He has published more than 10 scientific papers in high-impact journals.

Chuxuan Ding Postdoctoral Researcher at the University of Science and Technology of China. Her main research interests include molecular coupled transport dynamics in low-dimensional materials, the fitting of machine learning potential functions, and using machine learning to accelerate theoretical chemistry calculations.

Paradigm Shift and Application of Life Science Driven by Big Data and Artificial Intelligence

Xin Li, Haiping Jiang, Wenhao Liu, Chen Fang, Cong Li, Haorao Wang, Xusheng Ma, and Huanhuan Wu

Abstract Driven by advancements in artificial intelligence (AI) and big data, life science research is shifting from a traditional paradigm combining experimental and data science to a new paradigm based on experimental science but mainly driven by AI and big data. Emerging AI technology holds great potential in fundamental research, healthcare, agriculture and biomanufacturing. They have not only transformed research approaches but also fueled innovation and growth in bio-related industries. However, integrating AI with biological big data remains challenging, with obstacles including multimodal data integration, constructing life science specific big models, and limited simulation resolution of life systems. This paper outlines the development of life sciences and research paradigms, highlights the practical applications of the "Big Data + AI" framework, and discusses future directions for the field.

Keywords Research paradigm · Big data · Artificial intelligence · Life sciences

Introduction

The era of Artificial Intelligence has ushered in a transformative wave that is reshaping the landscape of scientific research, industrial development, and societal activities. Leveraging its unparalleled capabilities in data processing and pattern recognition, AI has fundamentally altered our cognitive frameworks and modes of interaction with the world. Among the diverse disciplines, life sciences, owing to their intrinsic connection with human health, have emerged as a leading frontier for AI-driven innovation. The integration of AI has not only expedited the pace of research in life sciences but has also redefined its underlying paradigms. By enabling sophisticated analyses of omics data, including genomics and proteomics, and demonstrating remarkable potential in drug discovery and precision medicine, AI has become a

X. Li (✉) · H. Jiang · W. Liu · C. Fang · C. Li · H. Wang · X. Ma · H. Wu
Institute of Zoology, Chinese Academy of Sciences, Beijing, China
e-mail: xinli@ioz.ac.cn

Beijing Institute for Stem Cell and Regenerative Medicine, Beijing, China

China's e-Science Blue Book 2025, https://doi.org/10.1007/978-981-96-8252-2_13

cornerstone of modern life science research. Its ability to decipher intricate patterns within vast biological datasets and unravel the fundamental mechanisms of life has catalyzed revolutionary advancements. As a pivotal player in global technological innovation, China has significantly contributed to this AI-powered transformation in life sciences. This article delves into the multifaceted impact of "Big Data + AI" on the evolution of life science research paradigms, highlights the latest breakthroughs and practical applications, and explores future trajectories in this dynamic field.

1 "Big Data + AI" Driving the Transformation of Life Science Research Paradigms

1.1 The Evolution of Life Science Research Paradigms

The history of life sciences is defined by a continuous journey toward a deeper understanding of living systems, marked by four distinct evolutionary stages. **First Stage: Phenotypic Observation and Description (Ancient Times to the 19th Century)**. In its infancy, life sciences were rooted in the observation of phenotypic traits and behaviors, which laid the groundwork for identifying macro-level patterns in living systems. For instance, the ancient Chinese agricultural text *Xia Xiaozheng* documented relationships between animal reproduction, migration, and climatic variations. Charles Darwin's *On the Origin of Species* revolutionized biology by proposing the theory of evolution through meticulous phenotypic analysis [1]. Yet, this era's insights were confined to surface-level observations, with the microscopic underpinnings of life remaining unexplored. **Second Stage: Cellular and Molecular Exploration (19th Century to the Early 21st Century)**. Advancements in physical and chemical analytical tools catalyzed a shift from macro to micro-level investigations. The advent of microscopy in the nineteenth century unveiled cells as the fundamental units of life, while mid-twentieth century breakthroughs in biochemistry and X-ray diffraction culminated in Watson and Crick's landmark discovery of the DNA double helix, heralding the dawn of molecular biology [2]. This era also witnessed the formulation of the central dogma, propelling advances in genetics, immunology, and embryology [3]. However, living systems are more than the sum of their microscopic components; their complexity arises from multi-scale interactions that defy reductionist explanations. **Third Stage: Holistic Profiling of Components (Early 21st Century to Present)**. The rapid evolution of sequencing technologies has enabled comprehensive characterization of living systems. The Human Genome Project, leveraging first-generation sequencing, decoded the human genome, accelerating the growth of omics technologies [4]. High-throughput second-generation sequencing of genomes, transcriptomes, and epigenomes has since allowed scientists to map the dynamic states of living systems across physiological and pathological contexts. Despite this progress, static, single-scale analyses remain inadequate for capturing the multi-dimensional complexity of life, as omics data, though voluminous, often fail to reveal the high-dimensional latent patterns governing biological processes. **Fourth Stage: System-Level Simulation and Reconstruction (Present to Future)**. Artificial Intelligence has emerged as a transformative force in decoding the intricate dynamics of living systems. AlphaFold's atomic-level protein structure predictions resolved a 50-year challenge in biology [5], while the EVO model, powered by vast genomics datasets, simulates genomes to decode the "encyclopedia of life," enabling the design and analysis of DNA, RNA, and protein sequences at unprecedented scales [6]. AI's ability to address multi-scale, multi-modal, and spatiotemporal complexities is poised to unlock the fundamental principles of life, paving the way for a new era of system-level understanding.

The four paradigms of scientific research, as articulated by Turing Award winner Jim Gray, have profoundly influenced the scientific community. Gray classified research into four paradigms: experimentation, theoretical modeling, computational simulation, and data-driven discovery. However, life sciences exhibit unique characteristics that challenge this framework. Historically, life sciences have been dominated by the experimental paradigm, from Andreas Vesalius' anatomical studies in the sixteenth century [7] to Watson and Crick's DNA model [2] and the discovery of the central dogma [3]. The absence of universal laws akin to Newton's principles or Maxwell's equations has hindered the development of a theory-driven second paradigm. Similarly, the lack of foundational laws has limited the application of computational simulation as a third paradigm. The surge in omics data since the early twenty-first century has ushered in a data-driven fourth paradigm, where high-throughput data acquisition and computational analysis uncover hidden correlations and patterns.

In conclusion, life sciences today navigate a dual paradigm of experimentation and data science, with limited reliance on theoretical and computational approaches. The exponential growth of biological data, while expansive, often yields low-dimensional insights that fail to capture the full complexity of living systems. AI technologies offer a promising solution by integrating multi-dimensional biological data to bridge the gap between low-level genomic information and high-level cellular and tissue mechanisms. This integration is essential for uncovering the underlying principles of biological complexity and advancing life sciences toward a new research paradigm driven by experiments, data, and artificial intelligence.

2 "Big Data + AI" Driving the Evolution of Life Science Paradigms

The rapid advancement of omics technologies has catalyzed an exponential surge in biological data, presenting both opportunities and challenges for life sciences. To manage this deluge of data, biologists have developed a multitude of biological databases for storage and organization. However, the inherent complexity of biological data continues to hinder its effective analysis and interpretation. In recent years, large AI models such as ChatGPT, ERNIE Bot, and Taichu have revolutionized fields like natural language processing and image generation. In life sciences, AI innovations like the AlphaFold series have sparked a paradigm shift, leveraging the vast repositories of biological big data to decode intricate biological processes. This fusion of data and intelligence has ushered in a new era of research, fundamentally transforming how life sciences explore biological phenomena and their underlying mechanisms.

2.1 Generation and Management of Biological Data

The accumulation of biological data has evolved in tandem with the progression of life sciences. During the eras of phenotypic observation and cellular-molecular analysis, researchers gathered data across diverse scales, spanning species classification, evolutionary patterns, organ structures, cellular morphology, and DNA replication mechanisms. While these data provided valuable insights into the phenotypic and microscopic aspects of life, their fragmented and limited scope hindered the development of comprehensive theories. The launch of the Human Genome Project in 1987 marked a turning point, not only unveiling the blueprint of human genetics but also laying the groundwork for genomics research [4]. China's participation in 1999, contributing 1% of the sequencing effort, underscored its growing role in global science. The advent of second-generation sequencing technologies dramatically reduced costs and accelerated sequencing speeds, leading to an exponential surge in data volume—from GB to TB and now PB levels. The integration of single-cell and spatial omics data has further diversified datasets, adding layers of complexity to their analysis. This big data era, while transformative, has also introduced challenges in data integration, necessitating robust storage and management platforms to ensure consistency, integrity, and global accessibility. Internationally, databases like GenBank and PubMed in the U.S., and ENA and UniProt in Europe, have become invaluable resources. China has also made significant contributions with platforms such as the Genome Sequence Archive (GSA), the China National GeneBank (CNGB), and the multi-species single-cell dataset scCompass [8]. These repositories have become cornerstones for global research, driving innovation and application in life sciences.

2.2 The Integration and Development of AI in Life Sciences Research

Since the inception of the term "artificial intelligence" in 1956, AI technologies have undergone transformative evolution—from symbolic logic reasoning and machine learning to deep learning and large model technologies—dramatically reshaping life sciences research. In the 1950s–1980s, AI revolved around symbolic logic and expert systems, which employed rule-based reasoning to tackle complex problems, such as simulating molecular reaction pathways or diagnosing diseases [9]. The 1990s to early twenty-first century witnessed a shift toward machine learning, driven by enhanced computational power. This era saw the emergence of deep learning, which emphasized pattern extraction from data, with supervised and unsupervised learning methods demonstrating significant potential in applications like genome analysis and disease prediction [10–14]. The twenty-first century marked the third wave of AI, characterized by the rise of deep learning, which revolutionized tools for protein structure prediction and gene regulatory network inference [15, 16]. Today, propelled by the rapid growth of big data and computational capabilities, AI has entered the "era of large models," achieving groundbreaking progress in life sciences. Pre-trained models such as EVO [6], Geneformer [17], GeneCompass [18], and AlphaFold [19] have catalyzed innovation in areas like precision medicine and drug development, redefining the boundaries of life sciences research. Looking ahead, the creation of cross-omics digital cell models represents a pivotal next step. By integrating multi-layered data from genomics, transcriptomics, proteomics, and metabolomics, a "digital cell" model could simulate and predict cellular behaviors and responses across diverse biological environments. This advancement will offer unprecedented precision and comprehensiveness, enabling deeper insights into complex biological processes, accelerating disease mechanism research, and advancing personalized treatment strategies.

3 Applications of "Big Data + AI" in Life Sciences

Under the dual impetus of big data and artificial intelligence, life sciences research is undergoing a paradigm shift. AI's ability to uncover complex patterns within biological big data empowers scientists to decode the cross-scale mechanisms of biological systems—from the central dogma to cellular behaviors and tissue-organ functions. Leveraging this synergy, "Big Data + AI" is transforming diverse biological applications, including healthcare, ecological agriculture, biomaterials, and environmental sciences, positioning itself as a cornerstone of innovation in the bioeconomy.

4 Empowering Basic Research in Life Sciences with "Big Data + AI"

4.1 Decoding the Central Dogma with "Big Data + AI"

Every cell in plants, animals, and human harbors billions of molecular machines, with the central dogma—comprising DNA, RNA, and proteins—serving as the core mechanism for life activities and genetic information flow. Accurately deciphering the structures and interactions of these macromolecules is pivotal for unraveling life's mechanisms, understanding disease pathology, and advancing drug design.

DNA, the genetic blueprint of life, lies at the heart of biological principles. Stanford University's EVO model [6], pre-trained on 2.7 million evolutionarily diverse prokaryotic and phage genomes, exemplifies this. EVO predicts the impact of nucleotide sequence changes on organismal fitness, generates long DNA sequences with functional architectures, and designs protein-RNA complexes. This demonstrates the ability of large models to decode the language of life from vast genomic data, bridging the gap from DNA to proteins. RNA, a pivotal intermediary in the central dogma, bridges DNA and proteins through its folding into secondary and tertiary structures. These structures underpin its roles in genetic encoding, translation, regulation, and gene expression. The RhoFold + model [20], developed by a consortium including the Chinese University of Hong Kong and Fudan University and pre-trained on 23.7 million RNA sequences, enables end-to-end prediction of RNA 3D structures. Such pre-trained RNA models accelerate insights into RNA structure and function, driving innovations in RNA-based drug design and synthetic biology. Proteins, as the executors of gene functions, are central to understanding and engineering life. DeepMind's AlphaFold3 [19], built on a Transformer + Diffusion architecture, has revolutionized the prediction of molecular structures and interactions. Chinese researchers have also made remarkable strides in this field. For instance, the HeliXonAI platform outperformed AlphaFold2 in a global protein structure prediction competition. The ESM-AA model [21], a collaborative effort by Tsinghua University AIR, Peking University, and Nanjing University, excels at processing biological structures across scales. Meanwhile, the CHOOSER model [22], a joint initiative by Tongji University, Nanjing Medical University, and Zhejiang University, is the first AI large model tailored for identifying self-processing Cas12 proteins. To integrate DNA, RNA, and proteins into a unified framework, Alibaba Cloud's Feitian Lab introduced the "LucaOne" model [23], the first integrated biological large model of its kind, offering a holistic view of the central dogma.

The central dogma drives life activities, transforming genetic information from DNA into functional tissues and organs. This process evolves from low-dimensional nucleotide sequences to high-dimensional biological activities, presenting a nonlinear and complex challenge. Large models, trained on extensive datasets of DNA, RNA, and proteins, exhibit exceptional generalization and scalability. These biological foundation models predict DNA coding and regulatory patterns, RNA and protein 3D structures, and enable the leap from low-dimensional data to high-dimensional biological functions. This foundational capability paves the way for predicting cellular and tissue-organ behaviors, offering unprecedented insights into life's complexity.

4.2 Decoding Cellular Characteristics and Behaviors with "Big Data + AI"

Cells, as the fundamental units of life, form the basis of physiological and pathological activities in tissues, organs, and organisms. With the rapid evolution of single-cell sequencing technologies, biologists can now efficiently capture multidimensional biological data, including gene expression profiles, epigenetic states, and spatial distributions. By leveraging AI to analyze these complex datasets and construct predictive models, researchers are unlocking deeper insights into the mechanisms of cellular identity and function.

In the realm of single-cell data analysis, AI models such as Phenograph [24], MAGIC, and Seurat [27] are instrumental in mapping intercellular relationships. The scGNN model employs a single-cell graph neural network architecture to express and aggregate cell relationships, utilizing a Gaussian mixture model to simulate heterogeneous gene expression patterns [25]. The scVI model, developed by Nir Yosef's team using hierarchical Bayesian and deep neural network frameworks, excels in tasks like batch effect correction, clustering, and visualization of single-cell data [26]. AI models now enable the simulation of gene perturbations and their phenotypic effects, providing invaluable guidance for wet-lab experiments. For instance, the scGen model, created by Fabian's team using generative adversarial networks, offers a computational tool for experimental design, screening perturbation responses in disease and drug treatment contexts [27]. The CellPolaris model, developed by China's Compass team, accurately identifies core factors in cell fate transitions and simulates transcriptional factor perturbations, proving highly effective in gene regulatory mechanism analysis and disease gene discovery [28].

Recent breakthroughs in natural language processing (NLP) have significantly influenced single-cell biology. Inspired by large language model frameworks, international research teams have utilized tens of millions of human single-cell transcriptomic datasets, combined with vast computational resources and biological knowledge, to develop foundational models capable of deciphering dynamic gene relationships. Notable examples include GeneCompass [18], scGPT [29], Geneformer [17], and scFoundation [30]. One standout is the model by Li Xin's team at the Institute of Zoology, Chinese Academy of Sciences, trained on 120 million human and mouse single-cell transcriptomic datasets with embedded biological knowledge. Using self-supervised learning, the model enhances understanding of gene regulatory mechanisms and identifies key cell fate regulators and potential drug targets [18]. The scFoundation model, developed by Professor Zhang Xuegong's team at Tsinghua University, processes 50 million single-cell gene expression datasets with 100 million parameters, capable of handling approximately 20,000 genes simultaneously [30]. Additionally, the LangCell model, a collaboration between Professor Nie Zaiqing's team at Tsinghua University and Shuimu Molecular, unifies single-cell data and natural language representations, enabling annotation of new cell types without prior labeling [31]. These models leverage advanced algorithms like Transformers to analyze gene expression and other fundamental biological data, using unsupervised or self-supervised learning to model complex biological processes. They have successfully uncovered critical aspects of gene regulatory networks, cell fate determination, and disease-related molecular mechanisms. Compared to traditional AI models, these foundational models enable global simulation of gene regulatory networks, prediction of cell fate transitions, and efficient transfer of insights across multiple downstream tasks.

4.3 Decoding Tissue and Organ Functions with "Big Data + AI"

Single-cell omics technologies, while powerful in their resolution, lose spatial information during tissue dissociation, limiting insights into cellular organization and interactions. Advances in spatial transcriptomics now generate large-scale datasets, enabling comprehensive analysis of tissue and organ functions across spatial and temporal scales. However, the rich biological knowledge embedded in these datasets often eludes direct human interpretation. AI's integration with spatial transcriptomics reveals gene expression patterns within tissue contexts and the characteristics of cellular microenvironments. Models such as SpaGCN [32], BASS [33], BayesSpace [34], and GraphST [35], trained using graph neural networks and Potts models, integrate spatial transcriptomic data to explore expression patterns and focus on cell–cell and cell-niche interactions. The shift from two-dimensional to three-dimensional analysis provides deeper spatial insights into gene and cell functions. Yet, existing 3D integration algorithms often fail to account for spatial distortions, resulting in discrepancies between reconstructed and actual cell positions, compromising downstream analysis reliability. Addressing this, China National GeneBank's ST-GEARS algorithm [36] integrates 3D spatial transcriptomic data, while their Spateo toolkit [37], developed in collaboration with multiple institutions, uses interdisciplinary mathematical models to reconstruct 3D organ structures and quantify spatiotemporal dynamics. The release of Spateo marks a transformative leap in spatiotemporal omics research, supporting studies in embryogenesis, brain science, and disease while advancing high-precision spatiotemporal life landscape analysis.

Foundational models for spatial transcriptomics, combining large-scale and cross-modal approaches, remain a key research frontier. The Nicheformer model [38], developed by Fabian's team, pre-trained on multi-platform spatial and single-cell transcriptomic data, predicts cellular microenvironment density and composition. Jiliang Tang's CellPLM model [39] explicitly encodes spatial locations into embeddings, using single-cell transcriptomic data for prediction and visualizing intercellular interactions through attention matrices. At the multicellular level, AI demonstrates immense potential, integrating multimodal data and complex algorithms to enable multiscale analysis from genes to tissues, driving advancements in tissue, organ, and system-level biological research.

5 Empowering the Biotech Industry with "Big Data + AI"

5.1 Revolutionizing Healthcare Applications with "Big Data + AI"

The convergence of "Big Data + AI" is catalyzing transformative breakthroughs in healthcare, particularly in disease diagnosis and treatment. AI has demonstrated remarkable progress in diverse areas, including pathological state simulation, disease diagnosis, metastasis detection, cancer subtype classification, and survival prediction. For instance, the Beijing Academy of Artificial Intelligence has developed a real-time cardiac electrophysiological simulation system. This system integrates 19 cellular physiological state variables and over 70 equations, enabling real-time 3D simulation of heart electrical activity and facilitating in-depth exploration of the impact of various physiological and pathological factors on cardiac function through parameter adjustments. Meanwhile, Gu Zhongze's team at Southeast University has leveraged deep learning algorithms to create an advanced human lung physiological system (Lung-MPS), which combines alveolar and bronchial compartments with immune cells to enable real-time monitoring of lung pathologies and inflammatory responses [40]. Hong Kong University of Science and Technology has pioneered four AI medical models: MOME (breast cancer diagnosis), mSTAR (pathology assistance tool), MedDr (general practice), and XAIM (explainable AI). These models assist general practitioners and specialists in diagnosing over 30 types of cancers and diseases, with some models achieving accuracy levels comparable to medical professionals with five or more years of experience. They are expected to reduce diagnostic time by 30% to 40%. In a groundbreaking move, BGI has fine-tuned large language models to identify pathogenic variants in rare genetic diseases, developing the Genetic Transformer model [41]. This model achieved impressive recall rates of 99% and 98% in simulated and real clinical samples, respectively, while boosting analysis efficiency by 20-fold. Collaborative efforts by the Chinese PLA General Hospital, the National Cancer Center, and Peking Union Medical College Hospital, in partnership with Insight Medical, have introduced the world's first AI system for clinical pathological diagnosis of complex organs [42]. The system demonstrated near-100% sensitivity and 80.6% specificity on over 3,000 real-world test slides at the PLA General Hospital. Additionally, Li Zhicheng's team at the Shenzhen Institute of Advanced Technology, Chinese Academy of Sciences, has developed an AI-powered integrated diagnostic system for glioma. Utilizing digital pathology images as input and adhering to the 2021 WHO Classification of Central Nervous System Tumors, the system delivers precise diagnostic results comparable to those of human pathologists [43].

In the realm of drug design, Wuhan University of Technology's Movable Type software [44] has significantly enhanced biomolecular free energy calculations by integrating molecular dynamics simulations. Huawei's PanGu model has improved drug design efficiency by 33% and optimized molecular binding energy by over 40% in key scenarios such as target pocket discovery and molecular docking. A notable achievement by the First Affiliated Hospital of Xi'an Jiaotong University is the discovery of the first new antibiotic target and class in nearly 40 years, breaking the stagnation in traditional antibiotic development. Gu Zhongze's team at Southeast University, in collaboration with Huawei, has developed the world's first pharmaceutical model for human organ chips. Meanwhile, Insilico Medicine's AI-discovered small-molecule drug, ISM001-055, has entered international multicenter Phase II clinical trials in China and the U.S., with the first batch of patients already dosed. This milestone marks the first AI-generated drug candidate, from novel target discovery to molecular design, advancing to Phase II clinical trials.

The integration of "Big Data + AI" is revolutionizing healthcare by enhancing diagnostic accuracy, efficiency, and personalized treatment possibilities. It enables a deeper understanding of physiological and pathological states, delivering innovative solutions that significantly elevate disease diagnosis, management, and treatment outcomes.

5.2 *Transforming Agriculture with "Big Data + AI"*

As the global population continues to grow, the demand for food and employment is escalating. Traditional agricultural practices are increasingly inadequate, and the integration of "Big Data + AI" is paving the way for a new era of agricultural development. The rise of "smart agriculture" holds immense promise for improving crop quality and yield. Crop breeding, a cornerstone of food security and economic growth, has seen significant advancements driven by AI technologies. For instance, Northwest A&F University has developed CropGPT [45], an AI-driven breeding system based on the IPDB strategy. This system integrates and analyzes diverse biological datasets to assist biologists in bulk cloning of functional genes, predicting key functional sites, and providing reference targets for breeders to develop new materials with gene aggregation. Weed control and soil nutrition management are critical to enhancing crop yield, as they directly influence crop growth cycles. In collaboration with China Agricultural University and the Shanghai AI Laboratory, the Yazhou Bay National Laboratory has launched SeedLLM, the first large language model dedicated to seed science. The model comprehensively analyzes and indexes breeding literature, technical resources, and online data to provide detailed insights and solutions for key applications such as variety selection, agronomic trait description, cultivation technique recommendations, and historical promotion region queries. In evaluations by domestic breeding experts, SeedLLM outperformed undergraduate agricultural students, achieving an overall score 4.87 times higher. Additionally, China Agricultural University has unveiled the Shennong Model 2.0, which offers capabilities such as agricultural knowledge Q&A, text understanding, summarization, and decision-making inference. Focused on critical segments of the agricultural value chain—from breeding, planting, and livestock farming to remote sensing and meteorological monitoring—the model is driving the transformation of modern agriculture toward smart and precision farming practices.

5.3 Empowering Biomanufacturing with "Big Data + AI"

Synthetic biology, leveraging engineering principles to redesign and optimize natural organisms, is a game-changing technology revolutionizing biomanufacturing and the bioeconomy. AI plays a pivotal role in identifying optimal solutions from the vast and complex array of modification strategies. A breakthrough by Wang Xiaowo's team at Tsinghua University involves the use of Generative Adversarial Networks (GANs) to design novel genetic promoters [46]. The GAN extracts feature from natural promoters, generates millions of new sequences, and filters them through promoter activity prediction models. Impressively, 70.8% of the AI-designed promoters demonstrated functionality, with some even surpassing the activity of the most potent natural promoters and their strongest mutants. These promoters exhibit low global sequence similarity to the E. coli genome and contain unconventional motifs, offering new theoretical insights for promoter design. Building on this, the team developed DeepSEED, an AI-driven flanking sequence optimization method. By integrating expert knowledge and employing techniques akin to image generation, they successfully designed constitutive promoters in E. coli. DeepSEED not only enhances understanding of sequence functionality but also provides a valuable framework for further research and optimization [47]. In another groundbreaking development, Wu Bian's team at the Institute of Microbiology, Chinese Academy of Sciences, utilized AI computational techniques to engineer a series of novel enzymes, enabling catalytic reactions unprecedented in nature. The team also achieved the world's first computationally guided, industrial-grade microbial strain, marking a significant milestone in AI-driven biomanufacturing for industrial applications [48]. The application of AI in synthetic biology continues to expand, driving progress from foundational innovations to the commercialization of practical products. This advancement not only lays a robust foundation for future research but also holds promise for addressing challenges in biomedicine, industrial production, and beyond.

5.4 Empowering Environmental Protection with "Big Data + AI"

Protecting Earth's ecosystems, ensuring sustainable resource utilization, and safeguarding biodiversity are central to sustainable development, underpinning the well-being of present and future generations. In biodiversity conservation, deep convolutional neural networks have been successfully deployed for wildlife identification, counting, and characterization, significantly advancing animal classification and conservation efforts [49, 50]. By aggregating extensive datasets on reptiles, birds, and mammals [51–53], researchers can construct models to analyze ecological patterns at regional and global scales, predict factors influencing biodiversity, and develop targeted conservation strategies. In environmental pollution monitoring and control, AI models built on large-scale environmental metagenomic data can identify microbial communities critical for pollutant degradation, facilitating the development of advanced bioremediation strategies [54]. Machine learning algorithms such as random forests and support vector machines are also employed to predict the presence of heavy metals and organic pollutants in soil, enabling proactive environmental management [55, 56]. Furthermore, AI optimizes soil microbiome composition, enhancing plant health, boosting crop yields, and reducing reliance on chemical fertilizers and pesticides [57]. The integration of "Big Data + AI" has revolutionized bioenvironmental monitoring and management, providing powerful tools for environmental treatment and restoration. This marks the dawn of a new era in environmental protection, paving the way for more efficient and sustainable practices.

6 Future Trends in Applying "Big Data + AI" in Life Sciences

6.1 *Developing High-Quality, Systematic Biological Big Data*

The intricacy and dynamic nature of biological big data present formidable challenges for data processing. These datasets span multi-omics domains—including genomics, transcriptomics, and proteomics—as well as diverse data types such as medical imaging and electronic health records. The exponential growth in data volume, coupled with its diverse sources, renders traditional analytical methods insufficient. The imperative for simulating life systems further underscores the need for high-quality, systematic, and modality-aligned biological big data. To effectively simulate and analyze multi-scale, multi-modal, and spatiotemporal dynamic life systems, it is critical to establish comprehensive, high-quality data resources. Only such data can be efficiently transformed into unified AI-ready biological datasets, enabling the development of robust biological large models. In China, proactive measures have been implemented to address these challenges. For instance, the Chinese Academy of Sciences initiated the "Core Technologies and System Development for Biological Big Data" program, a forward-looking strategic initiative aimed at advancing breakthroughs in biological big data technologies, algorithms, and database construction. This program emphasizes rigorous standards and precise objectives, contributing significantly to the national biological information infrastructure. Additionally, China is accelerating the development of a nationwide integrated big data center collaborative innovation system, enhancing capabilities in data storage, transmission, processing, and analysis. Looking ahead, the establishment of large-scale biological devices, such as the Human Organ Physiological and Pathological Simulation Device (HOPE), will enable the continuous generation of standardized biological big data across multiple scales—genomics, transcriptomics, metabolomics, and organoids. These standardized datasets will reduce variability, enhance the accuracy and reliability of analyses, and provide a robust foundation for artificial intelligence, minimizing noise and bias while uncovering intricate structures and patterns within complex life systems.

7 Building Large-Scale AI-Ready Biological Datasets

AI-ready data refers to meticulously curated and optimized datasets designed to meet the demands of AI model training and application. These datasets excel in quality and completeness while ensuring consistency, accessibility, interoperability, and reusability, thereby supporting complex AI-driven analysis and predictive tasks in life sciences. Biological data in life sciences span diverse types and scales, ranging from microscopic to macroscopic, representing multifaceted perspectives of life systems. They not only capture partial features but also exhibit biological causal relationships and complementary characteristics. Integrating these multi-modal and multi-scale biological datasets into systematic AI-ready formats is essential for constructing frameworks capable of comprehending the complexity of life systems.

However, transforming biological big data into AI-ready datasets is fraught with challenges, including high dimensionality, data sparsity, inconsistent feature representations across modalities, and difficulties in establishing correlations. Although pioneering studies like CZ CELLxGENE [58] and scCompass [8] have developed billion-level, high-quality single-cell transcriptomic datasets—used to build foundational models such as Geneformer [17] and geneCompass [18]—they remain limited to single modalities. The multi-scale, multi-modal, and interconnected nature of life systems necessitates AI-ready datasets that are both comprehensive and richly detailed to enable AI systems to fully interpret life system information. Future efforts must focus on generating and collecting vast quantities of high-quality multi-modal biological data to achieve comprehensive coverage of life system information; developing cross-modal feature alignment techniques to capture inter-modal relationships and achieve unified data representations; and creating efficient multi-modal data fusion methods to integrate AI-ready datasets across different scales and levels. These AI-ready biological big datasets will serve as a cornerstone for future intelligent simulations of life systems, driving advancements in life sciences and beyond.

8 Establishing Life Foundation Model Architectures

Currently, many AI architectures in life sciences are borrowed from computational science, but the unique characteristics of life systems necessitate the development of AI frameworks specifically tailored to this domain. The remarkable success of AlphaFold in predicting protein structures stems from its innovative architecture, which combines the powerful Transformer framework with modules such as multiple sequence alignments and spatial geometric relationships [5]. However, existing foundation models are predominantly designed for single-modal data and largely rely on Transformer architectures, making them ill-equipped to address the multi-modal, interactive, and spatiotemporal dynamic nature of life systems. This limitation results in subpar performance in terms of simulation accuracy and generalization. For instance, the genome foundation model EVO [6] integrates dense quadratic Transformer operators with sub-quadratic Hyena operators within the StripedHyena framework to enhance computational efficiency. Additionally, the model expands its context window, significantly improving its ability to identify connections between genes and other regulatory elements. Despite these advancements, EVO remains confined to genomic-level analysis and cannot capture higher-dimensional features such as transcription, translation, or cellular and tissue organization. Moving forward, it is crucial to develop technical frameworks capable of integrating multi-modal data and efficiently extracting biological insights from diverse data types. This will enable the establishment of a unified life foundation model that elucidates causal relationships and emergent properties across different scales of life systems.

9 Building a New Paradigm for Life Sciences Research

Traditional "wet experiments," which provide direct empirical insights into life processes, have long been the cornerstone of life sciences research. However, these experiments are often labor-intensive, costly, and limited in their data acquisition capabilities. In contrast, "dry experiments" leverage computational simulations and data analysis to efficiently process and analyze vast datasets, uncovering deep patterns and fundamental principles. While dry experiments excel in speed and predictive power, their reliability hinges on data quality and model accuracy. By constructing life foundation models, researchers can seamlessly integrate empirical data from wet experiments with computational insights from dry experiments within a unified framework. This integration will drive a paradigm shift in life sciences research, transitioning from the isolated operation of wet and dry experiments to a synergistic "dry-wet integration" approach driven by computation and theory. Furthermore, the incorporation of automated experimental technologies can streamline the entire process—from data collection to analysis—enhancing efficiency and precision while enabling large-scale, high-throughput, and systematic experimentation. This transformative research paradigm, which integrates "AI + wet experiment validation + automation," will foster the self-iterative evolution of life foundation models through a continuous "simulation-prediction-validation-feedback" loop. By doing so, it will empower researchers to delve deeper into the complexities of life phenomena and uncover the fundamental principles governing life processes, paving the way for groundbreaking discoveries in life sciences.

Conclusion

The convergence and widespread adoption of AI and big data technologies are catalyzing a profound transformation in life sciences research paradigms—evolving

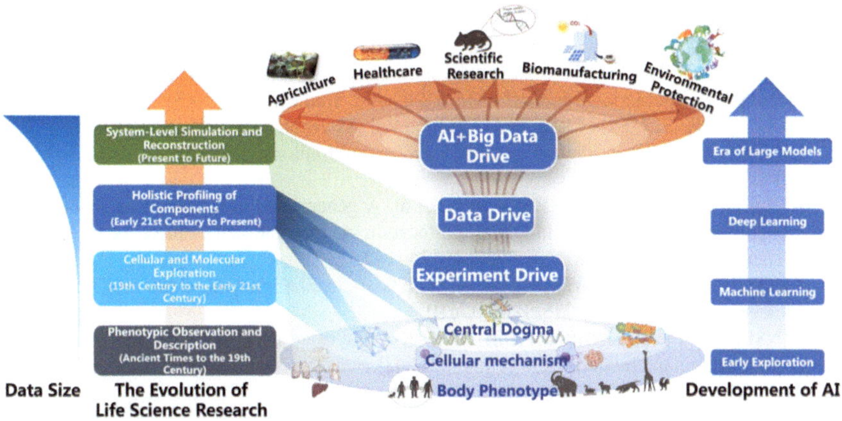

Fig. 1 Transformation and application of life sciences research paradigm

from the traditional integration of experimental science and data science to a groundbreaking framework anchored in experimentation and propelled by the synergistic forces of artificial intelligence and big data (Fig. 1). This paradigm shift not only significantly amplifies the efficiency and depth of life sciences research but also delivers innovative strategies and methodologies to tackle pressing challenges across healthcare, ecological agriculture, biomaterials, and environmental protection.

Amid the surge of "big data + AI" revolutionizing life sciences, China is emerging as a pivotal player in the global innovation landscape. The nation has prioritized scientific research and big data development, vigorously advancing the application of big data and AI technologies in life sciences. This strategic focus underscores China's commitment to contributing its unique wisdom and solutions to the sustainable development of human society, paving the way for transformative advancements and a brighter future.

References

1. CHARLES D. The Origin of Species[M]. London: John Murray, 1859.
2. WATSON J D, CRICK F H C. Molecular Structure of Nucleic Acids: A Structure for Deoxyribose Nucleic Acid[J]. Nature, 1953, 171(4356): 737–738.
3. CRICK F H. On protein synthesis[J]. Symposia of the Society for Experimental Biology, 1958, 12: 138–163.
4. LANDER E S, LINTON L M, BIRREN B, et al. Initial sequencing and analysis of the human genome[J]. Nature, 2001, 409(6822): 860–921.
5. JUMPER J, EVANS R, PRITZEL A, et al. Highly accurate protein structure prediction with AlphaFold[J]. Nature, 2021, 596(7873): 583–589.
6. NGUYEN E, POLI M, DURRANT M G, et al. Sequence modeling and design from molecular to genome scale with Evo[J]. Science, 2024, 386(6723): eado9336.
7. ANDREAS V. De Humani Corporis Fabrica[M]. Basel: Andreas Oporinus, 1543.

8. WANG P, LIU W, WANG J, et al. scCompass: An integrated cross-species scRNA-seq database for AI-ready[A]. bioRxiv, 2024: 2024.11.12.623138.
9. VAN MELLE W. MYCIN: a knowledge-based consultation program for infectious disease diagnosis[J]. International Journal of Man-Machine Studies, 1978, 10(3): 313–322.
10. GOLUB T R, SLONIM D K, TAMAYO P, et al. Molecular Classification of Cancer: Class Discovery and Class Prediction by Gene Expression Monitoring[J]. Science, 1999, 286(5439): 531–537.
11. KULP D, HAUSSLER D, REESE M G, et al. A generalized hidden Markov model for the recognition of human genes in DNA[J]. Proceedings. International Conference on Intelligent Systems for Molecular Biology, 1996, 4: 134–142.
12. ALTSCHUL S F, GISH W, MILLER W, et al. Basic local alignment search tool[J]. Journal of Molecular Biology, 1990, 215(3): 403–410.
13. JANSEN R, YU H, GREENBAUM D, et al. A Bayesian networks approach for predicting protein-protein interactions from genomic data[J]. Science (New York, N.Y.), 2003, 302(5644): 449–453.
14. 王化军, 陈润生, 倪向善, et al. 预测蛋白质二级结构的人工神经元网络方法[J]. 生物物理学报, 1989(4): 422–427.
15. XU Z, WANG W, YANG T, et al. STOmicsDB: a comprehensive database for spatial transcriptomics data sharing, analysis and visualization[J]. Nucleic Acids Research, 2024, 52(D1): D1053–D1061.
16. LI H, SUN Y, HONG H, et al. Inferring transcription factor regulatory networks from single-cell ATAC-seq data based on graph neural networks[J]. Nature Machine Intelligence, 2022, 4.
17. THEODORIS C V, XIAO L, CHOPRA A, et al. Transfer learning enables predictions in network biology[J]. Nature, 2023, 618(7965): 616–624.
18. YANG X, LIU G, FENG G, et al. GeneCompass: deciphering universal gene regulatory mechanisms with a knowledge-informed cross-species foundation model[J]. Cell Research, 2024.
19. ABRAMSON J, ADLER J, DUNGER J, et al. Accurate structure prediction of biomolecular interactions with AlphaFold 3[J]. Nature, 2024, 630(8016): 493–500.
20. SHEN T, HU Z, SUN S, et al. Accurate RNA 3D structure prediction using a language model-based deep learning approach[J]. Nature Methods, 2024: 1–12.
21. ZHENG K, LONG S, LU T, et al. ESM All-Atom: Multi-Scale Protein Language Model for Unified Molecular Modeling[C]//Forty-first International Conference on Machine Learning. 2024.
22. LI W, JIANG X, WANG W, et al. Discovering CRISPR-Cas system with self-processing pre-crRNA capability by foundation models[J]. Nature Communications, 2024, 15(1): 10024.
23. HE Y, FANG P, SHAN Y, et al. LucaOne: Generalized Biological Foundation Model with Unified Nucleic Acid and Protein Language[A]. Bioinformatics, 2024.
24. LEVINE J H, SIMONDS E F, BENDALL S C, et al. Data-Driven Phenotypic Dissection of AML Reveals Progenitor-like Cells that Correlate with Prognosis[J]. Cell, 2015, 162(1): 184–197.
25. WANG J, MA A, CHANG Y, et al. scGNN is a novel graph neural network framework for single-cell RNA-Seq analyses[J]. Nature Communications, 2021, 12(1): 1882.
26. LOPEZ R, REGIER J, COLE M B, et al. Deep generative modeling for single-cell transcriptomics[J]. Nature Methods, 2018, 15(12): 1053–1058.
27. LOTFOLLAHI M, WOLF F A, THEIS F J. scGen predicts single-cell perturbation responses[J]. Nature Methods, 2019, 16(8): 715–721.
28. FENG G, QIN X, ZHANG J, et al. CellPolaris: Decoding Cell Fate through Generalization Transfer Learning of Gene Regulatory Networks[A]. bioRxiv, 2023: 2023.09.25.559244.
29. CUI H, WANG C, MAAN H, et al. scGPT: Towards Building a Foundation Model for Single-Cell Multi-omics Using Generative AI[R]. Bioinformatics, 2023.
30. HAO M, GONG J, ZENG X, et al. Large-scale foundation model on single-cell transcriptomics[J]. Nature Methods, 2024, 21(8): 1481–1491.

31. ZHAO S, ZHANG J, WU Y, et al. LangCell: Language-Cell Pre-training for Cell Identity Understanding[A]. 2024.
32. HU J, LI X, COLEMAN K, et al. SpaGCN: Integrating gene expression, spatial location and histology to identify spatial domains and spatially variable genes by graph convolutional network[J]. Nature Methods, 2021, 18(11): 1342–1351.
33. LI Z, ZHOU X. BASS: multi-scale and multi-sample analysis enables accurate cell type clustering and spatial domain detection in spatial transcriptomic studies[J]. Genome Biology, 2022, 23(1): 168.
34. ZHAO E, STONE M R, REN X, et al. Spatial transcriptomics at subspot resolution with BayesSpace[J]. Nature Biotechnology, 2021, 39(11): 1375–1384.
35. LONG Y, ANG K S, LI M, et al. Spatially informed clustering, integration, and deconvolution of spatial transcriptomics with GraphST[J]. Nature Communications, 2023, 14(1): 1155.
36. XIA T, HU L, ZUO L, et al. ST-GEARS: Advancing 3D downstream research through accurate spatial information recovery[J]. Nature Communications, 2024, 15(1): 7806.
37. QIU X, ZHU D Y, LU Y, et al. Spatiotemporal modeling of molecular holograms[J]. Cell, 2024: S0092-8674(24)01159-0.
38. SCHAAR A C, TEJADA-LAPUERTA A, PALLA G, et al. Nicheformer: a foundation model for single-cell and spatial omics[A]. bioRxiv, 2024: 2024.04.15.589472.
39. WEN H, TANG W, DAI X, et al. CellPLM: Pre-training of Cell Language Model Beyond Single Cells[A]. Bioinformatics, 2023.
40. CHEN Z, HUANG J, ZHANG J, et al. A storm in a teacup -- A biomimetic lung microphysiological system in conjunction with a deep-learning algorithm to monitor lung pathological and inflammatory reactions[J]. Biosensors and Bioelectronics, 2023, 219: 114772.
41. LIANG L, CHEN Y, WANG T, et al. Genetic Transformer: An Innovative Large Language Model Driven Approach for Rapid and Accurate Identification of Causative Variants in Rare Genetic Diseases[A]. medRxiv, 2024: 2024.07.18.24310666.
42. SONG Z, ZOU S, ZHOU W, et al. Clinically applicable histopathological diagnosis system for gastric cancer detection using deep learning[J]. Nature Communications, 2020, 11(1): 4294.
43. WANG W, ZHAO Y, TENG L, et al. Neuropathologist-level integrated classification of adult-type diffuse gliomas using deep learning from whole-slide pathological images[J]. Nature Communications, 2023, 14(1): 6359.
44. LIU W, LIU Z, LIU H, et al. Free Energy Calculations Using the Movable Type Method with Molecular Dynamics Driven Protein–Ligand Sampling[J]. Journal of Chemical Information and Modeling, 2022, 62(22): 5645–5665.
45. ZHU W, HAN R, SHANG X, et al. The CropGPT project: Call for a global, coordinated effort in precision design breeding driven by AI using biological big data[J]. Molecular Plant, 2024, 17(2): 215–218.
46. WANG Y, WANG H, WEI L, et al. Synthetic promoter design in Escherichia coli based on a deep generative network[J]. Nucleic Acids Research, 2020, 48(12): 6403–6412.
47. ZHANG P, WANG H, XU H, et al. Deep flanking sequence engineering for efficient promoter design using DeepSEED[J]. Nature Communications, 2023, 14(1): 6309.
48. CUI Y, WANG Y, TIAN W, et al. Development of a versatile and efficient C–N lyase platform for asymmetric hydroamination via computational enzyme redesign[J]. Nature Catalysis, 2021, 4(5): 364–373.
49. THALOR M A, NAGABHYRAVA R, RAJKUMAR K, et al. Deep learning insights and methods for classifying wildlife[C]//2023 3rd International Conference on Advance Computing and Innovative Technologies in Engineering (ICACITE). 2023: 403–407.
50. NOROUZZADEH M S, NGUYEN A, KOSMALA M, et al. Automatically identifying, counting, and describing wild animals in camera-trap images with deep learning[J]. Proceedings of the National Academy of Sciences, 2018, 115(25): E5716–E5725.
51. OSKYRKO O, MI C, MEIRI S, et al. ReptTraits: a comprehensive dataset of ecological traits in reptiles[J]. Scientific Data, 2024, 11(1): 243.
52. CHIA S Y, FANG Y T, SU Y T, et al. A global database of bird nest traits[J]. Scientific Data, 2023, 10(1): 923.

53. DING C, LIANG D, XIN W, et al.A dataset on the morphological, life-history and ecological traits of the mammals in China[J]. Biodiversity Science, 2022, 30(2): 21520.
54. GAO Y Z, LIU H, CHAO H J, et al. Constitutive Expression of a Nag-Like Dioxygenase Gene through an Internal Promoter in the 2-Chloronitrobenzene Catabolism Gene Cluster of Pseudomonas stutzeri ZWLR2-1[J]. Applied and Environmental Microbiology, 2016, 82(12): 3461–3470.
55. PALANSOORIYA K N, LI J, DISSANAYAKE P D, et al. Prediction of Soil Heavy Metal Immobilization by Biochar Using Machine Learning[J]. Environmental Science & Technology, 2022, 56(7): 4187–4198.
56. WU G, KECHAVARZI C, LI X, et al. Machine learning models for predicting PAHs bioavailability in compost amended soils[J]. Chemical Engineering Journal, 2013, 223: 747–754.
57. MO Y, BIER R, LI X, et al. Agricultural practices influence soil microbiome assembly and interactions at different depths identified by machine learning[J]. Communications Biology, 2024, 7(1): 1–16.
58. ABDULLA S, AEVERMANN B, ASSIS P, et al. CZ CELLxGENE Discover: a single-cell data platform for scalable exploration, analysis and modeling of aggregated data[J]. Nucleic Acids Research, 2024: gkae1142.

Xin Li a dual-appointed researcher at the Institute of Zoology, Chinese Academy of Sciences, and the Beijing Institute of Stem Cell and Regenerative Medicine, is also a distinguished doctoral supervisor and a 2022 recipient of the National Overseas Outstanding Young Scholars Program. Dr. Li earned his Ph.D. in Developmental Biology from Northeast Agricultural University's "National Science Base Class" program (jointly trained by the Institute of Zoology, Chinese Academy of Sciences) in 2015. He furthered his expertise through postdoctoral research at the University of California, San Diego, and the laboratory of renowned cancer biologist Robert Weinberg at the Massachusetts Institute of Technology. Since joining the Institute of Zoology, Chinese Academy of Sciences, in December 2021, Dr. Li has spearheaded groundbreaking research in AI-driven biology, stem cell and developmental biology, aging, and cancer metastasis, with significant translational applications. He has authored or co-authored multiple high-impact papers in leading journals such as *Cell*, *Science*, *Cell Research*,*Nature Cell Biology*, and *Cell Stem Cell*. Among his notable contributions are the establishment of the first rat haploid and mouse-rat cross-species heterologous diploid embryonic stem cells, which advanced our understanding of stem cell pluripotency and evolutionary gene regulation mechanisms (*Cell Stem Cell* 2014, *Cell* 2016); the discovery of transgenerational inheritance of acquired metabolic traits via tsRNA and the epigenetic regulation of lipid metabolism in aging (*Science* 2016, *Nature Cell Biology* 2018, STTT 2022), recognized as one of China's Top 10 Scientific Advances in 2016; and the development of the world's first cross-species life foundation model (*Cell Research* 2024, cover story) and a transfer learning-based gene regulatory network generation model. Dr. Li also serves as a doctoral supervisor at Northeast Agricultural University and holds memberships in the

Anti-Aging Committee of the Chinese Association of Gerontology and Geriatrics and the Aging Genetics Committee of the Chinese Genetic Society.

Haiping Jiang a postdoctoral researcher at the Institute of Zoology, Chinese Academy of Sciences, focuses on aging, cancer, and artificial intelligence.

Wenhao Liu a Ph.D. candidate jointly trained by the Institute of Zoology, Chinese Academy of Sciences, and Northeast Agricultural University, specializes in life data science, bioinformatics, and artificial intelligence.

Chen Fang a Ph.D. candidate at the Institute of Zoology, Chinese Academy of Sciences, is dedicated to foundational models for AI in life sciences.

Cong Li a Ph.D. candidate at the Institute of Zoology, Chinese Academy of Sciences, is committed to foundational models for AI in life sciences.

Haorao Wang a Ph.D. candidate at the School of Frontier Interdisciplinary Sciences, University of Chinese Academy of Sciences (jointly trained by the Institute of Zoology, Chinese Academy of Sciences), focuses on deep learning, cancer, and aging.

Xusheng Ma a master's candidate at the Institute of Zoology, Chinese Academy of Sciences, explores life sciences and artificial intelligence.

Huanhuan Wu a Ph.D. candidate at the School of Interdisciplinary Sciences, University of Chinese Academy of Sciences (jointly trained by the Institute of Zoology, Chinese Academy of Sciences), investigates artificial intelligence and aging.

Development and Innovative Applications of Data and AI Models in Materials Science

Miao Liu, Zongguo Wang, Yangang Wang, Sheng Meng, Tenglong Lu, Meng Wan, Ziyi Chen, and Yang Yuan

Abstract Materials are a fundamental cornerstone of economic and social development, and the integration of artificial intelligence (AI) technologies has opened new opportunities for research in materials science. This paper reviews the advancements and innovative applications of data and large AI models in the field of materials. It begins by discussing the progress and trends in materials science databases, highlighting internationally renowned databases and their applications, along with case studies showcasing data-driven materials research and development. The paper then explores prediction models for material properties, force fields, and Hamiltonians, emphasizing the critical roles of various AI models and platforms in materials research. Furthermore, it examines the applications of language models, knowledge graph technology, and enhanced search capabilities in materials science, such as vertical language models supporting research, knowledge graphs uncovering latent connections, and enhanced search optimizing information retrieval. Detailed examples include innovative applications like MatChat, developed by the Chinese Academy of Sciences, for predicting compound synthesis pathways, and GPTFF, a large AI model for force field predictions. Finally, the paper reflects on the future development of data and AI technologies in materials science. Although challenges remain in advancing these technologies, the field holds immense potential, and their applications are set to drive further progress in materials science research and industrial development.

Keywords Materials science · Fundamental models · Database · AI

M. Liu (✉) · S. Meng · T. Lu
Institute of Physics, Chinese Academy of Sciences, Beijing, China
e-mail: mliu@iphy.ac.cn

Z. Wang · Y. Wang · M. Wan · Z. Chen · Y. Yuan
Computer Network Information Center, Chinese Academy of Sciences, Beijing, China

China's e-Science Blue Book 2025, https://doi.org/10.1007/978-981-96-8252-2_14

Introduction

Materials science is the underlying power of social development and the material foundation of scientific and technological progress. With the continuous progress of science and technology, research in the field of materials is also deepening, and the Artificial Intelligence (AI) technology emerging in recent years is deeply empowering the development of materials science from multiple levels, bringing new opportunities for the innovative development of new materials.

In today's era, information technology is at the forefront of empowering materials science and opening a new chapter. The precise mining of material properties and performance through big data has accelerated the development of new materials; the simulation of the evolution of complex material structures with the help of Artificial Intelligence has optimized materials design; and efficient integration of dispersed materials science and technology resources via cloud computing has promoted scientific research collaboration and sharing. These integrations not only break through the time and space limitations of traditional materials science research but also make the microscopic world of materials clearly visible from a digital perspective. With its powerful data processing and learning capabilities, the AI Large Model can quickly analyze the complex relationship between various performance parameters, structural features, and synthesis paths of materials, which greatly shortens the research and development cycle of new materials and improves the accuracy of research. The application of AI in materials science is maturing, and it has become the most dynamic and promising frontiers of current materials science research, driving the field toward a digital and intelligent future. It is expected to generate breakthrough innovations in many key technological fields, reshaping the competitive landscape of global materials science and technology.

In the journey of materials science digitization, many pioneering works have been carried out by the global research community. Lawrence Berkeley National Laboratory's Materials Project [1], like a treasure trove of data, pioneered the large-scale integration and mining of materials data, lighting the way for subsequent development. As the wave of digitization advances, tech giants are following suit, coming on strong. DeepMind's GNoME [2], using advanced artificial intelligence algorithms and fueled by vast amounts of data, deeply predicts new material structures, demonstrating impressive innovative capabilities; Microsoft leveraging its technological strength, launched MaterSim and Mattergen [3, 4], making significant strides in the field of materials simulation and generation [3, 4]; Meta's OMat24 dataset [5] has contributed a rich and diverse resource to the entire field of materials science, promoting the sharing and exchange of knowledge.

In the global wave of materials science digitization, our nation has demonstrated strong capabilities and vigorous innovation. Institute of Physics Chinese Academy of Sciences and Computer Network Information Center of Chinese Academy of Sciences have made a series of remarkable breakthroughs in this field. The collaborative team constructed the Atomly Materials Science Database [6, 7], which collects, organizes, and deeply analyzes various material data through high-throughput computation, building a comprehensive and detailed materials information database, providing extensive data support for materials research. The launched MatChat inorganic material synthesis path prediction model, utilizing advanced algorithms and extensive data training, can accurately predict the synthesis routes of inorganic materials, greatly improves the efficiency and success rate of material synthesis while reducing unnecessary trial and error in experiments. The development of GPTFF General AI Force Field for Materials Science [8] provides a powerful tool for the simulation of materials' microstructures and performance calculations, enabling researchers to explore materials at the atomic scale and opening new paths for the design and optimization of new materials. These achievements not only highlight our nation's profound expertise and innovation in the field of materials science digitization but also serve as the key cornerstone and indispensable core force driving the industry towards a higher level of development.

Of course, the application of large models in the field of materials science still faces some challenges. For example, while the potential of pituitary large models in driving paradigm shifts in basic scientific research is evident, the industry as a whole is still in the early exploration phase and has not yet fully addressed the research and development pain points. In the future, it is still necessary to focus on issues such as erroneous data, missing data, and model hallucinations, continuing iterative upgrades.

This paper reviews the development history, from the early stages of scientific exploration to the involvement of tech giants today, discussing conceptual breakthroughs and technological innovations that have reshaped the research paradigm of materials science. It also summarizes the existing rich datasets in the materials field, analyzes various cutting-edge technologies, and showcases key infrastructures. Finally, the paper looks forward to the future development trend of the deep integration of AI and materials science, offering valuable insights for researchers, policymakers and practitioners.

1 Advances and Trends in Materials Science Databases

In today's digital era, the rise of materials science databases has brought new opportunities and challenges for materials research, development and application. Understanding the rise of materials science databases and their international background is of great significance in promoting the development of materials science. With the rapid development of information technology, materials science research is gradually changing from traditional experiment-driven to data-driven. The emergence of materials science databases has provided materials researchers with a large amount of experimental data, theoretical calculations and literature, enabling them to carry out materials design, performance prediction and optimization more efficiently. The resulting materials informatics, an emerging discipline that combines information science and materials science, aims to collect, organize, analyze and apply materials data using information technology tools. Materials science database is an important part of materials informatics, which provides data base and technical support for the development of materials informatics. Materials research and development is a complex and long process, which requires a large number of experiments and tests. The emergence of materials science database can provide borrow and reference for materials research and development personnel, reduce the number of experiments and costs, and improve the efficiency and success rate of research and development.

1.1 Background on the Development of Materials Science Databases

In the field of materials science, with the development of advanced technologies such as Big Data, Artificial Intelligence, and Machine Learning (ML), materials science databases have become an important force in promoting materials research. These databases provide abundant experimental and computational data, which serve as a solid foundation for materials design, performance prediction, and application innovation.

Materials science research often relies on a large amount of experimental data and theoretical calculations. However, the properties of materials are multidimensional, involving various factors such as chemical composition, structure, properties, environmental influences, etc., and the data sources are scattered and vast. In order to achieve efficient materials research and performance optimization, scholars have gradually realizd the need for a systematic and standardized data management platform that integrates various types of data for rapid query and analysis. With the rapid growth of data volume, materials science databases have gradually transformed from simple experimental records to integrated large-scale databases, covering information on the composition, structure, properties, manufacturing processes and other aspects of materials. Materials science databases not only store traditional experimental data but also include predictive data obtained from new technologies such as high-throughput computing and machine learning. One of the directions for the development of materials science databases is to achieve open sharing and promote global data exchange and cooperation. Through open databases, researchers can conduct secondary development and research based on existing data, thereby promoting interdisciplinary innovation.

The Materials Genome Initiative (MGI) [9] in the United States was launched in 201, aiming to enhance the efficiency and innovation of materials research by accelerating the discovery and optimization of materials. The core idea of MGI is to accelerate the process of materials from research to application through big data and computational modeling. To this end, MGI has promoted the construction of many materials databases, including Materials Project [1], a database based on high-throughput computing that provides data on the structure, properties, and behavior of materials. Additionally, databases such as AFLOWlib and Open Quantum Materials Database (OQMD) [10, 11] provide important data support for researchers. In

Europe, several collaborative projects are also dedicated to the sharing and integration of materials data, with the European Horizon 2020 project providing financial and technical support for the development of materials databases, thereby fostering multidisciplinary cooperation.

China has also made significant progress in construction of materials science databases. China launched its own Materials Genome Program, aiming to accelerate the research and development of new materials through the establishment of databases. The Ministry of Science and Technology has set up a special project, with the University of Science and Technology Beijing leading the establishment of relevant materials science database. Beijing and the Chinese Academy of Sciences jointly built the Huairou Materials Genome Research Platform, which is an advanced platform for materials exploration, basic and applied research, integrating materials computation, big data processing, high-throughput materials synthesis and characterization, and high-throughput technology research and development, thereby vigorously promoting the development of materials science research and application. This platform has strongly promoted the development of materials science research and applications.. The Scientific Data Bank (ScienceDB) [12] established by the Chinese Academy of Sciences, as a new scientific data management and computation platform, with its unique data modeling, storage and computation capabilities, as well as powerful parallel computing and elastic expansion functions, is of great significance in promoting efficient management, analysis and sharing of scientific data. It plays a crucial role in accelerating scientific research innovation and developmen in materials and other disciplines.

1.2 Introduction to World-Renowned Materials Science Databases

Data resources are the foundation for the training and application of large models. Materials databases and data quality are the key elements in the construction of material large models. Currently, data is not only a carrier of information but also an important driving force for scientific and technological innovation and competition. Especially in the field of materials science, with the continuous emergence of new materials, how to efficiently utilize and manage data resources has become an major challenge for researchers and enterprises.

Notable authoritative international databases include: SpringerMaterials, one of the world's largest high-quality numerical databases on material properties, provided by Springer Nature Group; Total Materia, one of the world's most comprehensive materials databases for materials data and cross-country standard comparisons; MatNavi [13], a free database organized by the National Institute for Materials Science (NIMS) in Japan; The Materials Project [1], a materials computational data analysis library jointly developed and maintained by Lawrence Berkeley National Laboratory and the Massachusetts Institute of Technology; OQMD (Open Quantum

Materials Database) [11], an open-source quantum materials database developed by Northwestern University; AFLOW [10], co-developed by Duke University (Duke) and the AFLOW Consortium; as well as NOMAD [14], a simple database migration tool for developers, AiiDA [15], a high-throughput computational database and workflow management system, and the Inorganic Crystal Structure Database (ICSD) [16], the world's largest database of inorganic crystal structures, provided by FIZ Karlsruhe in Germany. Tech giants are also making their presence felt in this field, with Meta releasing the OC20, OC24 and OMat24 datasets, a series of important open-source datasets in materials science [5, 17].

The development of domestic materials science databases in China is relatively recent, but there are typical cases of "catching up." Currently, one of the most internationally recognized examples is Atomly [6], a world-class materials science data platform jointly developed by the Institute of Physics of the Chinese Academy of Sciences and the Songshan Lake Materials Laboratory, which holds independent intellectual property rights. This platform contains more than 300,000 inorganic compounds, atomic structure, electronic structure, thermodynamic stability, X-ray diffraction patterns, and other information.

It is evident that the current data production model, based on first-principle calculations, has fully dominated this field. This has driven computational power and highly automated workflows to become the core productivity of the materials field. Data is the most important asset, and the primary ways of accumulating data in materials science currently include experimental and theoretical calculations. However, the magnitude of the experimental database in materials science currently cannot compared with computational database, and there is still a lot of room for development, as well as greater challenges. Constructing experimental database is not simply about collecting numerous experimental data; it requires a longer period of accumulation and manual processing to unified and standardized the data. Therefore, the data accumulated through the experimental methods cannot match the theoretical computational methods in the short term. After decades of development, theoretical computational methods, such as those based on density functional theory, have matured significantly and can provide large amounts of high-precision, highly standardized data with high efficiency, making them naturally compatible with the requirements of machine learning models for datasets. Currently, work on establishing a high-quality, large-scale computational materials database is is progressing rapidly on the international stage, and our country still needs to build its strength in this field to vigorously develop computational databases with independent intellectual property rights that are at the forefront of the world.

2 Artificial Intelligence Models and Platforms in Materials Science

The rapid development of artificial intelligence technology has brought about profound changes in the way materials science research is conducted. By applying advanced algorithms such as deep learning and neural networks, AI plays an important role in material prediction, design, optimization and experimentation. Materials science not only requires a massive amount of data, but also relies on powerful computational capabilities and highly accurate models, which makes AI a key driver for materials research and development. AI models in materials science rely on large-scale training datasets, and by analyzing the structure, composition, properties of materials, they can effectively predict the properties of new materials and their potential applications. At the same time, open platforms for materials science are accelerating the practical application of AI models. These platforms integrate computational resources, automated experimental equipment, and intelligent tools, pushing materials science research into a new era of efficiency, precision, and intelligence. Below, we will delve into several major AI models used in materials science and analyze their potential to enhance the efficiency and precision of materials research as well as how they are applied.

2.1 Prediction Models for Physical Properties, Force Fields, and Hamiltonian Quantities in Materials Science

2.1.1 Material Property Prediction Model

One of the keys to material property prediction is the accurate acquisition of material descriptors. Extensive innovative work has emerged from the persistent efforts of researchers in this field. The representative networks and models mainly focus on the following:

(1) CGCNN (Crystal Graph Convolutional Neural Network), a graph convolutional neural network based on crystal graphs [18], constructs a graph representation of the crystal structure, encoding the information of atoms and their chemical bonds as graph data. It then extracts the intrinsic features of the crystal structure through graph convolutional operations at multiple depths.
(2) CrabNet (Compositionally Restricted Attention-Based Network), is a model based on a unique component-restricted attention mechanism [19]. It effectively captures interactions between different elements and directly predicts material properties from chemical formulas with high accuracy and interpretability
(3) MEGNet (MatErials Graph Network), a general-purpose material graph network model for accurate prediction of molecular and crystalline properties [20].
(4) TiraCGCNN (Tripartite interaction representation algorithm-enhanced Crystal Graph Neural Networks), a crystal graph convolutional neural network model

[21] that explicitly encodes the features of three-body interactions, integrates atomic information, bond length and bond angle information, and clearly defines the vector relationship between atoms and edges.

(5) The Universal formation energy model, a model [22] trained on a large amount of data to predict the formation energy of inorganic compounds,, is innovative in its use of a set of structurally relevant descriptors. These descriptors emphasize key factors such as electronegativity differences and coordination numbers, allowing the model to accurately capture the fundamental characteristics of atomic interactions in materials, leading to outstanding predictive performance.

(6) SteelBERT, an end-to-end strategy [23] for quantitatively predicting material properties from large-scale historical textual data, consists of a materials-specific natural language encoder and a multimodal deep learning framework. It takes composition and processing texts as input to quantitatively predict mechanical properties.

2.1.2 Atomic-Scale Force-Field Models

The core problem of materials science is to understand the interactions between atoms, and atomic-scale general-purpose force field models provide an important tool for simulating atomic interactions. With the development of artificial intelligence technology, high-precision atomic-scale force field models have brought about a profound transformation in computational simulations in the field of materials science. The representative force field models widely used in materials simulation both domestically and internationally include the following:

(1) DeepMD/DPA1/DPA2 [24], a deep learning method for representing multi-body potentials, calculates atomic energy by decomposing the system potential into a sum of atomic energy contributions and constructing atomic environment descriptors to input into a deep neural network. It can be used for molecular dynamics simulations, since it relies on open-source datasets, its modeling capabilities are limited.

(2) M3GNet [25], a general-purpose interatomic potential function model based on graph neural networks that considers three-body interactions, is used to describe the interatomic potential energy surface. Its architecture incorporates many-body interactions by constructing a material graph containing information such as atomic coordinates and lattice matrices. It extends three-body angular interactions using an efficient base set of spherical Bessel and spherical harmonic functions, and is optimized iteratively with multi-step updates. Trained on ten years of large-scale structural relaxation data from the Materials Project, it can accurately predict the energy, force, and stress of various materials, and has been widely applied in structural relaxation, dynamics simulation, and property prediction.

(3) CHGNET [26], a pre-trained, general-purpose machine-learning interatomic potential model, is designed for atomic-scale modeling. It is based on graph neural networks and describes interatomic interactions by constructing graph

structures for atoms, chemical bonds, and other information in material structures. A key characteristic of CHGNet is its ability to leverage information on energy, force, stress, and magnetic moments from the Materials Project Trajectory Dataset, which contains approximately 1.5 million inorganic structures' DFT calculation data. In particular, it infers atomic charges from magnetic moments, which helps to better capture electronic interactions and distinguish between ions of different valence states in atomic-scale simulations.

(4) MACE [27], a novel neural network model for the fields of computational chemistry and materials science, is based on equivariant message-passing neural networks (MPNNs). It employs a hierarchical higher-body-order extension to pass higher-order messages through an innovative message construction mechanism, addressing the limitations of traditional MPNNs, which only pass two-body messages, thereby limiting model expressivity and increasing computational cost.

(5) SevenNet [28], a highly innovative development in molecular dynamics (MD) simulations, focuses on solving the parallelization challenges faced by graph neural network interatomic potentials (GNN-IPs). Its core spatial decomposition parallelization algorithm is built on the NequIP architecture. Within this architecture, the simulation unit is cleverly divided into multiple subdomains, each assigned to a different GPU for processing. During message passing, SevenNet efficiently exchanges atomic positions, node features, and gradient information, reducing inter-processor communication overhead and significantly improving computational efficiency.

(6) GPTFF [8], jointly developed by Institute of Physics of Chinese Academy of Sciences and Songshan Lake Materials Laboratory, is constructed based on Graph Neural Networks (GNNs). It projects atoms in a crystal structure into a high-dimensional space based on their element types, representing atoms using embedding vectors and describing their geometric structures (e.g., bond lengths) with edge vectors. These atom and edge vectors are then concatenatedto form comprehensive representations. Additionally, three-body interaction information is incorporated by mapping the cosine values of bond angles into higher-dimensional space, enabling the model to learn interactions between different atomic nodes and edge vectors, thereby enhancing prediction accuracy. As a general-purpose force field, GPTFF can rapidly optimize given crystal structures and is well-suited for the screening and relaxation of unknown structures. With high accuracy and broad applicability, GPTFF is among China's leading contributions.

2.1.3 Hamiltonian Prediction Model

The application of deep learning techniques in materials science is rapidly advancing. Among them, Hamiltonian prediction models offer a novel approach to accelerating electronic structure calculations of materials. These models demonstrate significant advantages, particularly in handling complex material systems and circumventing the high computational cost of traditional methods. Their successful application highlights the tremendous potential of integrating deep learning with materials science, providing powerful tools for predicting and designing new materials in academic research.

(1) Deep Learning Density Functional Theory Hamiltonian (DeepH) [29], developed by the research group of Yong Xu and Wenhui Duan at the Department of Physics, Tsinghua University, is a deep learning-based first-principles calculation method. This approach learns from density generalized functional (DFT) data, given a material structure, to accurately predicts its DFT Hamiltonian, enabling precise calculations of various material properties. The core innovation of DeepH lies in leveraging neural networks that are covariant under the three-dimensional Euclidean group (E(3) group) to predict the DFT Hamiltonian corresponding to microscopic atomic structures. This significantly accelerates first-principles electronic structure calculations, especially in predicting the electronic structures of complex material systems.
(2) HamGNN (Hamiltonian Graph Neural Network) [30], developed by the research team of Professors Hongjun Xiang and Xinguo Gong at Fudan University, is a graph neural network model designed to efficiently predict the electronic Hamiltonian of materials. By integrating machine learning techniques, HamGNN rapidly constructs electronic Hamiltonian matrices for arbitrary compositions and crystal structures, eliminating the expensive self-consistency process required in traditional electronic structure calculations. A key feature of HamGNN is its E(3) symmetry, which ensures high accuracy and generalization across different materials. Utilizing a two-step training approach consisting of pre-training and fine-tuning, HamGNN achieves accuracy comparable to first-principles methods.

2.2 *Applications of Language Models, Knowledge Graphs, and Enhanced Search in Materials Science*

2.2.1 Development and Application of Pendant Language Models for Materials Science

The pendant language models for materials science are designed to process materials-related textual data, including academic papers, patents, and experimental reports. Their development typically involves constructing large-scale materials science corpora by collecting textual data from various authoritative sources, such as classical textbooks, cutting-edge research journals, and industry reports. These rich and specialized datasets serve as the foundation for training high-performance language models in the field.

The MatChat model, developed by the Artificial Intelligence Department of the Computer Network Information Center of the Chinese Academy of Sciences (CNIC) in collaboration with the Institute of Physics, is an early heuristic case of such models, based on the LLaMA2-7B fine-tuned training [31] of a large-language model focusing on the task of path prediction for synthesis paths of inorganic materials, which is already online and open for use. Chem-GPT, developed at the University of Science and Technology of China, is a chemical data-driven model trained by incorporating chemists' knowledge so as to provide preliminary experimental suggestions. The model is trained based on the GPT architecture and focuses on chemical development [32].

2.2.2 Knowledge Graph Technology Enables Materials Science Discovery

Knowledge graph integrates and structurally represents materials science knowledge scattered in a large amount of literature and data by constructing a network of relationships between entities (e.g., materials, elements, properties, preparation processes, etc.) and entities. When constructing a materials science knowledge graph, it is first necessary to extract information from various data sources and identify the key entities and relationships in them. For example, information such as material names, constituents, crystal structures, mechanical properties, and other information and authors' interrelationships, presentations, and interrelationships between them are extracted from material science papers. Then the corresponding knowledge system is constructed to obtain deeper relationships. For example, important milestones and trends in the development of materials are studied through the correlation between material information and authors.

2.2.3 Development and Application of Enhanced Search in Materials Science

The application of enhanced search technology in materials science is dedicated to improving the ability of researchers to access accurate materials science information. On the one hand, semantic understanding technology is introduced so that the search system can deeply understand the semantics of the search terms and is not limited to literal matching of keywords. On the other hand, enhanced search combines user behavior analysis and personalized recommendation technology. By analyzing the user's historical search records, browsing behavior, and feedback on search results, the system is able to gradually understand the user's research interests and preferences, and provide the user with personalized search results sorting and recommended content.

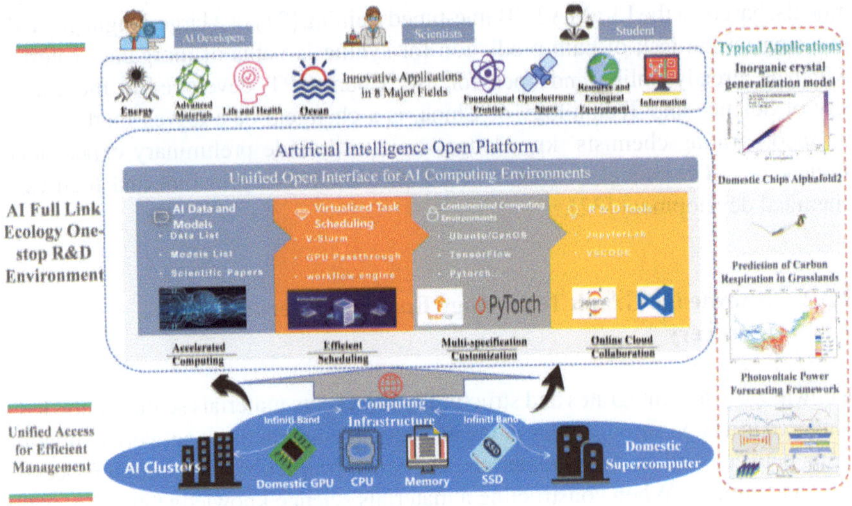

Fig. 1 Architecture diagram of AI open platform

2.3 *Artificial Intelligence Open Platform*

The application of materials models relies on the support of powerful computational resources. To overcome the limitations of heterogeneous resource scheduling in supercomputing and cloud computing, the Computer Network Information Center of the Chinese Academy of Sciences has developed VenusAI, an open artificial intelligence platform [33]. VenusAI provides comprehensive and diverse intelligent services for materials science research, covering the entire process from materials synthesis and performance prediction to design. This platform significantly enhances research efficiency and accelerates the transformation of scientific discoveries into practical applications (see Fig. 1). As an important part of China Science and Technology Cloud, VenusAI has been serving more than 200 research teams in the field of materials science, including universities, research institutes and enterprises. By seamlessly integrating advanced AI tools with large-scale computing resources, this platform is propelling materials science research into a new era of efficiency, precision, and intelligence.

3 Examples of Innovative Applications in Materials Science

3.1 *"Big Data + Artificial Intelligence" Empowering New Material Discovery*

(1) Discovery of Ternary Nitride Materials [34]

Over the past 70 years, humans have discovered an average of only 3.3 nitride crystals per year. However, Ceder's team at the University of California, Berkeley, has significantly accelerated this process by constructing a large interactive database of ternary nitride stability using high-throughput first-principles calculations. This database leverages computational materials discovery techniques and informatics tools to systematically cluster ternary nitrides into chemical families based on their stability and metastability. Within just one year, the team identified 92 new ternary nitride materials, seven of which have been successfully synthesized experimentally. This breakthrough demonstrates how data-driven approaches can dramatically enhance the efficiency of new material discovery.

(2) Discovery of New Materials for Energy Storage [35]

The Joint Center for Energy Storage Research (JCESR), established by the U.S. Department of Energy, is dedicated to advancing next-generation energy storage technologies to support the future energy transition and sustainable development. As part of its research on multivalent ion batteries, JCESR researchers leveraged the Materials Project database to screen tens of thousands of candidate materials. In just a year and a half, they successfully identified a spinel-type $MgTi_2S_4$ magnesium-ion cathode material, which was subsequently experimentally validated. Prior to this discovery, magnesium-ion cathode materials had suffered from low energy storage capacity, with no significant breakthroughs in nearly two decades. The emergence of spinel-type $MgTi_2S_4$ has doubled the previous world record for energy storage capacity in magnesium-ion cathodes, highlighting the remarkable potential of data-driven materials discovery.

(3) New Materials Discovered by GNoME in Collaboration with A-Lab [2, 36]

GNoME, a deep learning tool for the materials domain developed by DeepMind, was trained using workflows and datasets accumulated over more than a decade in the Materials Project. Its algorithms are continuously refined through active learning, enhancing its predictive capabilities. In parallel, the Joint Center for Energy Storage Research (JCESR), established by the U.S. Department of Energy, has been deeply engaged in energy storage research, aiming to support future energy transitions and sustainable development. To further advance autonomous materials discovery, UC Berkeley's Ceder team developed A-Lab, an innovative synthesis system that integrates robotics and artificial intelligence to automate material formulation and synthesis. Leveraging GNoME's predictions, A-Lab conducted experimental validation, forming a collaborative system for materials discovery. In just 17 days, A-Lab attempted to synthesize 58 target compounds, successfully obtaining 41 previously unknown materials, demonstrating the power of AI-driven materials discovery.

(4) Discovery of Quantum, Superconducting, and Other Materials [37–39]

With the help of the Atomly materials science database and first-principles calculations, a team from the Institute of Physics of the Chinese Academy of Sciences conducted a comprehensive search for superconducting materials. They systematically screened through 180,000 inorganic materials and discovered several superconducting materials with MgB2-like structures. Theoretical predictions indicate that

CaB2 exhibits a high superconducting transition temperature (Tc = 9.4–28.6 K). Additionally, several new superconducting materials were identified, such as $SrGa_2$ and $BaGa_2$. The collaborators experimentally confirmed the superconductivity in BaGa2, thus realizing an "end-to-end" model from material search to experimental verification.

Based on the Atomly database, the institute team also systematically explored the thermodynamic stability and electronic structure characteristics of 1,386 "CsV_3Sb_5-like"cages and identified 25 thermodynamically stable novel materials. $CsTi_3Bi_5$ and $RuTi_3Bi_5$ in the predicted structures have been experimentally synthesized, confirming the validity of the method.

In addition, with the help of high-throughput calculations, the research team quickly searched the phase diagrams of compounds in the Lu-H-N system and found that there is no stable ternary phase within 10 GPa; the comparison revealed that the XRD of Dias et al. is a mixture of LuH_2 and LuN phases [40]. It provides theoretical evidence to clarify the false papers in superconducting materials.

(5) High-throughput Screening of Coating Materials for All-solid-state batteries [41]

All-solid-state batteries are a future trend, significantly enhancing the battery's safety performance and energy storage density, with the "solid-solid interface" being the core scientific challenge in such applications. To address the sulfide-lithium iron phosphate interface stability issue, the team from the Institute of Physics of the Chinese Academy of Sciences explored the lithium iron phosphate all-solid-state battery system using the Atomly database. The team evaluated five key dimensions: material electronic conductivity, thermodynamic stability, electrochemical window, interfacial chemical stability, and lithium ion conductivity. After screening through 54,005 specific materials in the Atomly database, they identified 41 promising coating materials. These materials not only effectively prevent direct contact between sulfide electrolyte and lithium iron phosphate but also maintain lithium ion transport performance, providing valuable guidance for the further optimization of all-solid-state batteries.

3.2 *Matchat Large Model Application*

To promote the innovative application of large-scale language models in materials science, the Institute of Physics of the Chinese Academy of Sciences and the Computer Network Information Center collaborated to carefully construct the MatChat inorganic materials synthesis path prediction model [42]. The research team used the LLaMA2-7B model as the pre-training model [31, 42], and the dataset was derived from 35,675 inorganic material solid-phase reaction synthesis processes extracted from more than 4 million papers, and 13,878 high-confidence synthesis path descriptions were obtained after further screening, de-emphasizing, and cleaning to serve as the training set, which was then integrated into the form of instruction question and answer after pre-processing the dataset (see Fig. 2). MatChat enables users to query the synthesis methods of specific materials and provides chemical expressions of material reactions, synthesis conditions, and other relevant information.

Currently, the model is already running online and providing services to researchers in the field of materials (see Fig. 3). The model is basically capable of generating and reasoning knowledge in the field of materials synthesis, and has been experimentally verified to have a performance that exceeds that of ChatGPT in predicting the synthesis of complex inorganic materials.

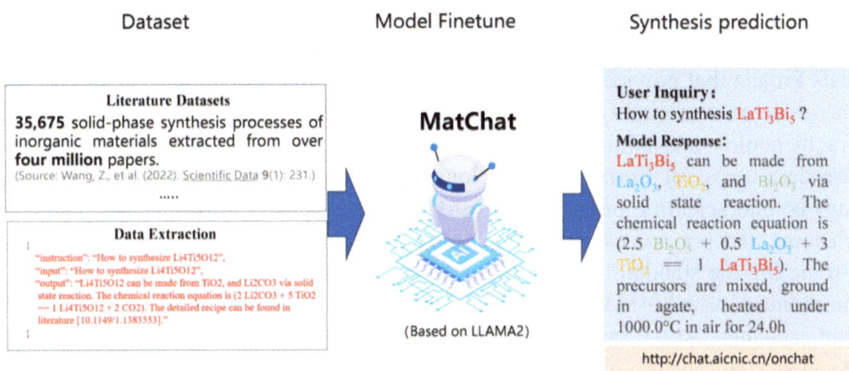

Fig. 2 Schematic diagram of Matchat's large model work

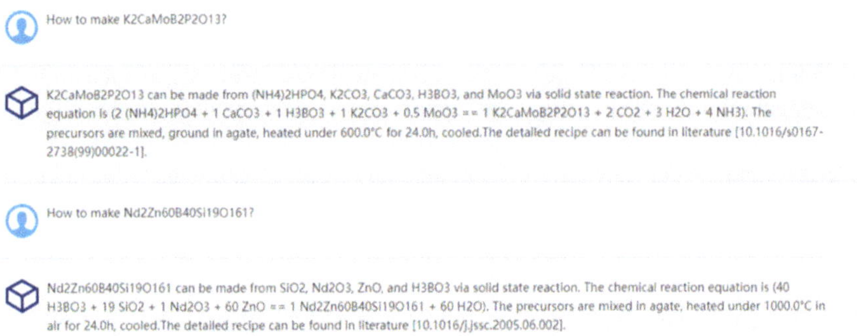

Fig. 3 Matchat large model usage interface

4 Summary and Outlook

In the field of materials science, the process of materials discovery and optimized design is becoming faster and more precise with the continuous development of big data and artificial intelligence technologies. However, the popularization and application of these advanced technologies still face many challenges. In order to further advance materials science, in-depth research and development of new technologies in the following areas are still needed to bring about more innovative material discoveries and applications.

(1) Continuing to Promote High-quality Databases in Materials Science

Although there are high-quality computational databases such as Atomly and Materials Project that provide a wealth of materials data, many materials databases have inconsistent data quality and lack uniform standards, leading to significant data fragmentation and siloing. This phenomenon makes it difficult to compare and use data across platforms, and the long update cycle of experimental data often lags behind scientific progress. Therefore, there is a need to further improve data cleaning and pre-processing technologies and develop unified industry standards. Promoting the standardization of data formats and the establishment of validation mechanisms, combined with automated updating mechanisms, will help data sharing and exchange. At the same time, open platforms should be encouraged to promote data contribution and sharing by more researchers.

(2) Building Efficient Materials Artificial Intelligence Prediction Models

Currently, techniques such as deep learning and graph neural networks have been widely applied to materials science, but they still suffer from the "black box" problem and lack sufficient interpretability. In addition, building efficient material prediction models usually requires a large amount of computational resources, and the limitation of computational power restricts the application of such models. In the future, AI models should be developed in the direction of being more transparent and easy to interpret, and the dependence on computational resources should be reduced through cutting-edge technologies such as optimization algorithms, distributed computing, and quantum computing, so as to enhance the usefulness and generalizability of the models.

(3) Deepening Automation Labs and AI Agent Applications

The automated materials science laboratories of the future will not be limited to the automation of individual devices, but will aim to achieve integrated automation across the entire experimental process. From material synthesis, processing, characterization, performance testing, to data analysis, full automation and integration will be realized. By combining advanced AI-driven models (such as MatChat and GPTFF) with high-throughput experimental equipment, a closed loop of data generation, analysis, and model validation can be established. Furthermore, as AI technology continues to advance, the application of AI Agents in materials science shows great promise, especially in materials design, performance prediction, experimental automation, and manufacturing process optimization. In traditional materials research and development, researchers often rely on experience and manual trial-and-error methods, which are inefficient and prone to bias. AI Agents, through machine learning and deep learning, can extract underlying patterns from vast amounts of existing materials data and predict which materials may exhibit specific physical, chemical, or mechanical properties. In the future, AI Agents are bound to play a crucial role in efficient material screening, automated material design, and reverse material design.

Acknowledgements Special Application Demonstration Project on Network Security and Informatization of Chinese Academy of Sciences(CAS-WX2022SF-0101).

References

1. JAIN A, ONG S P, HAUTIER G, et al. Commentary: The Materials Project: A materials genome approach to accelerating materials innovation[J]. APL Materials, 2013, 1(1): 011002.
2. MERCHANT A, BATZNER S, SCHOENHOLZ S S, et al. Scaling deep learning for materials discovery[J]. Nature, 2023, 624(7990): 80–85.
3. YANG H, HU C, ZHOU Y, et al. MatterSim: A Deep Learning Atomistic Model Across Elements, Temperatures and Pressures[J]. arXiv preprint, 2024, arXiv:2405.04967.
4. ZENI C, PINSLER R, ZÜGNER D, et al. MatterGen: a generative model for inorganic materials design[J]. arXiv preprint, 2023, arXiv:2312.03687.
5. BARROSO-LUQUE L, SHUAIBI M, FU X, et al. Open Materials 2024 (OMat24) Inorganic Materials Dataset and Models[J]. arXiv preprint, 2024, arXiv:2410.12771.
6. 刘淼,孟胜. Atomly.net 数据平台及其在无机化学中的应用[J]. SCIENTIA SINICA Chimica, 2022, 53(1): 19–25.
7. LU T, WANG Y, CAI G, et al. Synthesizability of transition-metal dichalcogenides: a systematic first-principles evaluation[J]. Materials Futures, 2023, 2(1): 015001.
8. XIE F, LU T, MENG S, et al. GPTFF: A high-accuracy out-of-the-box universal AI force field for arbitrary inorganic materials[J]. Science Bulletin, 2024, 69(22): 3525–3532.
9. DE PABLO J J, JACKSON N E, WEBB M A, et al. New frontiers for the materials genome initiative[J]. npj Computational Materials, 2019, 5(1): 41.
10. CURTAROLO S, SETYAWAN W, HART G L W, et al. AFLOW: An automatic framework for high-throughput materials discovery[J]. Computational Materials Science, 2012, 58: 218–226.
11. KIRKLIN S, SAAL J E, MEREDIG B, et al. The Open Quantum Materials Database (OQMD): assessing the accuracy of DFT formation energies[J]. npj Computational Materials, 2015, 1(1): 15010.
12. L. Chengzan, H. Yanfei, L. Jianhui and Z. Lili. ScienceDB: A Public Multidisciplinary Research Data Repository for eScience[C]. Auckland, New Zealand, IEEE, 2017.
13. TANIFUJI M, MATSUDA A, YOSHIKAWA H. Materials data platform-a fair system for data-driven materials science[C]//2019 8th International congress on advanced applied informatics (IIAI-AAI). IEEE, 2019: 1021–1022.
14. SCHEIDGEN M, HIMANEN L, LADINES A N, et al. NOMAD: A distributed web-based platform for managing materials science research data[J]. Journal of Open Source Software, 2023, 8(90): 5388.
15. PIZZI G, CEPELLOTTI A, SABATINI R, et al. AiiDA: automated interactive infrastructure and database for computational science[J]. Computational Materials Science, 2016, 111: 218–230.

16. ZAGORAC D, MÜLLER H, RUehL S, et al. Recent developments in the Inorganic Crystal Structure Database: theoretical crystal structure data and related features[J]. Journal of Applied Crystallography. 2019, 52(5): 918–925.
17. CHANUSSOT L, DAS A, GOYAL S, et al. Open Catalyst 2020 (OC20) Dataset and Community Challenges[J]. ACS Catalysis, 2021, 11(10): 6059–6072.
18. XIE T, GROSSMAN J C. Crystal Graph Convolutional Neural Networks for an Accurate and Interpretable Prediction of Material Properties[J]. Physical Review Letters, 2018, 120(14): 145301.
19. WANG A Y T, KAUWE S K, MURDOCK R J, et al. Compositionally restricted attention-based network for materials property predictions[J]. npj Computational Materials, 2021, 7(1): 77.
20. CHEN C, YE W, ZUO Y, et al. Graph Networks as a Universal Machine Learning Framework for Molecules and Crystals[J]. Chemistry of Materials, 2019, 31(9): 3564–3572.
21. YUAN Y, CHEN Z, FENG T, et al. Tripartite interaction representation algorithm for crystal graph neural networks[J]. Scientific Reports, 2024, 14(1): 24881.
22. LIANG Y, CHEN M, WANG Y, et al. A universal model for accurately predicting the formation energy of inorganic compounds[J]. Science China Materials, 2023, 66(1): 343–351.
23. TIAN S, JIANG X, WANG W, et al. Steel design based on a large language model[J]. Acta Materialia, 2024, 285: 120663.
24. WANG H, ZHANG L, HAN J, et al. DeePMD-kit: A deep learning package for many-body potential energy representation and molecular dynamics[J]. Computer Physics Communications, 2018, 228: 178–184.
25. CHEN C, ONG S P. A universal graph deep learning interatomic potential for the periodic table[J]. Nature Computational Science, 2022, 2(11): 718–728.
26. DENG B, ZHONG P, JUN K, et al. CHGNet as a pretrained universal neural network potential for charge-informed atomistic modelling[J]. Nature Machine Intelligence, 2023, 5(9): 1031–1041.
27. BATATIA I, KOVÁCS D P, SIMM G N C, et al. MACE: Higher Order Equivariant Message Passing Neural Networks for Fast and Accurate Force Fields[J]. Advances in Neural Information Processing Systems, 2022, 35:11423–11436
28. PARK Y, KIM J, HWANG S, et al. Scalable Parallel Algorithm for Graph Neural Network Interatomic Potentials in Molecular Dynamics Simulations[J]. Journal of Chemical Theory and Computation, 2024, 20(11): 4857–4868.
29. LI H, WANG Z, ZOU N, et al. Deep-learning density functional theory Hamiltonian for efficient ab initio electronic-structure calculation[J]. Nature Computational Science, 2022, 2(6): 367–377.
30. ZHONG Y, YU H, SU M, et al. Transferable equivariant graph neural networks for the Hamiltonians of molecules and solids[J]. npj Computational Materials, 2023, 9(1): 182.
31. TOUVRON H, MARTIN L, STONE K, et al. Llama 2: Open Foundation and Fine-Tuned Chat Models[J]. arXiv preprint, 2023, arXiv:2307.09288.
32. ZHAO H, CHEN W, HUANG H, et al. A robotic platform for the synthesis of colloidal nanocrystals[J]. Nature synthesis, 2023, 2(6): 505–514.
33. YAO T, WANG J, WAN M, et al. VenusAI: An artificial intelligence platform for scientific discovery on supercomputers[J]. Journal of Systems Architecture, 2022, 128: 102550.
34. SUN W, BARTEL C J, ARCA E, et al. A map of the inorganic ternary metal nitrides[J]. Nature Materials, 2019, 18(7): 732–739.
35. LIU M, JAIN A, RONG Z, et al. Evaluation of sulfur spinel compounds for multivalent battery cathode applications[J]. Energy & Environmental Science, 2016, 9(10): 3201–3209.
36. SZYMANSKI N J, RENDY B, FEI Y, et al. An autonomous laboratory for the accelerated synthesis of novel materials[J]. Nature, 2023, 624:86–91.
37. YU Z, BO T, LIU B, et al. Superconductive materials with MgB_2-like structures from data-driven screening[J]. Phys. Rev. B, 2022, 105(21): 214517.
38. JIANG Y, YU Z, WANG Y, et al. Screening Promising CsV_3Sb_5-Like Kagome Materials from Systematic First-Principles Evaluation[J]. Chinese Physics Letters, 2022, 39(4): 047402.

39. XIE F, LU T, YU Z, et al. Lu–H–N Phase Diagram from First-Principles Calculations[J]. Chinese Physics Letters, 2023, 40(5): 057401.
40. DASENBROCK-GAMMON N, SNIDER E, MCBRIDE R, et al. RETRACTED ARTICLE: Evidence of near-ambient superconductivity in a N-doped lutetium hydride[J]. Nature, 2023, 615(7951): 244–250.
41. LU T, MENG S, LIU M. Electrochemically and Chemically Stable Electrolyte-Electrode Interfaces for Lithium Iron Phosphate All-Solid-State Batteries with Sulfide Electrolytes[J]. Journal of Materials Chemistry A, 2024, 12(7): 3954–3966.
42. CHEN Z Y, XIE F K, WAN M, et al. MatChat: A large language model and application service platform for materials science[J]. Chinese Physics B, 2023, 32(11): 118104.

Miao Liu is a researcher and doctoral supervisor at the Institute of Physics of the Chinese Academy of Sciences and Songshan Lake Materials Laboratory. He is committed to the development of “Materials Big Data + Artificial Intelligence” research, using high-throughput first principle computation to build the industry’s top database for the field, and bring massive scientific data into the hands of researchers. His Atomly.net materials science database is one of the best databases in the world in terms of data quality and quantity.

Zongguo Wang Ph.D., is an Associate Researcher and Deputy Director of the Applied Software Development Laboratory in the Artificial Intelligence Technology and Application Development Department at the Computer Network Information Center, Chinese Academy of Sciences. She is also a member of the Chinese Academy of Sciences Youth Innovation Promotion Association and a representative of the Chinese delegation at the 8th BRICS Young Scientist Forum. Her research focuses on the application of artificial intelligence in various scientific disciplines. Dr. Wang has led more than ten research projects at the national, academy, and regional levels. In the field of materials science, she developed MatChat, a large language model for inorganic material synthesis. Additionally, her research team designed ANT-MOC, a large-scale neutron transport equation solver, which was nominated for the Best Paper Award at the 35th International Conference for High-Performance Computing, Networking, Storage, and Analysis (SC23), a prestigious CCF Class-A conference.

Yangang Wang Ph.D., is a Research Professor and Director of the Artificial Intelligence Technology and Application Development Department at the Computer Network Information Center, Chinese Academy of Sciences. His research focuses on AI computing and data service platform development, as well as AI-driven scientific discovery applications and parallel computing software. Dr. Wang has published over 80 research papers in leading international conferences and journals, including PPoPP and SC. He holds more than 30 patents and has authored three academic monographs. He has also led several national research projects, including the National Key Research and Development Program, the Chinese Academy of Sciences Strategic Priority Research Program (Category B), and the Chinese Academy of Sciences Informatization Special Project.

Sheng Meng is a research fellow at the Institute of Physics, Chinese Academy of Sciences, and director of the State Key Laboratory of Surface Physics. His research focuses on excited-state quantum dynamics, the microscopic mechanisms of energy conversion and storage, and surface quantum effects. He has developed novel first-principles computational methods that incorporate excited-state and nuclear quantum effects, along with related algorithms and software. His work has led to significant achievements across multiple scientific fields, with over 200 papers published in leading international journals such as Nature and Physical Review Letters. He is a recipient of the National Science Fund for Distinguished Young Scholars and a Fellow of the American Physical Society.

Tenglong Lu is a postdoctoral researcher at the Institute of Physics, Chinese Academy of Sciences, where he also received his Ph.D. in 2024. His research focuses on data-driven studies of energy and functional materials, as well as the application of machine learning in materials science. He earned his bachelor's degree from the Department of Physics at Beijing Normal University. He is a core developer of Atomly.net, a major inorganic materials database known for its scale and data quality in the field.

Meng Wan Engineer at the Computer Network Information Center, Chinese Academy of Sciences. His research interests include heterogeneous resource scheduling, AI cloud platforms, and time series forecasting. He received his B.Eng. degree from the School of Computer Science, Beijing University of Posts and Telecommunications, and his M.Sc. degree from the School of Electronics and Computer Science (ECS), University of Southampton, UK. He is also a core developer of MatChat, a foundation model for materials science.

Ziyi Chen Master‘s student at the Computer Network Information Center, Chinese Academy of Sciences, Bachelor‘s degree. His primary research focuses on generative model-based algorithms for inverse materials design. He graduated from the Software College at Dalian University of Technology and is one of the core developers of the MatChat materials large model.

Yang Yuan Master‘s student at the Computer Network Information Center, Chinese Academy of Sciences, Bachelor‘s degree. His primary research focuses on AI application technologies for scientific discovery. He graduated from the School of Computer Science and Technology at Tiangong University.

Exploration and Practice of ZiDongTaiChu Multimodal Large Model

Jinqiao Wang and Beiying Yang

Abstract In the new era of AI-driven technological innovation, large language models (LLMs) are leading a revolutionary transformation in scientific research paradigms. As a significant initiative in China's LLM sector, the development trajectory of the ZiDongTaiChu multimodal large model demonstrates substantial progress in domestic innovation within China's artificial intelligence domain. This paper presents a comprehensive overview of the technological evolution and innovative achievements of the ZiDongTaiChu multimodal large model. It elaborates on the breakthrough advancements in cross-modal self-supervised learning, adaptation of domestically developed operators, and the evolution of multimodal technologies, with particular emphasis on the technological progression from tri-modal to comprehensive modal capabilities. The study provides an in-depth analysis of representative applications across the industrial manufacturing, healthcare, and education sectors. Finally, it examines the collaborative innovation ecosystem model centered around the ZiDongTaiChu multimodal large model, elucidating the practical implementation pathway for intelligent transformation of traditional industries through large model technology.

Keywords Multimodal large model · Self-supervised learning · Cross-modal learning

Introduction

In the new era of technological innovation driven by artificial intelligence (AI), large-scale model technology is revolutionizing research paradigms. President Xi

J. Wang (✉) · B. Yang
Foundation Model Research Center, Institute of Automation, Chinese Academy of Sciences, Beijing, China
e-mail: jqwang@nlpr.ia.ac.cn

B. Yang
e-mail: beiyingyang15@gmail.com

J. Wang
School of Artificial Intelligence, University of Chinese Academy of Sciences, Beijing, China

China's e-Science Blue Book 2025, https://doi.org/10.1007/978-981-96-8252-2_15

Jinping has emphasized that accelerating the development of next-generation AI is a pivotal strategic initiative for securing global technological leadership, serving as a vital resource for advancing China's scientific progress, industrial optimization, and productivity transformation. Globally, competition in AI has intensified, with major nations elevating AI to a national strategic priority to vie for dominance in leadership and standard-setting within the field.

As a cutting-edge research institution focused on cross-modal intelligence, the Institute of Automation, Chinese Academy of Sciences (CASIA) has leveraged more than two decades of AI expertise to pursue systematic innovation in multimodal large-scale models. The ZiDongTaiChu Large Model Center team pioneered a novel multi-task multimodal self-supervised learning framework, achieving deep integration and cross-modal alignment across token-level, modality-level, and sample-level dimensions. Architecturally, ZiDongTaiChu Multimodal Large Model has evolved from its 1.0 multimodal model to the 2.0 omnimodal model and further to the 3.0 multimodal agent-based model. This progression introduced a unified native encoding mechanism, enabling natural emergence of multimodal capabilities without explicit modality fusion. At the data level, the team developed a semantic unification method for weakly correlated multimodal data, supporting hybrid training across three or arbitrary pairwise modalities. By adopting a behavior-aligned learning paradigm, the framework transcends traditional single-task cross-modal approaches (e.g., speech recognition or image captioning), enabling simultaneous cross-modal task generalization and hybrid modality comprehension. These innovations not only demonstrate China's pioneering capabilities in AI but also represent critical advancements toward general artificial intelligence.

This paper systematically examines the technological evolution and breakthroughs of the ZiDongTaiChu, analyzing its contributions to foundational theory, architectural innovation, and industrial applications. Technically, it details transformative achievements in cross-modal self-supervised learning, domestically optimized operator adaptation, and multimodal architectural evolution, emphasizing leaps from trimodal to omnimodal and agent-based paradigms. Practically, it evaluates case studies in industrial manufacturing, healthcare, and education, while exploring the collaborative ecosystem centered on ZiDongTaiChu. The study further elucidates actionable pathways for large-scale models to drive intelligent transformation across traditional industries.

1 Development Status of Multimodal Artificial Intelligence

1.1 Multimodal Artificial Intelligence Technologies

In recent years, next-generation artificial intelligence (AI) has become a focal point of global competition, with significant strategic implications for securing technological leadership in the future. Against this backdrop, multimodal pre-trained large models, as a cutting-edge exploration in AI, have emerged as a key area of focus for both academia and industry. As research advances, the field of AI is undergoing a critical transition from single-modal supervised learning to multimodal unsupervised learning.

Currently, over 90% of online data consists of images and videos, which contain vast knowledge resources. Humans inherently employ cross-modal interactions for information acquisition, environmental perception, knowledge learning, and expression. However, existing single-modal pre-trained models face substantial challenges in processing large-scale datasets and enhancing model performance. These models typically handle only unimodal information (e.g., text, audio, or visual data) from internet sources, failing to fully leverage or learn from multimodal data that integrates text, speech, images, and videos. Furthermore, human cognition and communication are inherently multimodal, necessitating AI systems capable of processing and interpreting multimodal information.

To develop more generalizable AI models, pre-trained models are evolving from unimodal to multimodal frameworks. This requires models to integrate information from diverse modalities—such as text, speech, images, and videos—and to investigate intermodal correlations and cross-modal information transformation. The advancement of multimodal AI has thus become a pivotal driver of innovation in AI technologies.

Multimodal pre-trained large models share architectural similarities with GPT (Generative Pre-Trained Transformer) [1] and BERT (Bidirectional Encoder Representations from Transformers) [2], as they are also based on Transformer deep learning architectures with self-attention mechanisms. A defining feature of these models is their capacity to process multiple modalities (e.g., text, speech, images, videos) as simultaneous inputs, unlike their unimodal counterparts that handle only one modality (e.g., text, speech, or images). Multimodal large models typically refer to deep neural networks with parameter sizes exceeding hundreds of millions and inputs incorporating two or more modalities. A key scientific challenge lies in designing neural architectures with robust unsupervised learning and cross-domain knowledge transfer capabilities, enabling performance improvements across diverse tasks under a unified framework with minimal annotation requirements.

A promising approach involves leveraging cross-modal semantic associations to enhance multimodal fusion comprehension, cross-modal transformation, and generative capabilities. Multimodal pre-trained models are typically trained on large-scale, unlabeled datasets sourced from internet-based multimodal content (e.g., web pages, videos), without requiring manual annotation, thereby achieving good scalability and generalizability. Without fine-tuning or with minimal fine-tuning, these models can directly address diverse multimodal data processing tasks, such as automatic video captioning, audio-visual synchronization, and generating images or video clips from textual or auditory inputs.

1.2 The Technical Status of Multimodal Large Models at Home and Abroad

1.2.1 Transformer Encoder-Based Multimodal Understanding Models

The emergence of BERT [2], a Transformer encoder-based architecture, has demonstrated the exceptional performance of large-scale self-supervised pre-training models. Beyond their widespread adoption in natural language processing, these models have progressively been extended to multimodal domains. Transformer encoder-based multimodal understanding models primarily employ the encoder component of Transformer architectures to learn semantic representations and intermodal relationships. Current research methodologies can be categorized into single-stream and dual-stream approaches. Single-stream methods jointly process visual and textual modalities through a unified encoder. Representative works include VL-BERT (Visual-Linguistic BERT) [3], VideoBERT [4], and UNITER (UNiversal Image-TExt Representation) [5]. VL-BERT introduces a multimodal pre-training framework that integrates image captions and object regions as BERT inputs, employing random masking of textual tokens and image regions to enhance model robustness. VideoBERT pioneers video-text pre-training by fusing textual information with video sequence inputs. UNITER advances the field through multi-task pre-training, incorporating an image-text matching module to strengthen cross-modal alignment. Dual-stream approaches employ separate encoders for individual modality feature extraction followed by cross-modal interaction. Notable implementations include ViLBERT (Visual-Language BERT) [6], LXMERT (Learning Cross-Modality Encoder Representations from Transformers) [7], and CLIP (Contrastive Language–Image Pre-training) [8]. ViLBERT features an alternating architecture of single-modality encoding and cross-modal interaction using Transformer blocks, trained through masked prediction and image-text matching tasks. LXMERT extends this paradigm by incorporating visual question answering as a pre-training objective, achieving superior performance on downstream VQA tasks. CLIP adopts a dual-tower structure with ResNet/ViT-based image encoding and BERT-based text encoding, leveraging contrastive loss for semantic alignment. Pre-trained on hundreds of millions of image-text pairs, CLIP demonstrates remarkable zero-shot transfer capabilities in retrieval tasks.

1.2.2 Transformer Decoder-Based Multimodal Generation Models

The evolution of GPT series models [9, 10] has drawn significant attention to their powerful text generation capabilities using Transformer decoder architectures. GPT-3 demonstrated that language models can guide neural networks in performing diverse text generation tasks. Inspired by this, researchers have explored large-scale pre-training for multimodal generation. Notably, OpenAI's DALL-E (Digital Artist's Lifelike Likenesses & Everything) [11] employs a 12-billion parameter GPT-3 variant for text-to-image synthesis, dubbed as the visual counterpart of GPT. DALL-E exhibits capabilities in manipulating object arrangements within generated images and creating novel conceptual hybrids (e.g., avocado-shaped armchairs). While offering partial controllability over object attributes and positioning, its success rate remains sensitive to linguistic formulations and degrades significantly with increased object complexity. Concurrently, Tsinghua University and BAAI proposed CogView [12], adopting a similar VQVAE + GPT architecture. This model achieves superior FID and IS metrics compared to DALL-E while enabling flexible task adaptation through fine-tuning, including super-resolution and style transfer.

1.2.3 Transformer-Based Encoder-Decoder Multimodal Models

Transformer-based encoder-decoder multimodal models enhance discriminative understanding by incorporating decoder architectures for generative pre-training, enabling superior learning of cross-modal correlations. Representative implementations include VL-T5 [13], E2E-VLP [14], M6 [15]. VL-T5 adopts an encoder-decoder framework to unify diverse multimodal tasks as text generation problems. E2E-VLP employs conventional masked language modeling and image-text matching as encoder-side pre-training objectives, while integrating object detection with semantic captioning as decoder-side tasks. The M6 model, developed by Alibaba DAMO Academy, shares parameters between encoder and decoder, utilizing attention masking mechanisms to regulate task-specific pre-training.

From a technological evolution perspective, multimodal AI is undergoing profound transformation. Traditional isolated encoding architectures prove inadequate for extracting deep semantic inter-modal relationships [8, 11], making the establishment of unified native encoding mechanisms and implementation of end-to-end multimodal data learning critical pathways to overcome technical bottlenecks. Concurrently, self-supervised learning methodologies [5, 7] have gained prominence to address data annotation scarcity, enabling models to learn universal feature representations from unlabeled data through multi-task pre-training objectives. Building on these foundations, multimodal generative capabilities have expanded beyond text synthesis to encompass "text-image" and "speech-text" conversions [5, 7]. The adoption of unified sequence modeling frameworks and multimodal embedding spaces has transcended traditional single-task cross-modal learning limitations. As model parameters continue to scale exponentially, developing efficient general-purpose multimodal architectures presents key technical challenges, driving intensive research into unified frameworks and model optimization. These evolutionary trajectories are collectively steering multimodal AI toward greater efficiency and universal applicability.

2 Technological Innovations and Implementation of the ZiDongTaiChu Large Model

2.1 Cross-Modal Self-supervised Learning

In artificial intelligence, cross-modal understanding and generation capabilities have become core competencies for large models. Self-supervised learning represents an emerging deep learning technique that eliminates reliance on manual data annotation by deriving supervisory signals directly from data itself. This approach enables automatic information completion and learns universal feature representations for downstream tasks, thereby addressing limitations of traditional supervised learning such as high data dependency, labor-intensive labeling, and poor generalization.

The ZiDongTaiChu model has made significant strides in cross-modal self-supervised learning by proposing a systematic technical framework that achieves deep semantic understanding and interactive generation across modalities. This groundbreaking advancement not only demonstrates China's robust R&D capabilities in AI but also marks a critical breakthrough in core AI technologies. Compared to existing models like CLIP [8] and DALL-E [11], The ZiDongTaiChu model pioneers a novel multi-task multimodal self-supervised learning architecture [16], as shown in Fig. 1, establishing a comprehensive theoretical system through three-level deep integration at token-level, modality-level, and sample-level hierarchies.

At the data processing level, ZiDongTaiChu has developed a semantic unification methodology for weakly correlated multimodal data. This approach enables hybrid training with three or arbitrary two-modal data combinations, significantly reducing data collection and cleaning costs. The tri-modal pre-training framework incorporates dedicated text, visual, and audio encoders to achieve unified representation of heterogeneous data. In application scope, ZiDongTaiChu innovatively realizes unified representation and inter-modal generation capabilities across image, text, and speech modalities. These capabilities span diverse understanding and generation tasks including cross-modal retrieval, multimodal classification, speech recognition,

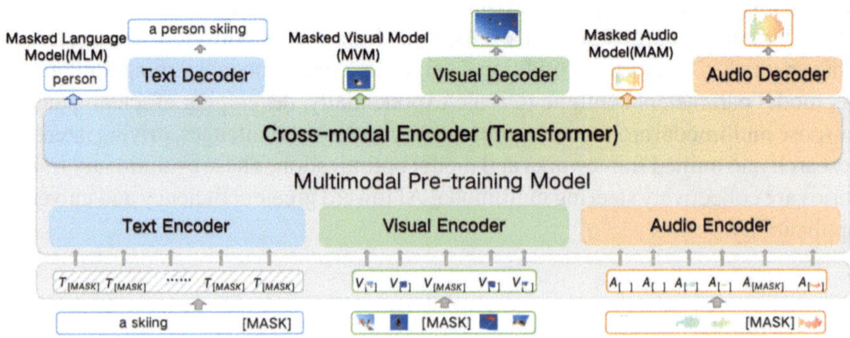

Fig. 1 Schematic diagram of the ZiDongTaiChu model

text generation, image synthesis, and audio production. Through its cross-modal Transformer architecture, the model simultaneously processes multi-scale features at sample-level, modality-level, and token-level hierarchies, supporting sophisticated multimodal operations. To enhance synergistic capabilities, ZiDongTaiChu implements an efficient inter-modal coordination mechanism enabling deep interactions among text, visual, and auditory modalities. This framework supports complex cross-modal conversion tasks such as machine translation, text-to-image generation, speech blending, visual captioning, and image-to-audio synthesis. By integrating sample-level alignment learning with modality-level feature fusion, the model achieves more natural and accurate multimodal understanding and generation.

These technological breakthroughs establish ZidongTaiChu as the industry's first large-scale system realizing unified multimodal representation and generation, charting a strategic direction for next-generation AI development.

2.2 Localization-Optimized Operators and Model Adaptation

In advancing the autonomous development of artificial intelligence technologies, the ZiDongTaiChu multimodal large model has conducted systematic exploration in operator localization adaptation and training optimization. Through independent innovation and refinement, efficient deployment and training of trillion-parameter multimodal models on domestic computing platforms have been achieved.

In constructing the foundational software-hardware architecture, the research team established a fully localized stack through deep collaboration with Huawei's Ascend AI training processors and the MindSpore framework. This process involved completing localization adaptation and optimization for over 200 operators, laying a robust foundation for large-scale model training. By developing innovative ultra-large-scale cross-modal distributed training acceleration technologies that integrate data parallelism, model parallelism, and other hardware-software co-design acceleration mechanisms, training efficiency was significantly enhanced. These technological breakthroughs overcame hardware limitations in domestic platforms, reducing the training time for trillion-parameter models from four months to four weeks while maintaining model quality, thereby dramatically improving iteration efficiency. For training optimization, the team proposed a novel Periodic Momentum Decay LAMB (PMD-LAMB) optimizer [17], which mitigates the adverse effects of historical gradient lag and accelerates model convergence. This method innovatively scales single-batch detection to over 1,000 samples without compromising accuracy. Practical applications demonstrate remarkable efficiency improvements: MS COCO object detection training time was reduced to 12 min and instance segmentation to 17 min, achieving minute-level model iteration.

These technological advancements enable the ZiDongTaiChu multimodal large model to operate efficiently on domestic computing platforms, resolving critical challenges in large model localization while establishing technical foundations for scalable applications. These innovations mark significant progress in

China's autonomous development of large model technologies and hold substantial implications for promoting AI technology indigenization.

2.3 Evolution of Multimodal Large Model Technology

From a macro development perspective, artificial intelligence is transitioning from perceptual understanding to generative creation, with large-scale models emerging as the dominant paradigm for advancing general-purpose intelligence. Within this trajectory, the technological evolution of multimodal large models has exhibited a unique developmental pathway.

In terms of architectural design, early AI algorithms adopted a "one-scenario-one-model" approach, where distinct small models were tailored for specific tasks. These models, characterized by limited parameter scales (ranging from 1,000 to 20,000 parameters) and poor generalization capabilities, necessitated laborious processes—from scratch development, isolated optimization, iterative refinement, to occasional complete overhauls—during the 2008–2020 period.

In 2021, the team transitioned to collective research breakthroughs: ZiDongTaiChu1.0 completed the development of a billion-parameter model in January, successfully debugged a hundred-billion-parameter model in July, and made its global debut at the HC Conference in September. By June 2023, the full-modal large model ZiDongTaiChu2.0 achieved multimodal data processing capabilities covering text, speech, images, videos, 3D, and signal data, supporting multimodal input and output. This version successfully enabled the model to assimilate massive knowledge, breaking previous limitations of one-model-per-scenario and evolving into a shared-model-for-multiple-scenarios paradigm, significantly enhancing generalization capabilities. In November 2024, the multimodal agent-based large model ZiDongTaiChu3.0 introduced architectural innovations (as shown in Fig. 2). This version upgraded from modality-specific encoding to unified native encoding, innovatively incorporating a behavior-aligned multimodal learning paradigm that enabled natural emergence of multimodal capabilities. Compared to version 2.0, version 3.0 achieved a paradigm shift in intelligent agents: from tool-assisted processing to integrated memory, planning, and reasoning capabilities; and from basic perception/cognition to collaborative reasoning and problem-solving for complex multimodal tasks, closely approximating human-level comprehension and reasoning abilities. This architecture not only supports processing of audio, image, text, video, 3D, and signal inputs, but also delivers multimodal output capabilities including music generation, image generation, text generation, video generation, 3D generation, and signal generation. Concurrently, the team established the ZiDongTaiChu Large Model Open Service Platform, achieving significant progress in industrial implementation. Leveraging the Beijing-Wuhan computing power center infrastructure, the ZiDongTaiChu large model has been deployed at scale across over 10 industrial domains including automotive manufacturing quality inspection, media services,

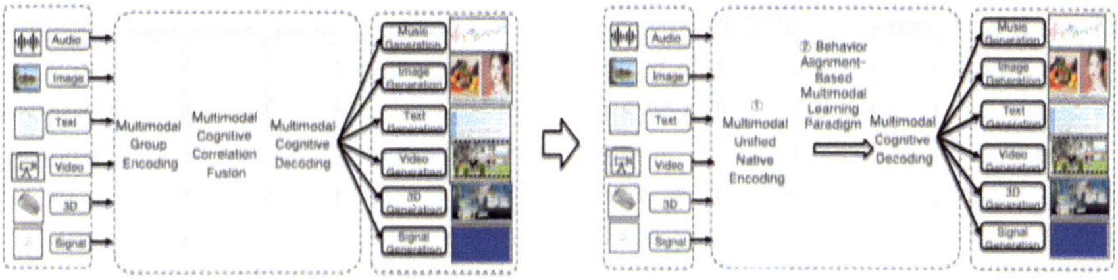

Fig. 2 ZiDongTaiChu large model base

cultural tourism, sign language education, textile production quality control, and medical device manufacturing.

This series of technological advancements not only reflects the rapid development of multimodal artificial intelligence but also charts a clear direction for the future of general intelligent technology. The practice of ZiDongTaiChu demonstrates that, through continuous innovation in model architecture and refinement of computing power infrastructure, multimodal large models are becoming a key driver in advancing artificial intelligence technologies.

3 Case Studies of ZiDongTaiChu's Industrial Applications

3.1 Intelligent Transformation in Industrial Manufacturing

Under the new paradigm of manufacturing transformation and upgrading, AI technologies are accelerating deep integration with traditional manufacturing industries. Leveraging its robust multimodal comprehension and generation capabilities, the Zidong Taichu Multimodal Large Model has played a pivotal role in driving digital transformation within the manufacturing sector. Through in-depth collaborations with multiple manufacturing enterprises, this model has achieved remarkable results in intelligent manufacturing, quality control, and production efficiency enhancement.

The "Intelligent Weld Seam Guidance and Recognition System Based on Industrial Large Models," jointly developed by the ZiDongTaiChu team and Huagong Tech, was selected as a representative case in the Ministry of Industry and Information Technology (MIIT)'s electronic information field. This milestone signifies substantial progress in implementing large model technologies in manufacturing. Targeting challenges such as low accuracy, instability, and inefficiency in welding processes within complex equipment production lines, the system innovatively adopts industrial large model-powered weld seam tracking technology. Replacing traditional computer vision models, the ZiDongTaiChu Multimodal Large Model serves as an intelligent infrastructure that enables high-precision, high-efficiency real-time weld seam tracking. It captures weld seam data in real time with millisecond-level responsiveness, unaffected by external interference, and automatically generates optimal paths through algorithms to guide robotic welding. Compared to conventional computer vision methods, the system achieves 50% higher recognition efficiency, superior precision, and enhanced stability. China's welding materials market is projected to exceed 40 billion CNY by 2025. As laser welding constitutes a core technology for intelligent manufacturing, intelligent weld seam recognition, tracking, and detection are critical. The deployment of this system significantly improves production efficiency across downstream industries including construction, shipbuilding, equipment manufacturing, automotive, and chemical sectors, demonstrating extensive applicability.

Through continuous industry-academia-research collaboration, ZiDongTaiChu facilitates deep integration between large model technologies and traditional manufacturing, providing robust technical support for intelligent industrial transformation.

3.2 Precision Diagnosis and Treatment in Healthcare

Amid rapid advancements in AI, intelligent transformation in healthcare has become inevitable. TheZiDongTaiChu Large Model exhibits unique advantages in this domain through its powerful multimodal reasoning capabilities, offering strong support for precision medicine. The model assists medical professionals in knowledge dissemination and intelligent diagnostics by parallel processing diverse medical data modalities—textual reports, medical images, and clinical speech—thereby delivering comprehensive knowledge support and diagnostic insights to address complex medical challenges.

In orthopedic instrument management, ZiDongTaiChu collaborated with Jiuzhoutong Medical Equipment Group to develop the "Orthopedic Chang'e" Intelligent System. Utilizing advanced multimodal recognition, this system achieves intelligent identification and full-process tracking of complex implants and surgical tools, covering preoperative planning, intraoperative logistics, and postoperative management. Post-implementation, the system reduced surgical order processing time from 30 min to under five minutes, resolving efficiency bottlenecks in orthopedic services and demonstrating AI's potential in addressing healthcare pain points.

For surgical assistance, Peking Union Medical College Hospital developed an Intracranial Surgery Multimodal Intelligent Assistant based on the ZiDongTaiChu Large Model. This system innovatively integrates visual and tactile data with real-time voice interaction to provide surgical teams with precise guidance. It effectively warns of potential neurovascular injury risks and delivers phase-aware decision support, significantly enhancing surgical safety. This breakthrough application has received high acclaim from stakeholders including Hong Kong Chief Executive John Lee, underscoring AI's value in improving medical safety.

ZiDongTaiChu's healthcare applications exhibit promising prospects. Through deep partnerships with medical institutions, the model is becoming a cornerstone technology for intelligent healthcare transformation.

3.3 Educational Large Models and Personalized Learning

As AI evolves rapidly, large models demonstrate immense potential in education. The ZiDongTaiChu Large Model utilizes parallel processing and fusion of multimodal data (speech, images, text) to achieve comprehensive educational knowledge integration. Combined with digital human technology, it enables automated lesson planning, virtual teaching, problem-solving, and personalized assignment grading, realizing holistic and accurate educational intelligence.

In the specific practice of education, the ZiDongTaiChu team has carried out in-depth cooperation with Central China Normal University to explore innovative ways to empower education with artificial intelligence based on the ZiDongTaiChu large model. The two parties have cooperated around core scenarios such as lesson plan generation, homework evaluation, and diversified tutoring, and reshaped the education and teaching process through big model technology to improve teaching efficiency and learning effects. In the teaching scenario, the ZiDongTaiChu large model can intelligently generate personalized lesson plan content based on specific teaching objectives and student characteristics; in the homework evaluation link, it can realize intelligent correction and accurate comments, effectively reducing the workload of teachers; at the same time, with the help of digital human virtual teacher technology, it provides students with all-weather personalized learning tutoring services.

To advance educational transformation, ZiDongTaiChu and ChuangErXin (Beijing) Tech co-developed an education-specific application. Through deep learning and semantic analysis of educational big data, the model identifies individual learning patterns and knowledge mastery levels to deliver personalized recommendations. This bidirectional empowerment between AI and education fosters pedagogical innovation while refining AI capabilities through real-world use cases.

Empirical results show that educational large models substantially enhance teaching quality and learning efficiency. Teachers benefit from reduced administrative burdens, enabling greater focus on pedagogical innovation, while students gain flexible, personalized learning support that improves engagement and outcomes. This AI-driven model is catalyzing profound shifts in educational paradigms.

ZiDongTaiChu's educational practices offer valuable insights for informatization and intelligent education development. By deeply integrating AI with educational workflows, the model advances educational equity, quality, and personalized learning—key milestones toward modernizing education.

3.4 Explorations in Emerging Domains

With the rapid development of artificial intelligence technology, the ZiDongTaiChu Multimodal Large Model has not only achieved breakthroughs in traditional strongholds such as industrial manufacturing, healthcare, and smart education, but also demonstrated powerful innovation potential in emerging fields including smart cities, legal services, and land resource management. Through continuous exploration and practice, the model is expanding the application boundaries of large-scale AI technologies, injecting new momentum into industrial digital transformation.

In the field of smart city construction, the ZiDongTaiChu team conducted systematic exploration in the Wuhan Airport New City area. The Wuhan Artificial Intelligence Research Institute developed a Smart City Integrated Management Platform based on the ZiDongTaiChu Large Model, integrating advanced technologies such as large models, digital humans, and multimodal interaction. By consolidating multidimensional data spanning urban governance, smart transportation, smart construction sites, industrial parks, environmental sanitation, bridge-tunnel-slope monitoring, and hazardous source surveillance, the platform established a novel "AI + Unified Governance" paradigm. This system replaces the traditional inefficient practice of developing scenario-specific models with a "Large Model + Targeted Fine-tuning" development framework, enabling "real-time detection, precise localization, and rapid resolution" of urban management challenges. This innovative practice provides replicable experience for smart city development.

In the field of legal service innovation, addressing industry pain points such as China's scarcity of legal resources and high service costs, the ZiDongTaiChu team developed a Legal Case Intelligent Analysis System powered by its large model. By training on legal domain-specific data—including case precedents, laws, regulations, and litigation materials—the model enhances its capability to "communicate in professional legal language." The system reduces case analysis time by 90% through intelligent decomposition and key event extraction. For example, inputting traffic accident case materials into the system enables completion of complex analyses such as liability determination and legal relationship assessment within 0.546 s, demonstrating AI's transformative potential in legal services.

In the domain of intelligent land resource management, focusing on the supervision of permanent basic farmland and high-standard farmland, the team developed an Early-Warning System for Illegal Land Occupation. By integrating multi-source data from video surveillance, satellite imagery, and UAV surveys, the system identifies 15 types of illegal land occupation scenarios, achieving a regulatory transition from "single-dimensional to comprehensive," "static to dynamic," and "planar to 3D monitoring," forming a closed-loop smart governance mechanism of "early detection, timely intervention, and prompt resolution."

In the field of supply chain services, through deep collaboration with Hubei Port Group, the team established a multimodal data-integrated AI foundation with strong generalization capabilities and operational synergy. Leveraging fine-tuning solutions based on the pre-trained ZiDongTaiChu Large Model, domain-specific document recognition algorithms can be rapidly developed with only 20+ annotated samples, enabling accurate identification of over 100 types of logistics certificates, effectively overcoming the high costs and inefficiencies of traditional training methods.

In the domain of government services, the team's AI-powered Government Document Proofreading Agent processes approximately 30 error types, including semantic misuse, grammatical errors, formatting deviations, and logical inconsistencies, providing robust technical support for enhancing administrative efficiency and document quality.

These innovations fully demonstrate the ZiDongTaiChu Multimodal Large Model's potential in emerging domains. As technology evolves and applications deepen, large models will play an increasingly vital role in driving digital transformation and improving governance efficacy. Moving forward, ZiDongTaiChu will continue to advance technological innovation, expand application scenarios, and inject new vitality into high-quality socioeconomic development.

4 Conclusions and Prospects

This paper comprehensively presents the exploration journey and practical achievements of the ZiDongTaiChu multimodal large model in artificial intelligence. As a significant exploration in large model development, ZiDongTaiChu demonstrates China's independent innovation capabilities in AI. Through deep integration of theoretical innovation and industrial practice, the model has achieved remarkable progress in cross-modal intelligence and technological applications. Technologically, ZiDongTaiChu has accomplished three critical breakthroughs: from trimodal to full-modality capabilities, and further to intelligent agent systems, injecting new momentum into AI innovation. The model demonstrates comprehensive capabilities in multimodal comprehension, generation, and correlation, supporting diverse tasks including multi-turn Q&A, text creation, image generation, 3D understanding, and signal analysis with enhanced cognitive and creative capacities. In industrial applications, ZiDongTaiChu's open service platform effectively addresses enterprise challenges in development efficiency, training costs, generalization capabilities, and practical implementation. For ecosystem development, it has established a Multimodal AI Industrial Consortium incorporating over a hundred member organizations, creating a comprehensive innovation ecosystem. Innovative products like the "Jiangcheng Luoshen" AI creation platform and digital human platform demonstrate its potential in cultural innovation and government services.

As multimodal large models reshape scientific innovation paradigms, the ZiDongTaiChu team commits to advancing AI development through localized resource integration. Technologically, it will enhance fundamental capabilities through cross-modal self-supervised learning and domestic operator optimization. Industrially, it will establish benchmark applications in smart manufacturing, healthcare, urban management, autonomous driving, cultural tourism, and education. Practical implementations will expand to automotive quality inspection, media services, sign language education, textile production, and medical device applications.

Looking ahead, ZiDongTaiChu will persist in technological innovation through full-stack localization while strengthening industry-academia collaboration. By cultivating an open ecosystem and promoting AI-entity economy integration, it aims to provide sustained momentum for regional economic development. Through synergistic innovation and industrial implementation, this initiative will drive the deep convergence of AI technologies with real-world applications.

References

1. RADFORD A, NARASIMHAN K. Improving Language Understanding by Generative Pre-Training[EB/OL]. OpenAI blog, 2018.
2. DEVLIN J, CHANG M W, LEE K, et al. BERT: Pre-training of deep bidirectional Transformers for language understanding[J]. arXiv:1810.04805, 2018.
3. SU W, ZHU X, CAO Y, et al. VL-BERT: Pre-training of Generic Visual-Linguistic Representations[J]. arXiv:1908.08530, 2019.
4. SUN C, MYERS A, VONDRICK C, et al. Videobert: A joint model for video and language representation learning[C]//Proceedings of the IEEE/CVF international conference on computer vision. 2019: 7464–7473.
5. CHEN Y C, LI L, YU L, et al. Uniter: Universal image-text representation learning[C]//Proceedings of the European Conference on Computer Vision (ECCV). Springer, Cham, 2020: 104–120.
6. LU J, BATRA D, PARIKH D, et al. Vilbert: Pretraining task-agnostic visiolinguistic representations for vision-and-language tasks[J]. arXiv:1908.02265, 2019.
7. TAN H, BANSAL M. LXMERT: Learning cross-modality encoder representations from Transformers[J]. arXiv:1908.07490, 2019.
8. RADFORD A, KIM J W, HALLACY C, et al. Learning transferable visual models from natural language supervision[C]//International conference on machine learning. PMLR, 2021: 8748–8763.
9. RADFORD A, WU J, AMODEI D, et al. Better language models and their implications[J]. OpenAI Blog, 2019, 1(2).
10. BROWN T B, MANN B, RYDERN, et al. Language models are few-shot learners[J]. arXiv: 2005.14165, 2020.
11. RAMESH A, PAVLOV M, GOH G, et al. Zero-shot text-to-image generation[J]. arXiv:2102.12092, 2021.
12. DING M, YANG Z, HONG W, et al. CogView: Mastering Text-to-Image Generation via Transformers[J]. arXiv:2105.13290, 2021.
13. CHO J, LEI J, TAN H, et al. Unifying vision-and-language tasks via text generation[J]. arXiv: 2102.02779, 2021.
14. XU H, YAN M, LI C, et al. E2E-VLP: End-to-End Vision-Language Pre-training Enhanced by Visual Learning[J]. arXiv:2106.01804, 2021.
15. LIN J, MEN R, YANG A, et al. M6: A chinese multimodal pretrainer[J]. arXiv:2103.00823, 2021.
16. OPT: Omni-Perception Pre-Trainer for Cross-Modal Understanding and Generation. Jing Liu, Xinxin Zhu, Fei Liu, Longteng Guo, Zijia Zhao, Mingzhen Sun, Hanqing Lu, Weining Wang, Jiajun Zhang, Shiyu Zhou and Jinqiao Wang. http://arxiv.org/abs/2107.00249
17. WANG T, ZHU Y, ZHAO C, et al. Large batch optimization for object detection: Training COCO in 12 minutes[C]//Proceedings of the 16th European Conference on Computer Vision (ECCV). Glasgow, UK: Springer International Publishing, 2020: 481–496.

Jinqiao Wang Deputy Chief Engineer at the Institute of Automation, Chinese Academy of Sciences, Executive Deputy Director of the ZidongTaichu Large Model Research Center, Researcher, Doctoral Supervisor, Dean of Wuhan Artificial Intelligence Research Institute, Professor at the School of Artificial Intelligence, University of Chinese Academy of Sciences, Secretary-General of the Multimodal Artificial Intelligence Industry Alliance, mainly engaged in research on multimodal large models, video analysis and retrieval, large-scale object recognition, etc. He has published over 300 papers in international authoritative journals and conferences, including more than 50 in international journals and over 220 in international conferences. He has participated in the formulation of 16 national, industry, and group standards, applied for more than 50 invention patents, and won 15 international algorithm competition championships. He has been honored with the first prize of Beijing Science and Technology Progress Award, the second prize of Beijing Natural Science Award, Wu Wenjun Artificial Intelligence Science and Technology Progress Award, Huawei Genius Youth Peach and Plum Award, China Invention Innovation Silver Award; he has also received titles such as New Era Chinese Economic Innovation Figure, CAS Keyan Master Team, Wuhan Model, Beijing High-Gather Leading Talent, Guangzhou Innovative Team Leading Talent, Shandong Taishan Leading Talent, and Optics Valley 3551 Leading Talent.

Beiying Yang Foundation Model Research Center, Institute of Automation, Chinese Academy of Sciences. Research interests: multimodal large models, video analysis and retrieval.

Research on the Application of Knowledge Graph Technology in the Discovery of Crop Gene Knowledge

Ruixue Zhao, Tan Sun, and Dandan Zhang

Abstract In the scientific research of crop breeding, traits are mainly controlled by key genes, and the cultivation of new crop varieties with the aggregation of a variety of elite traits has always been the direction of breeders. Therefore, the discovery of crop gene knowledge, that is, the knowledge related to crop genes, can provide new research ideas for the analysis of the molecular regulation mechanism of crop traits, and will effectively help the cultivation of new crop varieties. In the context of the scientific research paradigm of AI for Science, artificial intelligence technology has become the engine to help the innovation and development of crop breeding scientific research. As a knowledge organization technology in the field of artificial intelligence, knowledge graph technology can capture the complex correlation between genes and traits across species, and trigger the discovery of knowledge related to crop genes. Based on the in-depth analysis of the demand for crop gene knowledge discovery services, combined with the correlation characteristics of multi-dimensional crop breeding scientific data, selected the PubMed scientific literature database and eight field scientific databases as data acquisition sources, and uses multi-path knowledge extraction to construct a knowledge graph of cross-species and multi-dimensional scientific data fusion. Based on the knowledge graph, a crop gene knowledge discovery model was constructed, and the discovery of elite genes and gene function knowledge was realized. Finally, from the perspective of building a knowledge base supporting AI4S, building a technical system for intelligent computing, and building a knowledge service model driven by "data + AI", the empowerment of artificial intelligence technology in agricultural scientific research is prospected.

R. Zhao · D. Zhang (✉)
Agricultural Information Institute, Chinese Academy of Agricultural Sciences, Beijing, China
e-mail: zhangdandan01@caas.cn

The Chinese Academy of Agricultural Sciences, Beijing, China

T. Sun
Key Laboratory of Agricultural Big Data, Ministry of Agriculture and Rural Affairs, Beijing, China

Key Laboratory of Knowledge Mining and Knowledge Services in Agricultural Converging Publishing, National Press and Publication Administration, Beijing, China

China's e-Science Blue Book 2025, https://doi.org/10.1007/978-981-96-8252-2_16

Keywords Subject knowledge discovery · Crop gene knowledge · Knowledge graph · Link prediction

Introduction

As an emerging form of technology with deep integration of "AI" and "scientific research", intelligent scientific research (AI for Science) (AI4S) can break the cognitive limitations of scientists and has become an important driving force to improve scientific research efficiency and promote scientific discovery and scientific and technological innovation [1]. In the scientific research of crop breeding, the cultivation of new crop varieties with the aggregation of a variety of excellent traits has always been the direction of breeders. Crop breeding has gone through the development stage from domestication breeding, hybrid breeding, molecular marker-assisted breeding to computational breeding, and the traditional scientific research paradigm has been difficult to reveal the essential law of the molecular regulation mechanism of crop traits in the growing biological big data. The logic of scientists' acquisition of knowledge has also shifted from hypothesis-driven passive exploration to data-driven active knowledge discovery. However, the hypothesis of association between genes and traits often requires the analysis of multi-dimensional scientific data. The existing bulk segregant analysis (BSA) methods [2], quantitative trait locus (QTL) [3], and genome wide association study (GWAS) [4] cannot simultaneously correlate and analyze multi-dimensional scientific data between genes and traits across species. It hinders the migration and reuse of existing crop breeding knowledge and maximizes the value of crop breeding scientific data, and does not form a joint force for crop breeding scientific research, so that it is difficult to find crop genetic knowledge. As a knowledge organization technology in the field of artificial intelligence based on graph for data association and knowledge fusion of scientific data in multi-source heterogeneous fields, knowledge graph can scientifically represent the complex association of multi-dimensional crop breeding scientific data at the ontology layer, and the data layer can associate and integrate multi-source crop breeding knowledge, and can build a gene knowledge discovery model on the basis of it to capture the complex association between cross-species genes and traits, and trigger the discovery of crop gene-related subject knowledge. It can be seen that knowledge graph technology can break through the limitation that existing methods can only perform correlation analysis for single-dimensional scientific data, and provides a new opportunity for the discovery of crop gene knowledge.

In recent years, scholars at home and abroad have carried out a large number of research on the application of knowledge graph in the field of life sciences. Alshahrani et al. developed a semantic disease gene embeddings (SmuDGE) semantic disease gene embeddings (SmuDGE) using the feature learning embedding entity vector method, aiming to scientifically rank disease-related genes [5]. Based on the constructed disease knowledge graph (CADA), Peng et al. used a combination of network representation learning and link prediction to predict the priority of pathogenic genes [6]. Choi et al. proposed a knowledge graph-embedding model

(KGED) based on convolutional neural networks, which used the gene–gene relationships inferred by KGED to generate gene interaction networks for each cancer type to mine highly correlated cancer genes [7]. The Lausanne Institute in the United Kingdom is based on the domain knowledge graph KnetMiner to realize the knowledge system query of gene regulatory networks [8]. The French Agricultural Research Centre for International Development (CIRAD) has constructed an Agronomic Linked Data (AgroLD) that integrates multiple scientific datasets in the field of plants to support the development of scientific hypotheses related to plant complex traits [9]. It can be seen that knowledge graph, as a knowledge organization technology based on graph structure to associate and integrate multi-source and multi-type scientific data, can apply knowledge reasoning, link prediction and other methods to obtain the intrinsic relevance of knowledge to realize the discovery of new knowledge in the discipline. However, the existing crop domain knowledge graph lacks the hierarchical knowledge structure of multi-dimensional scientific data organization and association between cross-species genes and traits, which makes the cross-species multi-dimensional scientific data lack effective association and fusion. It hinders the migration and reuse of existing crop breeding knowledge, and does not form a synergy for crop breeding research, which brings challenges to the discovery of crop gene knowledge [10].

In order to meet the needs of crop gene knowledge discovery, this paper analyzes the demand for crop gene knowledge discovery services based on the framework of crop trait regulation gene knowledge discovery method [10], and elaborates the demand-oriented domain knowledge graph ontology layer, data layer and knowledge graph-driven gene knowledge discovery model construction method. In this paper, the application scenarios of knowledge discovery embedded in the scientific research process of crop breeding are selected, and the results of deep crop gene knowledge discovery are expounded, in order to provide reference for personnel in related fields.

1 Crop Gene Knowledge Discovery Service Demand Analysis

In the scientific research of crop breeding, the cultivation of new crop varieties with the aggregation of a variety of excellent traits has always been the direction of breeders. Crop traits are mainly controlled by key functional genes, however, there is often a trade-off effect between genes regulating different traits, which makes it difficult to combine these superior traits. Exploring the excellent genes that regulate multiple traits such as salt resistance and insect resistance at the same time and analyzing the molecular regulation mechanism of traits can balance this trade-off effect, which is the key to cultivating new crop varieties with high yield and high quality. Therefore, the discovery of crop gene knowledge, that is, the subject knowledge related to crop genes, including the efficient association discovery of cross-species gene-level knowledge, the discovery of excellent genes and the efficient prediction of cross-species gene functions, will effectively help the cultivation of new crop varieties. With the promotion of technology, it is difficult for traditional scientific research paradigms to reveal the essential laws of life in the growing biological big data. Facing the service needs of crop gene knowledge discovery, this paper deeply analyzes the knowledge discovery service needs of efficient cross-species gene-level knowledge association discovery, excellent gene discovery, and efficient prediction of cross-species gene function, aiming to provide researchers with crop gene knowledge discovery services embedded in the whole process of scientific research [10].

(1) Efficient association discovery of cross-species gene-level knowledge. Today, researchers need to access gene-centric, hierarchical disciplinary knowledge across multiple scientific databases. In addition, it is necessary to obtain relevant subject knowledge of homologous genes in other species through step-by-step search of homologous genes across species, which is a time-consuming and cumbersome process [10]. In order to meet the service needs of efficient association discovery of cross-species gene-level knowledge, it is necessary to realize the association retrieval of multi-dimensional scientific data between species, so as to improve the discovery efficiency of cross-species interdisciplinary knowledge. Specifically, using a certain gene ID as the search term, we can not only obtain the knowledge of the gene at the gene level, but also obtain the subject knowledge of different scientific data dimensions at the gene level, protein level and enrichment pathway level of genes homologous in other species through the association of genes between species, so as to realize the efficient association discovery of gene-centered hierarchical knowledge.
(2) Discovery of excellent genes for target traits. In the scientific research of crop breeding, breeders have been committed to cultivating new crop varieties with the aggregation of a variety of excellent traits. Therefore, the discovery of excellent genes that can regulate multiple traits such as disease resistance and salt resistance at the same time is crucial for the development of new varieties of high-yield and high-quality crops. In order to meet the service needs for the discovery of superior genes for target traits, it is necessary to discover functional genes that regulate multiple superior traits at the same time, so as to provide researchers with potential discovery of superior genes for target traits [10]. Specifically, according to the target traits, the search keywords (such as "drought resistance" and "plant height"), through the multi-dimensional scientific data association between cross-species genes and traits, the excellent genes that regulate drought resistance and plant height at the same time were found, and the found excellent genes were assigned specific weight scores to carry out the visual interpretation of the excellent gene recommendation results.
(3) Efficient prediction of cross-species gene function. The analysis of the molecular regulation mechanism of crop traits is a complex and dynamic process, involving the whole process of transcription, translation and modification. As a functional unit of regulatory traits, an in-depth understanding of the molecular

biological function of genes not only helps researchers to carry out scientific gene editing in a targeted manner, but also has important guiding significance for promoting the precise improvement of target traits in crop breeding. In order to meet the service demand for efficient prediction of gene function, it is necessary to predict and analyze the molecular biological function of target genes to provide researchers with efficient prediction of gene function across species [10]. Specifically, using the target gene ID as the search term, through the hierarchical knowledge organization system between cross-species genes and traits, it provides researchers with molecular biological functional information of existing known homologous genes in cross-species, and provides scientific reference and experimental guidance for the research of new genes (target gene ID).

2 Ontology Construction of Crop Gene Knowledge Graph

The analysis of the molecular regulation mechanism of crop traits involves various types of scientific data and complex interrelationships. Therefore, after in-depth analysis of the correlation between multi-dimensional scientific data, the use of ontology technology to scientifically express their logical relationships can lay a solid semantic support for knowledge discovery in the discipline. As the core semantic framework of the knowledge graph, the ontology layer mainly includes entity types, data attributes, and object attributes between entities. As an important semantic framework for subsequent discipline knowledge discovery, the ontology layer can significantly improve the construction efficiency of knowledge graph and provide a key semantic foundation for the subsequent construction of knowledge graph.

2.1 Ontology Summary Model Design

The ontology summary model describes the logic of complex correlations between multi-dimensional crop breeding scientific data. Based on the discovery of crop gene-related knowledge, our research team takes genes, proteins and traits as the core entities on the basis of full reference to the main data sources, and integrates scientific data from different data types and dimensions by defining the scientific object attribute association. The constructed ontology summary model is shown in Fig. 1, and a multi-dimensional scientific data association semantic model covering 13 entities and 14 object attributes is finally constructed [11].

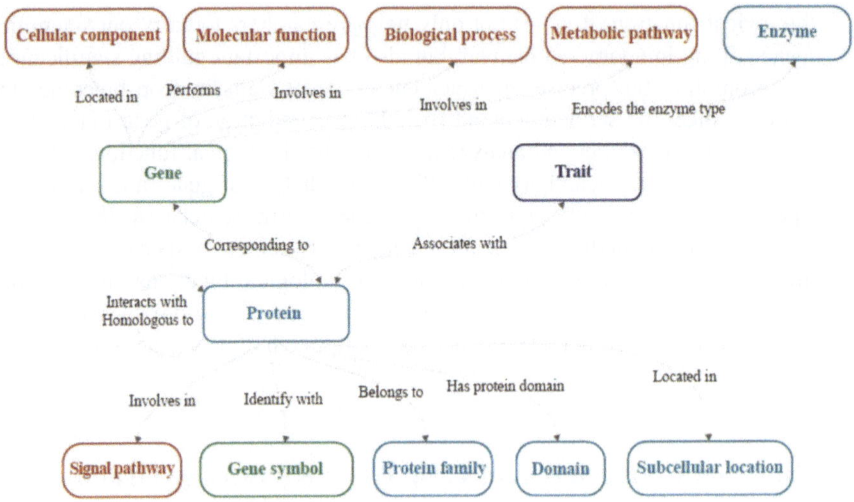

Fig. 1 A semantic model of multidimensional scientific data association

2.2 *Core Entity and Hierarchy Definitions*

In the knowledge graph ontology layer, entities are used to describe a collection of a class of concepts with the same attributes, which is an important part of the ontology semantic model. On the basis of determining the core entity types in the ontology semantic model, it is necessary to further define the entity hierarchy in the ontology semantic model. The research team defined 10 other associated entities with "trait", "gene" and "protein" as the core entities. includes "gene symbol" entities at the gene level; "protein family", "domain", "subcellular localization" and "enzyme" entities at the protein level; At the level of metabolic enrichment pathways, the entities "cellular component", "molecular function", "biological process", "metabolic pathway", and "signal pathway" [11].

2.3 *Entity Core Property Definitions*

In order to standardize the construction and management process of the ontology layer, after establishing the entity type and its hierarchy, it is necessary to further refine the intrinsic data attributes of the entities and the interrelated object attributes of the entities. Together, these attribute information constitute a comprehensive description of the essential characteristics of the entity and its interrelationships, so as to achieve a scientific representation of the knowledge organization system of the professional domain.

(1) Definition of entity data attributes

An entity's data attribute is the characteristic of the entity itself, i.e., if an entity has a property, all entities in the entity type have that property. Data attributes provide a more comprehensive description of an entity's own characteristics. For the 13 entity types defined in the ontology summary model, combined with the application requirements of crop gene knowledge discovery, the data attribute retention associated with other entity types was preferentially selected. For example, for gene classes, the genes we obtain and their corresponding unique gene identifiers, species, sequence length, etc., can be used as data attributes of the gene class. However, adhering to the scientific screening of data attributes, it is necessary to remove some irrelevant data attributes. In the process of screening the data attributes of each entity, on the basis of combining the needs of crop gene knowledge discovery services, and considering the quickness of the construction of the knowledge graph data layer, the data attributes associated with other entities were preferentially retained. Finally, data attributes such as the time when the protein was first discovered, gene identifiers, functional descriptions, and influencing phenotypic descriptions were retained [11].

(2) Definition of entity object attributes

The object properties of an entity can scientifically represent the semantic relationship between two entities, which determines the richness and application effect of the domain knowledge graph. According to the needs of crop gene knowledge discovery services, genes, proteins and traits are the core entities, firstly, focusing on the trait entities, and linking the trait entities with the protein entities through the "associates with" object attributes, so as to realize the establishment of the association between traits and proteins. In addition, the association between cross-species proteins is established through the "homologous to" object attribute, which is a key object attribute in the ontology layer, and is also an important basis for realizing the association and integration of multi-dimensional scientific data between species. In addition, the "located in", "belongs to", and "has protein domain" object attributes are associated with subcellular localization locations, protein families, and domains, respectively. At the same time, the relationship between the interacting proteins is established through the "interacts with" object property. In addition, the establishment of the association between proteins and gene symbols through "identify with" object attributes has become the key to the discovery of gene function knowledge across species. The association between proteins and genes is constructed through the "corresponding to" object attributes. In the same way, focusing on gene entities, the associations between gene and molecular functions, biological processes, and cellular components are established through the attributes of "performs", "involves in", and "located in", respectively. The association between genes and metabolic pathways is established through the attribute of "involves in", and the association between genes and enzymes is established through "encodes the enzyme type" [10].

3 Construction of Crop Gene Knowledge Graph Data Layer

The core task of the construction of the knowledge graph data layer is to extract the entities and their relationships based on the semantic framework of the entities and their relationships defined by the ontology layer and store them in the graph database. Compared with the general knowledge graph, the construction of the domain knowledge graph has more stringent requirements for the scientific rigor and accuracy of the data source, and the authority of the data source and the accuracy of knowledge extraction must be fully considered. As the main channel for researchers to obtain cutting-edge knowledge of the discipline, scientific and technological literature contains the latest research results in the discipline, and the knowledge extracted from it reflects the latest development trends of the discipline. At the same time, the domain science database constructed by experts also contains a large number of systematic and specialized subject knowledge. Therefore, the effective extraction, reintegration and fusion of these multi-source and heterogeneous multi-dimensional crop breeding scientific data can help stimulate new disciplinary knowledge discovery. However, in the current construction of crop domain knowledge graph data layer, there is a lack of effective correlation between crop breeding knowledge in scientific literature and related knowledge in domain science databases, which brings difficulties to the discovery of crop gene knowledge. In view of the limitations of the current methods of constructing the knowledge graph data layer in the crop field, in order to realize the migration and reuse of crop breeding knowledge and maximize the value of multi-dimensional crop breeding scientific data, scientific literature and domain science databases were selected as the key data acquisition sources, and the knowledge graph data layer construction method of multi-path knowledge extraction and multi-source knowledge association and fusion was adopted.

3.1 Select the Source of Data Acquisition

After a comprehensive analysis of the distribution of crop breeding scientific data and the authority of the domain science database, the PubMed literature database was selected as the data source for the construction of domain knowledge graph, taking the model plant Arabidopsis thaliana and the staple food crops rice, maize and wheat as the data collection objects. In addition, eight other field scientific databases were selected as data sources, including Ensembl (European Molecular Biology Laboratory's European Bioinformatics Institute) plants (genomic information of 4 species), Phytozome (genomic information of 4 species), and rice database RGAP (Rice Genome). Annotation Project), UniProt (Universal Protein) (protein annotation information of 4 species), Pfam (protein family analysis and modeling) (protein family information of 4 species), STRING (protein interaction information of 4 species), GO (Gene Ontology) (pathway annotation information for 4 species) and KEGG

(Kyoto Encyclopedia of Genes and Genomes) (pathway annotation information for 4 species) [10].

3.2 Multipath Knowledge Extraction

On the basis of selecting the data acquisition sources, the correlation between the data sources was further analyzed, and the scientific data structures in each data source were interpreted and analyzed. The types of data formats selected in the data sources are further clarified, including structured data in Excel, TXT, and TSV formats, semi-structured data in XML and FASTA formats, and unstructured data in Text format. Considering the efficiency and accuracy of knowledge extraction, the multi-path knowledge extraction method is used to extract entities and relationships from multi-source scientific data in different data formats [10].

(1) Knowledge extraction for structured data. Structured knowledge mostly comes from data in relational databases, and in this field, researchers mainly focus on how to convert database data into RDF data and OWL ontologies [12]. Typical languages are: DM (Direct Mapping) and R2RML languages. These two languages are used to define various rules for how data in a relational database is converted into RDF. Due to the particularity of domain scientific data naming, the knowledge extraction oriented to structured data is mainly based on the pandas tool to convert the format and clean the structured data, and the rule mapping method is further used to obtain the rule data that meets the storage format.
(2) Knowledge extraction for semi-structured data. Encyclopedia data and web page data are typical semi-structured data. Taking Wikipedia as an example, the entry page contains semi-structured data such as entry titles and entry summaries, which can provide data sources for the construction of large-scale knowledge graphs, such as DBpedia [13] and Yago [14]. For the knowledge extraction of semi-structured data, BLAST is mainly used to obtain the knowledge of protein homology relationship of protein sequence data in FASTA format by using BLAST, a sequence similarity calculation tool in the field of biology. In addition, Kettle, an ETL process tool, was used to parse the semi-structured data in XML format, and the relevant subject knowledge was extracted based on the rules.
(3) Knowledge extraction for unstructured data. Due to the scattered and complex agricultural data and the heterogeneity of multiple sources, it is extremely complex to extract relevant agricultural subject knowledge from unstructured data. In recent years, some typical machine learning methods, such as decision trees, support vector machines, and conditional random fields, have been gradually applied to agricultural knowledge extraction, among which the BiLSTM-CRF model has been used to extract crop diseases, insect pests, and livestock and poultry disease data, and has achieved good results [15]. In recent years, the

prevalence of attention mechanism has brought about a revolution in the field of deep learning, and the Transformer model has solved the problems of recurrent neural networks, which are difficult to deal with long-distance dependency and difficult to train in parallel [16]. Therefore, the BERT-LSTM-CRF model has also become a favored model for many researchers [17]. With the popularity of large models, generative knowledge extraction has become another research hotspot, such as InstructIE, UIE, etc., and the generative method based on large models and the ability to learn with fewer samples prompt further reduce the technical threshold of knowledge extraction, and can further improve the performance of model output. For knowledge extraction of unstructured data, our research team mainly adopts a combination of traditional deep learning models and generative knowledge extraction based on large models. The BERT-BiLSTM-CRF series model base was selected, and the NLTK package was used to process the literature in sentences, and then the weights of the pre-trained model of BERT were loaded to label the biological entities in the literature to achieve relevant knowledge extraction. In addition, the open-source Baichuan-7B, Qwen-7B and other large models were selected, and the prompt word engineering was used to endow the large model with prior knowledge in the subject area, and the model was instructed to identify the entities or triples in the text according to the predetermined entity and relationship types.

3.3 Multi-source Knowledge Association and Integration

Multi-source knowledge association fusion is the association and integration of the extraction results of the involved entities, entity data attributes and object attributes, which is crucial for the construction of knowledge graphs and provides important data support for subsequent subject knowledge discovery. The core challenge of the process is how to efficiently and accurately combine descriptive information from different sources about the same entity or relationships between entities, reducing redundancy. In order to achieve this goal, it is first necessary to clean the extracted data, distinguish and integrate the multi-dimensional scientific data entities extracted from the literature and the corresponding entities in the domain knowledge base, and eliminate duplicated, erroneous or incomplete information, so as to ensure the accuracy and credibility of scientific data. In terms of disambiguation and merging of entity concepts, we use mapping strategies to unify entity concepts from different sources according to the semantic associations in the ontology layer. In terms of the association and fusion of entity object attributes, based on the analysis of the association characteristics between different scientific data sources, we use the common protein identifier and gene ID of scientific databases as the mapping benchmark to map the relevant entities and their entity object characteristics of the scientific data obtained through multipath knowledge extraction, so as to realize the association and fusion of multiple types of triples in the knowledge graph data layer [10]. Finally, a knowledge graph covering 13 entities, 16 data attributes, and 14 object attributes, with a total of 125591 nodes and 547591 semantic relationships was formed [18].

4 Construction of Crop Gene Knowledge Discovery Model and Scenario Application

As a typical semantic representation model, knowledge graph can clearly display the knowledge in the subject domain through the relationship graph structure, and give scientific semantic associations and hierarchies to all kinds of knowledge. Among them, knowledge entities can construct complex indirect connections through one or more intermediate knowledge entities, and then generate multiple association paths. These association paths can reveal the "knowledge routes" of associations between knowledge entities, and provide strong support for the construction of knowledge discovery models based on knowledge reasoning, link prediction and other fields. Therefore, on the basis of the domain knowledge graph that has been constructed, some potential "knowledge routes" can be discovered by constructing a knowledge discovery model driven by the domain knowledge graph. Through the in-depth analysis of these "knowledge routes", the discovery of new knowledge in the discipline is realized.

In the process of generating new knowledge, expert prior knowledge can provide valuable constraints and feature relationships for knowledge discovery models, and help explain and understand the complex feature associations between multi-dimensional scientific data of crop breeding. Therefore, the construction of a crop gene knowledge discovery model that integrates the principles of knowledge mining methods with experts' prior knowledge can break the cognitive limitations of crop breeding scientists and effectively support the discovery of crop gene knowledge based on multi-dimensional scientific data forensic analysis. Based on the cross-species gene-level knowledge association model, gene regulatory trait prediction model [18] and gene function prediction model [11], this paper aims to realize the cross-species gene-level knowledge association discovery, excellent gene discovery and efficient prediction of gene function embedded in the scientific research process of crop breeding.

4.1 Cross-Species Gene-Level Knowledge Association Discovery

In the knowledge graph, path discovery refers to the method of discovering critical paths from complex networks, which can be used to analyze and discover complex paths between knowledge entities in the knowledge graph, and determine whether there is an implicit relationship between two seemingly unrelated knowledge entities within the specified path length [19, 20]. In the scientific research of crop breeding, the complex correlation between multi-dimensional scientific data between genes and traits provides an important data cornerstone for analyzing the molecular regulation mechanism of crop traits. Therefore, the discovery of cross-species gene-level knowledge associations can provide important experimental ideas for the study of the molecular regulation mechanism of crop traits. In this paper, a cross-species gene-level knowledge association model is constructed based on the principle of path discovery knowledge mining method, which can realize the efficient association discovery of gene-level knowledge. Based on this model, it can not only retrieve the content containing search keywords, but also display the search terms and related object attributes and related entities in detail, so as to obtain the hierarchical knowledge structure of multi-dimensional scientific data between genes and traits, and improve the efficiency of subject knowledge retrieval and discovery. The formula for the cross-species gene-level knowledge association model is as follows:

$$\begin{aligned}&\text{IF gene1} - [\text{corresponding}] - \text{protein1} - [\text{homologous}]\\&-\text{protein2} - [\text{corresponding}] - \text{gene2} - [] - \text{entity1}\\&\text{Then gene1} - [] - \text{entity1}\end{aligned}$$

Based on the cross-species gene-level knowledge association model, cross-species gene-level knowledge association discovery can be realized, that is, through cross-species gene-level knowledge association retrieval, the knowledge hierarchy system with gene entities as the core can be obtained, and the hierarchical knowledge of genes can be visualized from different molecular levels. In addition, the hierarchical knowledge of gene entities can be associated with retrieval and discovery from three different scientific data types: gene level, protein level, and enrichment pathway level. It can also support the search term with any entity name such as gene ID, protein ID, subcellular localization, etc., to realize the association and discovery of multi-dimensional scientific data between genes and traits, and obtain the hierarchical knowledge system related to the search term. As shown in Fig. 2, using the maize gene Zm00001d026139 as the search term, we can not only obtain the knowledge of the gene at the gene level, including the cellular components to which the gene belongs, the molecular function of the gene, the biological process involved in the gene and the type of enzyme encoded by the gene. At the same time, through the association of genes between species, the knowledge of different scientific data dimensions of rice gene LOC_Os04g47300 homologous at the gene level, protein level and enrichment pathway level can also be obtained, and then the association search and discovery of maize gene Zm00001d026139 (search term) related knowledge can be realized.

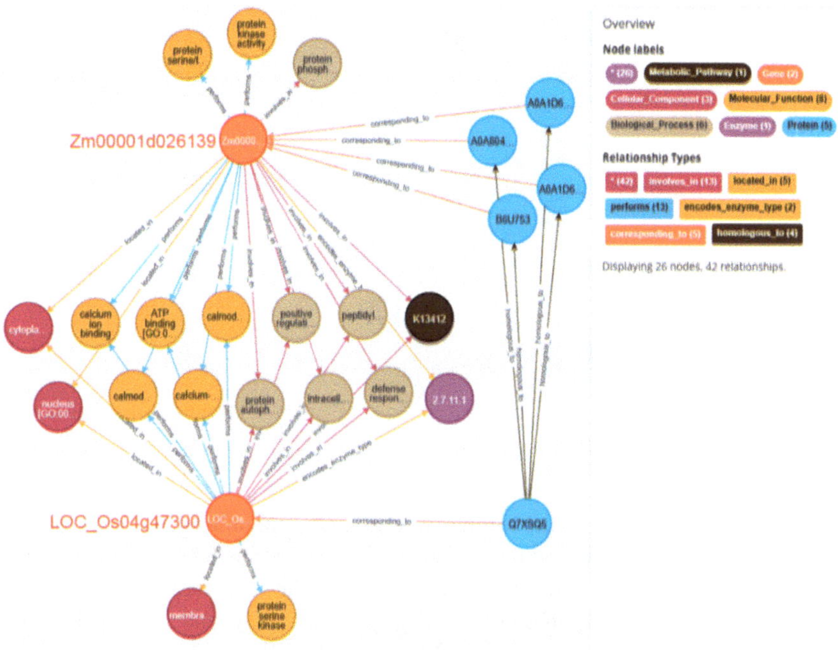

Fig. 2 Cross-species gene-level knowledge association retrieval

4.2 *Discovery of Excellent Genes for Target Traits*

In the knowledge graph, when the relationship between knowledge entities is missing or not discovered temporarily, link prediction can predict the likelihood of a "new link" between them by calculating the closeness between them. The core idea of link prediction is to measure the likelihood of a potential correlation between two entity nodes through a scoring function. The higher the score value, the more likely it is that there will be an association between the two entity nodes. In the scientific research of crop breeding, functional genes regulate crop traits by encoding proteins. The more similar the characteristics of two functional genes at the genome and proteomic levels, the more likely it is that the same traits will be regulated. In this paper, the prediction model of gene regulatory traits can be used to mine superior genes [18]. The core of the model is to use a similarity-based link prediction knowledge mining method to calculate the tightness between the entities of unknown functional genes and known functional genes (the relationship between genes and traits) for some unknown functional genes (loss of relationship between genes and traits), and predict the possibility of "new links" between unknown functional genes and traits according to the traits regulated by known functional genes [18]. The formula for calculating the similarity between gene entities defined in the model is:

$$\mathrm{S}(\text{``g1, g2"})'' = \mathrm{C}''(\mathrm{k}) \cdot \mathrm{D}(\mathrm{k}) \cdot \mathrm{S}(\text{``p1, p2"})$$

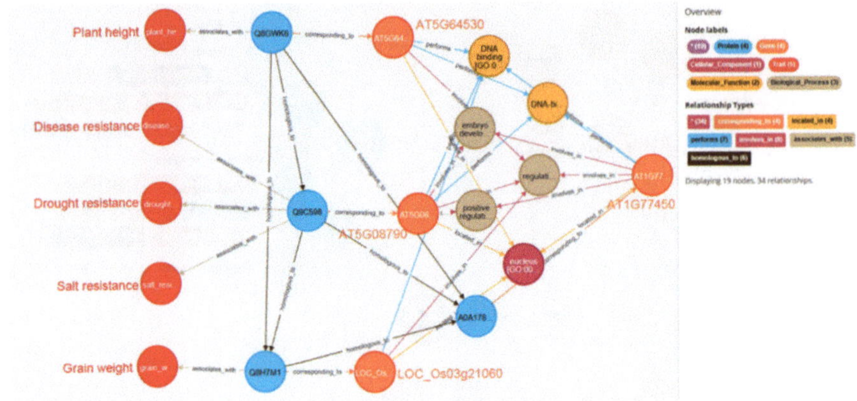

Fig. 3 Excellent gene discovery

where S("g1, g2") represents the similarity between the unknown gene and the known gene; "G1" is a known trait regulator, and "G2" is an unknown trait regulator. S(p1, p2) represents the sequence similarity score between the two protein entities, p1 and p2 are the proteins corresponding to "g1, g2", respectively, and N(x) is the set of nodes adjacent to the node x gene, $k = N(g1) \cap (g2)$; C(k) is the number of entities in the node set k, and D(k) is the number of entity classes in the node set k.

Based on the above prediction model of gene regulatory traits, the mining of excellent genes for the target traits can be realized. That is, by calculating the similarity score between the two gene entities, the unknown functional genes above the threshold were screened out and the association relationship with the traits was established. Finally, it is possible to achieve the potential discovery of excellent genes based on the needs of breeding goals. Taking the discovery of superior gene AT1G77450 as an example, the Arabidopsis gene AT5G64530 in Fig. 3 is associated with plant height traits, the Arabidopsis gene AT5G08790 is associated with disease resistance, drought resistance and salt resistance, and the rice gene LOC_Os03g21060 is related to grain weight traits. Based on the prediction model of gene regulatory traits [18], the similarity scores between the Arabidopsis thaliana gene AT1G77450 and the above three genes exceeded the threshold. Therefore, predictive genes AT1G77450 regulate five traits of plant height, disease resistance, drought resistance, salt resistance and grain weight.

4.3 *Efficient Prediction of Gene Function*

In the knowledge graph, rule reasoning can play a key role when the path between knowledge entities is short. Therefore, a series of knowledge reasoning rules can be constructed based on the semantic relationship between entities, aiming to explore the possible latent correlation between entities. This exploration process is highly dependent on the existing entity information and semantic relationships in the knowledge graph, and the correlation between entities is described by formulating accurate knowledge reasoning rules, so as to discover the new knowledge and new relationships hidden in the knowledge graph. In the scientific research of crop breeding, Arabidopsis thaliana plants have been regarded as typical model plants in plant research because of their short plant size and short growth cycle. Most of the genes in Arabidopsis thaliana have high homology with other complex crop genes, and are often used to provide high-quality gene function annotations for other crop functional gene studies to guide crop breeding scientific research. In order to realize the transfer and reuse of knowledge related to known functional genes and facilitate the retrieval between scientific databases in cross-species domains, gene symbols are often named according to the function of genes, that is, genes with the same or similar functions have the same gene symbols. This expert prior knowledge is also used as the key to the construction of gene function prediction models. In this paper, the gene function prediction model can be used to efficiently predict gene function across species [11]. The core logic of the association rule of the model is that if gene A corresponds to protein A, and protein A is defined as gene symbol S according to its function; At the same time, gene B corresponds to protein B, and protein B is also defined as the gene symbol S, so it can be inferred that gene A and gene B have the same or similar molecular biological functions [11]. The function identified with association inference rule is as follows:

$$
\begin{aligned}
&\text{IF gene1 } - \left[\text{corresponding to}\right] - \text{ protein1}\\
&- \left[\text{identify with}\right] - \text{ gene symbol S and}\\
&\text{gene2} - \left[\text{corresponding to}\right] - \text{protein2}\\
&-\left[\text{identify with}\right] - \text{gene symbol S}\\
&\text{Then gene1} - [\text{function identified with}] - \text{gene2}
\end{aligned}
$$

Based on the above-mentioned gene function prediction model, we can achieve efficient prediction of gene function across species. That is, based on the information of known functional genes of cross-species homology, the molecular biological function of target genes can be efficiently predicted. Taking the molecular biology function prediction of wheat gene TraesCS5B02G094600 as an example, the process of efficient cross-species gene function prediction was illustrated. As shown in Fig. 4, the wheat gene TraesCS5B02G094600 was used as the search term, and the association search was carried out in the constructed knowledge graph based on the gene function prediction model. The maize gene Zm00001d041606 and maize gene Zm00001d030614 have co-linked entity nodes in gene symbol (MCM7), cell composition, molecular function, biological process, metabolic pathway and enzyme type. In addition, the wheat gene TraesCS5B02G094600 with unknown function has the same gene symbol MCM7 as the above known functional genes. Combined with the established association inference rules, according to the functional annotation of known genes and the node information of co-association between known

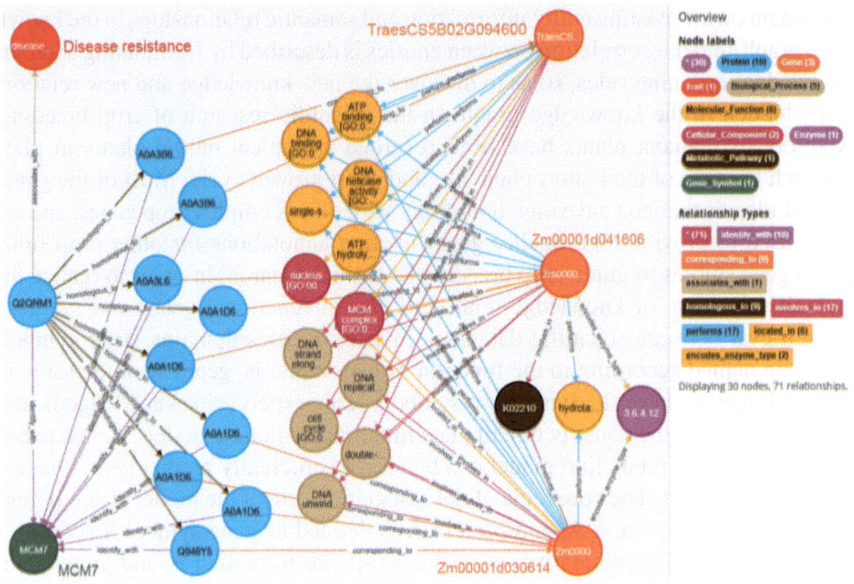

Fig. 4 Prediction of gene function across species

genes, multi-dimensional scientific data can be used to provide molecular biology function prediction supported by multi-dimensional scientific data for wheat gene TraesCS5B02G094600 with unknown function.

5 Summary and Outlook

Compared with the existing excellent gene mining methods, such as genome-wide association analysis, quantitative trait locus mapping, and mixed pool group analysis and mapping, the relevant gene loci can only be mapped into a region containing many candidate genes, and cannot provide numerical recommendations for excellent gene mining rankings [21–23]. The gene regulatory trait prediction model comprehensively considers the semantic association characteristics and graph structure characteristics of multiple types of scientific data between genes and traits, and integrates the attribute information of gene nodes themselves and the topological structure information of gene node relationship, which can provide the prediction results of the probability score of gene regulatory traits based on multi-dimensional scientific data analysis and calculation, which can realize excellent gene mining and recommendation. In this study, the knowledge graph technology was effectively applied to the discovery of crop gene knowledge, which realized the cross-integration of computer science and breeding science, transformed the existing phenotype-driven passive gene knowledge discovery into data-driven active gene knowledge discovery, and provided a new research idea for crop superior gene mining and functional analysis.

In the context of AI4S, the new paradigm of knowledge service, guided by scenario needs, with multi-dimensional scientific data as the knowledge base, and artificial intelligence technology empowerment as the driving force, will surely become a new battlefield to promote the innovation and development of scientific research in the field. The knowledge service model will also shift from the previous service frontier to actively explore frontier research, track the frontier of discipline to discover frontier knowledge of discipline, and understand the transformation of domain cognition to break through the limitations of domain cognition. In the future, the research team will continue to build a knowledge base that supports AI4S, build a technical system for intelligent computing, and build a "data + AI"-driven knowledge service model, aiming to provide new momentum for the improvement of agricultural research efficiency and scientific research innovation and breakthroughs.

Build a knowledge base that supports AI4S, provide an important data cornerstone for scientific and technological knowledge service innovation, and effectively support the comprehensiveness and accuracy of AI model training. Gather and integrate multi-source heterogeneous, multi-type (scientific and technological literature, scientific data, scientific and technological policies, etc.), multi-modal (pictures, videos, courseware, etc.) high-quality scientific and technological resources to build a comprehensive agricultural science and technology big data center; Establish a high-quality large-scale corpus knowledge base content system such as subject headings, classification systems, and encyclopedias, so as to provide an important corpus foundation for fine-grained knowledge discovery and knowledge computing. Construct ontologies and knowledge graphs for vertical disciplines, form an agricultural science and technology knowledge network, provide an important semantic foundation for knowledge reasoning and intelligent knowledge services, and become an important support for solving the illusion crisis of knowledge generation in the digital intelligence environment.

Break through the limitations and bottlenecks of existing emerging technologies such as knowledge graphs and large models, integrate new algorithms and discipline and professional domain models, and create a technical system for intelligent computing. Strengthen the research work of multi-source collection and aggregation of full-text resources with the most knowledge elements, domain characteristic data, deep fragmentation processing, semantic knowledge organization, local long-term storage, and annotated training data, so as to promote the improvement of the integration and processing efficiency of cross-domain and multi-modal scientific and technological resources; Research and development of a series of platforms such as text big data mining and knowledge computing, one-stop knowledge graph collaborative construction and management, etc., to achieve multi-granularity knowledge extraction and panoramic in-depth correlation integration, and help improve the intelligent level of knowledge organization and knowledge services; Promote the in-depth application of large models in vertical disciplines, tap high-quality potential knowledge, and inject new momentum into knowledge services.

Construct a "data + AI" driven knowledge service model, and promote a new paradigm of AI technology empowering the solution of major scientific problems in various disciplines. Give full play to the role of artificial intelligence technology in literature data acquisition, experimental prediction, result analysis, etc., realize the comprehensive empowerment of scientific research digitalization, networking and intelligence, and support the online iteration of theory and experiment. Build a new digital scientific research computing platform that is safe, reliable, open and collaborative, and provides computing power support, algorithm model operation, scientific research collaborative research, and intelligent knowledge services. For vertical domain knowledge discovery services, we will create a domain subject knowledge discovery service system that is embedded in the whole process of scientific research scenarios, and help new breakthroughs in field scientific research.

References

1. Sun Mengge, Huang Yuxin, Han Tao, et al. Challenges and opportunities of knowledge service under the new trend of intelligent scientific research[J]. Journal of Information, 2022, 41(6): 173–181, 107.
2. Yu X, Li Y, Cui X, et al. Simultaneously mapping loci related to two plant architecture traits by phenotypic recombination BSA/BSR in peanut (Arachis hypogaea L.) [J]. Theor Appl Genet, 2023, 136(6): 144.
3. Sonah H, O'Donoughue L, Cober E, et al. Identification of loci governing eight agronomic traits using a GBS-GWAS approach and validation by QTL mapping in soya bean [J]. Plant Biotechnol J, 2015, 13(2): 211.
4. Liu Y, Shen K, Yin C, et al. Genetic basis of geographical differentiation and breeding selection for wheat plant architecture traits [J]. Genome Biology, 2023, 24(1): 114.
5. Alshahrani M, and Hoehndorf R. Semantic Disease Gene Embeddings (SmuDGE): Phenotype-based disease gene prioritization without phenotypes [J]. Bioinformatics, 2018, 34: i901–i907.
6. Peng C, Dieck S, Schmid A,, et al. CADA: Phenotype-driven gene prioritization based on a case-enriched knowledge graph [J]. NAR Genomics Bioinf, 2021, 3: 1–7.
7. Choi W, and Lee H. Identifying disease-gene associations using a convolutional neural network-based model by embedding a biological knowledge graph with entity descriptions [J]. PloS One, 2021, 16(10): 1–27.
8. Hassani-Pak K, Singh A, Brandizi M,et al. KnetMiner: a comprehensive approach for supporting evidence-based gene discovery and complex trait analysis across species [J]. Plant Biotechnol. J, 2021, 19(8): 1670–1678.
9. Larmande P, Tagny Ngompe G, Venkatesan A, et al. AgroLD: a knowledge graph database for plant functional genomics [J] . Methods Mol. Biol, 2022, 2443: 527–540.
10. Zhang Dandan. Research on knowledge discovery of crop trait regulatory genes based on knowledge graph[D]. Beijing: Chinese Academy of Agricultural Sciences, 2024 (in Chinese).
11. Zhang Dandan, Zhao Ruixue, Xian Guojian, Xiong He. Construction and Application of Trait Regulatory Gene Ontology Model Integrating Cross-species Scientific Data[J]. Biotechnology Bulletin, 2024, 40(2): 313–324.
12. Cao Yuqing, Xian Guojian, Huang Yongwen, et al. Research on the construction of panoramic multi-path knowledge graph: A case study in the field of rice grain type gene[J]. Digital Library Forum, 2022 (4): 25–34.
13. Auer S, Bizer C, Kobilarov G, et al. Dbpedia: A nucleus for a web of open data[C]. Heidelberg: Springer Berlin Heidelberg, 2007: 722–735.

14. Suchanek F M, Kasneci G, Weikum G. Yago: a core of semantic knowledge[C]. Banff, Alberta, Canada: IEEE. 2007: 697–706.
15. Cheng M, Yu H, Feng Y, et al. Fishery Standard Named Entity Recognition Based on Attention Mechanism and BiLSTM+CRF[J]. Journal of Dalian Ocean University, 2020, 35(02): 296–301.
16. Vaswani A. Attention is all you need[C]. California: Curran Associates Inc, 2017.
17. Hu X, Zhang H, Hu S. Chinese named entity recognition based on BERT based-BiLSTM-CRF model[C]. Zhuhai. IEEE, 2022: 100–104.
18. Zhang D D, Zhao R X, Xian G J, et al. A new model construction based on the knowledge graph for mining elite polyphenotype genes in crops[J]. Front Plant Sci, 2024, 1: 361–716.
19. Chen C.Searching for Intellectual Turning Points:Progressive Knowledge Domain Visualization [J].Proceedings of the National Academy of Sciences,2004,101(S1): 5303–5310.
20. Song M,Heo G E,Ding Y.SemPathFinder: Semantic Path Analysis for Discovering Publicly Unknown Knowledge [J].Journal ofInformetrics,2015,9(4): 686–703.
21. Garcia M, Eckermann P, Haefele S, Satija S, Sznajder B, Timmins A, et al. Genome-wide association mapping of grain yield in a diverse collection of spring wheat (Triticum aestivum L.) evaluated in southern Australia [J]. PloS One, 2019, 14(2): 1–19.
22. Liu J, Qiao Y, Li C, et al. The NAC transcription factors play core roles in flowering and ripening fundamental to fruit yield and quality [J]. Front. Plant Science, 2023, 14.
23. Jiang T, Zhang C, Zhang Z, et al. QTL mapping of maize (Zea mays L.) kernel traits under low-phosphorus stress [J]. Physiol. Mol. Biol. Plants, 2023, 29(3): 435–445.

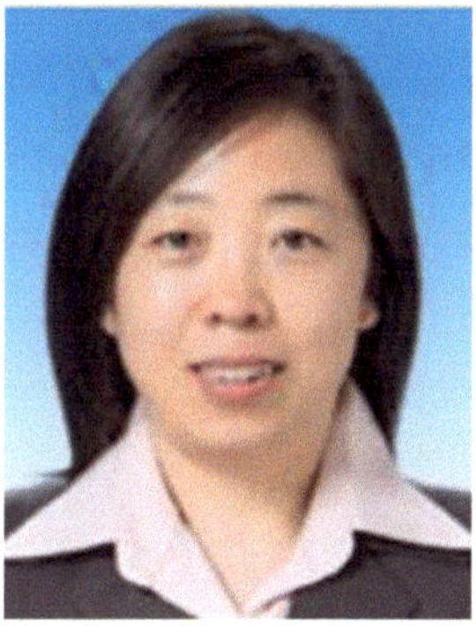

Ruixue Zhao Ph.D. in Management, Second-level Researcher, Doctoral Supervisor. He is currently the deputy director of the Agricultural Information Institute of the Chinese Academy of Agricultural Sciences, the chief of the big data and knowledge service innovation team of the Chinese Academy of Agricultural Sciences, and the director of the Key Laboratory of Knowledge Mining and Knowledge Service of Agricultural Integrated Publishing of the National Press and Publication Administration. He is also the director of the Library and Information Branch of the Chinese Agricultural Society, the executive director of the Chinese Society of Scientific and Technical Information, the vice chairman of the Professional Library Branch of the Chinese Library Association, a member of the National Library Standardization Technical Committee, and a member of the International Federation of International Federation of Science and Technology Libraries. He has long been engaged in scientific research and teaching in agricultural informatization, digital library, knowledge organization and knowledge service, agricultural scientific data management, etc. He has presided over or participated in more than 50 national science and technology support programs, national "863" programs, major projects of "new generation artificial intelligence", basic projects of the Ministry of Science and Technology, Chinese Academy of Engineering and other tasks, won 10 Beijing Science and Technology Progress and Ministerial Science and Technology Achievement Awards, and more than 40 computer software copyright registrations. He has published more than 100 scientific papers, authorized 3 invention patents, and published 15 monographs. She was awarded the "Women's Meritorious Service" model of the Chinese Academy of Agricultural Sciences.

Tan Sun in Management and doctoral supervisor, is currently the Vice President and Party Committee Member of the Chinese Academy of Agricultural Sciences. He has long been engaged in research on digital information description and organization, smart agriculture, and big data governance. In recent years, he has led more than 10 national and provincial-level projects, including the National Key R&D Program and the National Social Science Fund projects. He has published over 130 academic papers, authored or co-authored 8 books, and holds more than 10 authorized patents. In 2005, he was awarded the Special Government Allowance by the State Council, in 2014 he was recognized as a 'National Outstanding Science and Technology Worker,' in 2016 he was selected for the Chinese Academy of Agricultural Sciences Young Talent Program, and in 2018 he was selected as a Leading Talent, Category C, in the Agricultural Science Talent Program of the Chinese Academy of Agricultural Sciences.

Dandan Zhang Master of Science and Doctor of Management, mainly engaged in research on functional gene mining of crops, knowledge organization, and disciplinary knowledge discovery. Has participated in and led research projects including the National Key R&D Program project 'Formation and Regulation Mechanisms of Grain Crop Quality,' the Science and Technology Innovation 2030 Major Project for New Generation Artificial Intelligence 'Development and Application Demonstration of Agricultural Intelligent Knowledge Service Platform,' and sub-tasks on core algorithms and large model technology development and application for smart breeding. Has published 10 academic papers as the first author or corresponding author; holds 3 authorized invention patents; and has edited or contributed to 2 books.

Research on Social Science Issues in the Age of Artificial Intelligence

Xushu Peng, Litao Duan, and Yuanhao Ban

Abstract Artificial Intelligence (AI) is profoundly impacting not only the economic and social development but also the advancement of social sciences. Taking generative AI as an example, this paper explores the far-reaching influence of AI on social science research from the perspectives of theoretical logic, research subjects, paradigms, and the transformation of social science practitioners, and provides an outlook on future trends. The study posits that the impact of generative AI on social sciences tends to manifest in several ways: First, social science research is irreplaceable. Although AI may surpass human intelligence in certain domains, social sciences remain an embodiment of human wisdom, and AI-generated content cannot fully substitute for social science research; Second, the stratification of social science researchers is accelerating. The introduction of AI will lead to an "intelligence divide" among social science researchers, with those who can effectively utilize AI tools producing high-quality outcomes more rapidly; Third, the acceleration of interdisciplinary integration. The development of AI brings new topics to social science research, promoting the swift integration of different disciplines; Fourth, data-driven research. The rise of big data will make data-driven approaches a significant research paradigm in social sciences, altering traditional qualitative and quantitative methods and paving new paths for social science research.

Keywords Artificial intelligence · Social science research · Data-driven · Generative AI

Introduction

Artificial intelligence is regarded as one of the three most important technologies of the twenty-first century—alongside genetic engineering and nanoscience—and ranks among the most closely watched and exhilarating fields of recent technological development. In recent years, generative artificial intelligence, epitomized by

X. Peng (✉) · L. Duan · Y. Ban
Institute of Quantitative & Technological Economics, Chinese Academy of Social Sciences, Beijing, China
e-mail: pengxs@cass.org.cn

China's e-Science Blue Book 2025, https://doi.org/10.1007/978-981-96-8252-2_17

ChatGPT (Chat Generative Pre-trained Transformer), has achieved breakthrough advances. Numerous viewpoints suggest that AI may be approaching a "technological singularity," characterized by emergent intelligence [1], with some even arguing that general artificial intelligence could materialize within the next decade.

Unlike genetic engineering and nanotechnology, artificial intelligence is a prototypical general-purpose technology. It can be deeply integrated with economic and social development and widely applied across diverse facets of production and daily life—thereby influencing, transforming, and even upending human modes of production and living—while also exerting a profound impact on scientific research itself [2]. As academic disciplines and technological fields increasingly converge, AI is becoming not only the subject and tool of scientific inquiry but is also poised to revolutionize research objects, methodologies, and even the roles of researchers. The interplay between artificial intelligence and science has thus emerged as a central focus of scholarly attention, with "Artificial Intelligence for Research (AI4R)" coming to be viewed by some as the "fifth research paradigm" [3].

The knowledge-generation model of generative artificial intelligence bears a strong resemblance to the research paradigms in the social sciences, and the technical threshold for simulating content generation in social science research is comparatively lower. Consequently, it is imperative to analyze and accurately characterize the new features and emerging trends as the social sciences undergo a transformation. Against this backdrop, this paper examines the shift in the social sciences in the age of artificial intelligence from four perspectives: (1) the theoretical logic underlying the influence of generative AI on social science research, (2) the nature of research objects, (3) the evolution of research paradigms, and (4) the changing roles of social science scholars, and offers an outlook for the future.

1 The Theoretical Logic of Artificial Intelligence's Impact on Social Science Research

1.1 Technological Evolution of Artificial Intelligence

Since the advent of the concept of artificial intelligence in the 1950s, technological exploration has centered on understanding and emulating the mechanisms of the human brain. The first generation of AI—from the 1950s through the 1980s—primarily relied on constructing knowledge bases from extensive domain-specific expertise and employed artificial neural networks alongside heuristic reasoning models to simulate certain human cognitive behaviors, such as automated theorem proving, machine translation, and rudimentary human–computer dialogue. IBM's Watson serves as a quintessential example of this approach. This form of AI, often termed symbolic AI, builds its learning mechanisms predominantly through knowledge induction and logical reasoning. Although it has played an important role in understanding and applying natural language to address challenges in military, education, finance, healthcare, and other domains, its dependence on fixed and ample default information renders it effective mainly in simple task planning, scheduling, and even diagnostic applications, while leaving it ill-equipped to manage complex problems—especially under conditions of information scarcity [4].

Beginning in the 1980s, with advancements in computer technology, the development of deep learning models based on neural networks and back-propagation algorithms marked a significant shift. By emulating the connectivity patterns of human brain neurons and learning from vast datasets, connectionist AI gradually came to dominate the field. During this period, despite breakthrough progress in areas such as speech and image recognition and natural language processing, and despite acquiring certain learning and predictive capabilities, AI systems remained inefficient and suffered from issues of unreliability and insecurity, thereby limiting their large-scale applicability.

Entering the new century—particularly since 2011—continuous improvements in data mining techniques, multi-layer neural networks, and deep learning models, coupled with growing computational power and ever-expanding datasets for model training, have led behaviorist AI—grounded in the principles of logic, probability theory, and related foundations—to become mainstream, ushering in a new developmental surge. Notably, since 2017 the evolution of large language models has accelerated significantly. These models, trained on massive corpora of text, are capable of generating or comprehending natural language and, supported by data, sophisticated algorithms, and substantial computational resources, can produce a variety of outputs—including text, images, code, and video—based on their learned content. Generative AI models, exemplified by ChatGPT-4, have now achieved multimodal input and output capabilities (e.g., processing both text and images). This marks a significant improvement in AI's capacities for understanding, reasoning, generation efficiency, real-time performance, and the accuracy and innovativeness of its output, signaling that generative AI is rapidly advancing toward practical applicability. As its abilities in learning, perception, prediction, and decision-making continue to strengthen, the influence of generative AI is expected to expand further. This technological progress is widely regarded as a hallmark of a critical transitional phase in AI's evolution, heralding a new era of technological transformation and widespread application.

1.2 The Formation and Implementation of Generative Artificial Intelligence

Donald Hebb first discovered and explained the learning and memory mechanisms of brain neurons in 1994, arguing that the brain consists of numerous interconnected neurons and that learning occurs through repeated stimulation of neural cortices. Continuous simultaneous stimulation of neurons alters their synaptic connections, generating new information within the brain. Consequently, mimicking the operational principles of human neurons has been a natural research trajectory for artificial intelligence since its inception. Rosenblatt Frank [5] first proposed a single-layer neural network model with adjustable parameters, establishing a mapping relationship through data-driven learning. This was the first instance of expressing human learning functions via algorithmic models and endowing computer systems with the ability to learn from data [6].

The development of neural network models has been a major driving force behind AI's evolution, with the rise of generative artificial intelligence being largely attributable to breakthroughs in deep learning. Large language models (LLMs), such as ChatGPT, are fundamentally pre-trained models that leverage deep neural networks and vast datasets to learn and abstract probabilistic distributions and intrinsic patterns in data. This enables them to autonomously generate novel content that is related or analogous to given inputs, effectively replicating human cognitive and creative processes. The development and application of generative AI typically involve the following steps.

1. Developing or selecting an appropriate generative model. This often involves leveraging recurrent neural networks (RNNs) or transformers, combined with customized algorithms, to construct neural network architectures with varying parameter scales.
2. Preparing data and training the model. Text, images, audio, and video data are collected and preprocessed before being used to train the model. This allows the model to learn the sequential structure and probabilistic distribution of the input data, thereby acquiring an understanding of its features.
3. Training feedback and parameter optimization. Throughout the training process, the accuracy of randomly generated content is evaluated, and the model's parameters are fine-tuned accordingly. This process simulates human reinforcement learning mechanisms, iteratively improving the model's generative performance.
4. Model evaluation and refinement. Based on feedback from both model-generated outputs during training and real-world applications, the model architecture, neural network parameters, and training mechanisms are assessed and refined to ensure continuous performance improvement and practical effectiveness.
5. Prompting and content generation. During AI application, the system generates responses based on user inputs or its own perception of input data, including but not limited to natural language text, images, and audiovisual content. This process may be a one-time interaction or an iterative cycle of modifications and upgrades.

1.3 The Mechanisms of Generative AI's Impact on Social Science Research

Artificial intelligence was originally conceived as a means of simulating the human brain. The U.S. National Institute of Standards and Technology (NIST) defines AI as a complex system integrating software and/or hardware that possesses the capabilities to "learn, solve complex problems, make predictions, and perform tasks that require human sensory inputs such as vision, language, and touch—involving perception, cognition, planning, learning, communication, and physical movement." From the early expert systems to today's generative AI, the field has evolved into a multidisciplinary domain encompassing computer science, psychology, physics, mathematics, and both the natural and social sciences. This evolution has not only provided a robust scientific foundation for AI's capacity to simulate—and in some cases, surpass—certain human cognitive functions but has also offered profound opportunities for transformative changes in social science research.

At its essence, artificial intelligence refers to a set of theories and technologies aimed at replicating human intelligence, enabling computer systems to execute thought and decision-making processes analogous to those of humans. In contrast, social science research builds upon the cumulative body of human knowledge through preliminary study, supplemented by comprehensive literature reviews and empirical investigations to collect data. Researchers then analyze this data and existing literature using established methodologies, integrating theoretical insights to propose new frameworks or solutions. As evidenced in the implementation process of generative AI (see Table 1), training an AI system requires vast quantities of data—such as books, journals, and encyclopedic content—which are critical not only to constrain the adaptability and accuracy of AI but also constitute the essential data for social science research. In this respect, the data-driven training process of AI mirrors the learning, literature collection, and empirical research processes inherent in the social sciences.

Generative AI employs machine learning algorithms and deep learning mechanisms that emulate biological neural networks. By leveraging big data alongside advanced computational algorithms, such systems progressively achieve autonomous

Table 1 Training data sources and sizes for various AI large models

Large model type	Training data sources (GB)							Parameter scale
	Wikipedia	Books	Academic journals	Reddit	Common crawl	Others	Total	
GPT 1		4.6					4.6	
GPT 2				40			40	
GPT 3	11.4	21	101	50	570		753	175B
The Pile V1	6	118	244	63	227	167	825	
Megatron 11B	11.4	4.6		38	983	127	1374	
MT-NG	6.4	118	77	63	983	127	1374	530B
Gopher	12.5	2100	164.4		3450	4823	10550	280B

Source Yao, Qian. "Hosting and Governance of Training Data for ChatGPT-Like Large Models." China Finance 2023(06): 51–53 [7]

learning, pattern recognition, and decision-making capabilities, thereby generating new knowledge content. This process advances the automation of computer systems to unprecedented levels of sophistication and intelligence. Evidently, the knowledge-generation model of generative AI is highly analogous to the research paradigms of the social sciences in both logic and methodology. Moreover, because social science research generally does not depend on large-scale experiments or complex engineering validations, the technical threshold for AI to simulate content generation in this field is relatively low. Consequently, AI's impact on social science research is both direct and profound, not only introducing new tools and perspectives but also, to a significant degree, reshaping research processes and methodologies.

2 The Shift in Social Science Research Objects in the Age of Artificial Intelligence

2.1 The Expansion of Research Objects in the AI Era

In the age of artificial intelligence, the methods and tools of social science research are undergoing profound transformations, which in turn are expanding both the scope of inquiry and the subjects under investigation. Traditionally, social science has focused on classical domains such as human behavior, social structures, cultural mechanisms, political institutions, and economic activities. Its subjects have spanned multiple levels—from individual behavior to group relationships, and from social norms to the dynamics of economic operations—with an emphasis on uncovering the underlying logic and regularities of social systems. However, due to technological constraints and limitations in data acquisition, conventional social science research has encountered difficulties in addressing complex social phenomena, dynamic changes, and multidimensional interactions. Consequently, the objects of inquiry have gradually broadened beyond traditional social behaviors and economic phenomena to include the complex interactions between technology and individuals, as well as between technology and society, and the novel issues emerging from these interactions. New topics—such as algorithmic governance, platform economies, and intelligent social management—are not only reshaping the paradigms of social science research but are also propelling the diversification and increasing complexity of research subjects, thereby offering richer and more varied possibilities for inquiry.

In the field of economics, traditional research objects have primarily centered on supply–demand analysis, market equilibrium, and related topics, typically employing survey sampling and statistical analysis. In the AI era, however, research has expanded to include areas such as big data analytics, behavioral economics, fintech, and labor market dynamics. The advent of big data and machine learning has enabled economic research to shift from small sample studies to analyses involving massive datasets and large-scale models, thereby enhancing both the precision and depth of research and allowing for a more detailed characterization of complex economic systems [8]. For example, the use of machine learning algorithms to analyze vast quantities of consumer transaction data can yield more accurate market demand forecasts, optimize firms' production and marketing strategies, and improve the efficiency of resource allocation. Nevertheless, the extensive reliance on algorithms and big data for economic analysis may also lead to market inequalities and information monopolies. This is primarily because algorithms, in processing data, might inadvertently reinforce pre-existing social biases, resulting in unfair treatment of certain groups. Consequently, economic research has increasingly turned its attention to issues of algorithmic ethics, data privacy protection, and model interpretability, with a view to developing analytical tools that are both efficient and equitable.

In sociology, traditional research has focused on social structures, social relationships, and social change, typically employing questionnaires and in-depth interviews. With the rapid development of artificial intelligence and big data technologies, the availability of social media data, smart device data, and large-scale social network analyses now enables sociologists to investigate social interactions and structural changes with unprecedented breadth and depth. Machine learning algorithms can be applied to analyze group behaviors, the dynamics of social influence, and the formation of social capital within networks, thereby revealing latent patterns in social relations [9]. As a result, the subjects of sociological inquiry have expanded to include virtual social spaces, online behavioral patterns, and digital societies. For instance, analyzing data from platforms such as Weibo or WeChat can facilitate studies on the formation and dissemination of online public opinion, thus contributing to more effective social governance. However, the use of personal data in social research raises significant concerns regarding privacy and data security, with ethical issues in data collection and analysis becoming increasingly prominent. Although data analysis on social media platforms can uncover users' networks, interests, and behavioral patterns, inadequate privacy protection measures may result in the leakage or misuse of sensitive personal information. Moreover, algorithmic recommendation systems might create filter bubbles that reinforce users' existing biases, restrict exposure to diverse viewpoints, and exacerbate social fragmentation.

In the field of law, traditional research objects have included the interpretation of legal texts, analysis of judicial cases, and the development of legal institutions, primarily employing textual analysis and case studies. The advent of artificial intelligence has broadened these objects to encompass new areas such as AI-generated content, smart contracts, and legal technology (Legal Tech). Advances in natural language processing now allow for the automated analysis of legal texts, enabling legal scholars to systematically organize and compare vast collections of case precedents and statutes, thereby gaining deeper insights into patterns and deviations in legal applications. Additionally, AI-generated simulated scenarios can be employed in experimental legal research to examine the efficacy of different legal systems under specific conditions. As a result, legal research has expanded to address new challenges brought about by artificial intelligence, such as data privacy protection, algorithmic transparency and fairness, and the attribution of liability in autonomous driving incidents [10]. For instance, if an AI-assisted decision-making system exhibits bias or defects, it may lead to erroneous judicial outcomes, with accountability becoming difficult to determine. Similarly, in the event of an accident involving an autonomous vehicle, it remains unresolved whether liability should rest with the vehicle owner, the manufacturer, or the algorithm developer—issues that current legal frameworks have yet to fully address.

In philosophy, traditional research objects have encompassed fundamental issues such as ontology, epistemology, ethics, and aesthetics. The rise of artificial intelligence has prompted a reevaluation of the relationship between humans and machines, as well as the meaning of human existence. In creative domains such as art and literature, the application of AI has partially supplanted human intellectual activities, thereby posing new challenges to the notion of human subjectivity. Philosophical inquiry now must scrutinize the ethical dimensions of human–machine interactions and consider the impact of technological progress on free will, responsibility, and human well-being. Questions regarding whether artificial intelligence might possess consciousness and how to define the ethical status of machines have emerged as significant topics of debate. Furthermore, the rapid development of AI challenges traditional conceptions of human uniqueness, free will, and moral responsibility. If machines are capable of simulating or even surpassing human cognitive functions, what then remains of human distinctiveness? Should artificial intelligence be held accountable for moral responsibilities? Such questions have spurred profound reflections on technology ethics and the essence of humanity. In this context, philosophical research must reassess the relationship between humans and technology and explore how individuals can preserve their intrinsic value and dignity in a society where technological advancements are pervasive.

2.2 New Problems and Challenges for Economic and Social Development in the Age of Artificial Intelligence

The rapid development of artificial intelligence is reshaping the economic and social landscape at an unprecedented pace, ushering in profound transformations and entirely new developmental trends. From the perspective of economic growth models, key hallmarks of this era include intelligent manufacturing and personalized customization, the transformation of decision-making mechanisms driven by data, and the cross-industry integration and innovation. For example, the rise of intelligent manufacturing is gradually supplanting traditional mass production methods. The widespread deployment of robotics and automation not only enhances production efficiency and reduces costs but also enables more intelligent and refined production management [11]. By leveraging big data analytics and machine learning algorithms, firms can accurately predict consumer preferences and implement full-chain personalized customization, thereby boosting customer satisfaction and opening new avenues for growth. Concurrently, data-driven decision-making has emerged as a core feature of the digital economy era. The application of big data and AI technologies allows companies to extract insights from vast data resources, optimize decision-making processes, enhance efficiency, and foster innovations in business models [12]. Moreover, industry convergence has become one of the most conspicuous trends in the AI era. For instance, the integration of the Internet with manufacturing has given rise to the industrial Internet; the emergence of financial technology is reshaping financial services; and the combination of education and AI is promoting personalized learning and the proliferation of online education [13]. These trends not only generate new business formats and models but also accelerate technological iteration and the broader transformation of economic and social systems [14]. Clearly, the pervasive application of artificial intelligence is altering the fundamental logic of economic and social development, with data emerging as a key new production factor and algorithms coupled with computational power serving as core drivers of economic growth [15].

The AI-driven technological revolution has significantly boosted productivity, yet it is also accompanied by challenges related to labor market transformations, data security and privacy protection, as well as the lag in ethical and legal norms. First, the evolution of AI has instigated shifts in the labor market and heightened issues of social inequality. Automation and intelligent systems are progressively replacing traditional manual labor, particularly in sectors such as manufacturing, logistics, and finance, where many repetitive and low-skilled jobs are being supplanted by intelligent machines and algorithms. At the same time, AI development is giving rise to new occupations and skill demands, compelling workers to continually acquire new competencies to adapt to evolving technological environments. Given that technological dividends tend to preferentially benefit workers with higher education and skill levels, this structural imbalance in the labor market may further exacerbate social inequality [16].

Second, data security and privacy protection have emerged as central challenges in the AI era. Since AI relies heavily on the collection of vast amounts of data, companies increasingly mine user behavioral data for precision marketing, thereby amplifying the risks of data leakage and misuse. In the realm of public governance, large-scale data collection can infringe on citizens' privacy, and the improper use of such data not only undermines individual interests but also threatens social stability. This situation underscores the urgent need for enhanced data protection regulations and robust technological safeguards [17, 18].

Finally, the lag in ethical and legal frameworks represents another significant challenge in the AI era. The pace of AI technological advancement often outstrips the capacity of existing ethical norms and legal systems to keep pace. For example, issues such as determining liability in traffic accidents involving autonomous vehicles, ascertaining the copyright of AI-generated content, and addressing algorithmic biases in intelligent decision-making all entail complex ethical and legal considerations. Currently, existing legal frameworks and ethical standards fall short of fully addressing these emerging issues, resulting in legal gaps and ethical controversies in the application of these technologies.

In summary, the AI era presents unprecedented opportunities and challenges for social science. Research in the social sciences must adapt to these technological developments by broadening its research subjects and methods and by strengthening interdisciplinary collaboration. At the same time, it is imperative to remain vigilant about the potential ethical, privacy, and security issues that new technologies may engender. Only by upholding both scientific rigor and ethical principles can social science effectively contribute to the development and progress of human society.

3 Paradigm Shift in Social Science Research in the Age of Artificial Intelligence

The concept of a "paradigm" was first introduced by Thomas S. Kuhn, who described it as "the accepted exemplars of certain scientific practices that provide models for particular coherent traditions of scientific research." In this context, a paradigm refers to the basic norms and methodologies that researchers in a given field accept and adhere to [19]. The social science research paradigm encompasses the fundamental methods, theoretical frameworks, and academic norms used to study social phenomena and human behavior. This paradigm offers systematic modes of thought and methodological guidance, enabling researchers to comprehend and interpret complex social phenomena within established theoretical and methodological systems. However, constrained by strict theoretical assumptions, limited sample data, and singular research subjects, traditional paradigms in social science research have often been criticized for their lack of cognitive accuracy [20]. The advancement of artificial intelligence has once again challenged these traditional paradigms. The high dimensionality of data, the vast number of model parameters, and the intrinsic complexity of systems have produced qualitative leaps in both the breadth and depth of social science research. Consequently, the boundaries of theoretical foundations, research scope, complexity, and methodologies have expanded further. Social science research is increasingly evolving toward interdisciplinary integration, culminating in an AI-driven paradigm shift and the emergence of new avenues for computational social science research [21, 22].

3.1 From Theory-Driven to "Data-Model" Driven Research

Traditionally, social science research has been anchored in classical theoretical frameworks—such as market theory in economics or structural-functional theory in sociology—and, while empirical analyses (including those in economics) often employ data and mathematical models, limitations in data scale and model parameters have generally confined research to an analysis of social phenomena driven by theory. Social science is influenced by a multitude of factors—culture, religion, national context, political systems, and ethnicity—which often give rise to multiple, sometimes conflicting, theoretical explanations for the same phenomenon. For instance, debates between Keynesian and Austrian schools in economics illustrate how divergent theories can offer diametrically opposed interpretations of the same economic events. Such a theory-driven approach can lead to a myopic and self-referential mode of inquiry. The advent of big data and artificial intelligence offers a potential remedy to these shortcomings by shifting social science research from a purely theory-driven model to one driven by "data–model" integration [8, 20].

Following the emergence of large language models in artificial intelligence, the support provided by hundreds of billions—or even trillions—of parameters and massive datasets has enabled these models to offer more comprehensive explanations of complex issues, thereby significantly enhancing our ability to analyze and understand social problems [21]. First, the nature of research data has shifted from low-dimensional structured data to high-dimensional unstructured data. Traditional social science research has typically relied on small-scale, structured numerical data (such as economic or statistical data), whereas generative AI can harness vast quantities of multimodal unstructured data—encompassing text, images, audio, and video—from the Internet for information extraction and modeling analysis [23]. Second, the sources of data for social science research have been broadened. While traditional studies have been constrained by limited data processing capabilities—often relying on structured survey data—AI-based data processing relaxes these constraints, thereby expanding the channels for collecting survey data. Third, the availability of comprehensive literature data has increased. Literature reviews, which are indispensable in social science research, have been greatly enhanced by large language models that improve the capacity for reading, organizing, and even independently searching for and systematically cataloging literature on specific topics, thereby establishing a new paradigm for intelligent research [24].

Supported by big data and large-scale models, social science research is becoming increasingly data-driven. At the same time, researchers are engaging in simulated interactions with data, literature, and events via large models and social networks, thereby progressively transforming traditional social science into computational social science [25].

3.2 From Traditional Sociological Analysis to Interdisciplinary Analysis

As a novel form of productive force, the rapid development of artificial intelligence (AI) technology and its gradual penetration into various industries have given rise to new social phenomena [26], new modes of production [22], and new organizational forms [27] that span multiple fields. This evolution necessitates that social science research move beyond analyses confined solely to traditional sociological perspectives; it must also fully consider these emergent social phenomena and the technological foundations that underlie them. In particular, AI-based social science research relies extensively on various information technologies, inherently linking it with disciplines such as computer science, information economics, and physics. Consequently, in the AI era, social science research is bound to transition from a single-discipline focus to an interdisciplinary approach.

The boundaries among traditional social science disciplines have long been blurred—for example, between economics and political science, economics and sociology, sociology and history, and history and political science—and the advancement of AI further reinforces this ambiguity. Contemporary economics, for instance, is deepening its engagement with techno-economic analysis; it now encompasses not only traditional branches such as neoclassical economics, classical political economy, and macroeconomics, but also relies on insights from management studies, computer science, control theory, and other natural sciences. First, the new social phenomena introduced by AI require collaborative explanations from multiple disciplines. Each technological revolution inevitably brings forth entirely novel issues. For example, the Industrial Revolution precipitated Luddism [28], autonomous driving has sparked ethical debates [29], and the advent of ChatGPT has raised concerns about academic misconduct [30]. Such issues are often beyond the explanatory reach of any single discipline. For instance, addressing the question of copyright ownership in the context of AI necessitates integrating theoretical perspectives from computer science, law, and sociology. Second, the new modes of production engendered by AI also demand multidisciplinary explanations. In the AI era, intelligent technologies and algorithms have elevated data to the status of a key production factor in large-scale social production. This transformation not only reduces the reliance on human labor—potentially triggering an unemployment crisis and fostering social inequities such as algorithmic discrimination [28, 31]—but also accelerates the division between mental and manual labor. As a result, routine, low-creativity tasks are increasingly being supplanted by large-scale models and autonomous agents, gradually shifting the nature of labor toward greater creativity. Moreover, AI continues to create new value chains, as seen in emerging sectors such as online literature, short-form web dramas, digital broadcasting, and AI-generated video content. Consequently, research on modes of production must necessarily integrate insights from economics, sociology, philosophy, and other disciplines. Third, the new organizational forms facilitated by AI also require multidisciplinary elucidation. At the level of group organization, internet platforms and AI algorithms simultaneously promote interpersonal connections through interest alignment while exacerbating the fragmentation of communities [32], contributing to the polarization of online discourse. Issues such as "technological neutrality" and "algorithmic neutrality" must be considered with great care. At the level of organizational division of labor, internet platforms and AI algorithms are gradually transforming enterprise management from traditional hierarchical (pyramidal) structures to flatter organizational forms, thereby directly altering internal communication modes and decision-making processes. This shift not only impacts decision-making efficiency within firms but also reshapes the very logic of corporate management. Underpinning the continuous evolution of AI technology are advances in computer science, control theory, and related fields in the physical sciences.

3.3 From Linear, Single-Cause Research to Multivariate Complexity Research

Since its inception, social science has struggled to establish its scientific legitimacy under the Newtonian mechanistic paradigm. Within this framework, social science was built on linear thinking and deterministic causality, often neglecting the true complexity of social reality by focusing on reductive, fragmented analyses based on unidirectional causal relationships. With the development of AI technology, an increasing number of social scientists have come to acknowledge that social phenomena are products of complex systems characterized by multiple causes and effects and nonlinear dynamic evolution. This recognition necessitates a more comprehensive approach to studying social phenomena—one that draws on complexity theory to understand the intricate mechanisms at work. This paradigm shift represents not only a methodological advance but also a profound epistemological transformation, driving social science research toward a more integrated and dynamic orientation. It requires adjustments in both epistemological and methodological stances, moving beyond the traditional oscillation between social constructivism and social realism and avoiding the confines of strict methodological individualism or holism. Instead, researchers must employ a variety of theories and methods within a complexity framework to fully grasp and explain the multifaceted nature of social phenomena.

First, there is an increasing emphasis on nonlinear dynamic changes. Traditional social science research has often relied on linear causal models, whereas a complexity-oriented approach leverages the capabilities of large-scale models to focus on the interactions among multiple factors and their collective influence on social phenomena. This approach more accurately captures the dynamic and intricate nature of real-world society. For example, in financial markets, the interplay among individual trading behaviors, macroeconomic factors, and policy interventions gives rise to unpredictable fluctuation patterns. In such cases, a simple linear, single-cause explanation is insufficient to account for the complexity of market dynamics. Utilizing big data analytics and machine learning models allows researchers to capture these nonlinear relationships, thereby more closely simulating the true dynamics of market fluctuations.

Second, there is a shift from localized analysis to a more holistic, global analysis. The proliferation of AI technologies has led to new social phenomena that manifest as complex systems with emergent properties—where the overall characteristics of a system exceed the mere sum of its parts once a certain level of complexity is reached [33]. For instance, consider large-scale project management across multiple time zones. In such extensive, distributed teams, individual contributions may initially appear independent and dispersed. However, with the integration of AI-enabled collaborative tools—such as intelligent scheduling, real-time collaboration, and data analytics—the overall efficiency of the team can far exceed the simple aggregation of individual efforts. Specifically, AI technologies facilitate dynamic optimization of resource allocation, predict potential issues, and automatically generate solutions, thereby fostering an adaptive, synergistic mechanism within teams. This systemic collaborative effect not only enhances overall performance but also reveals innovative capacities that individual contributions alone cannot achieve.

Third, social science research is increasingly reliant on massive datasets. With the advent of AI and big data technologies, real-time data analysis and dynamic decision support systems have become integral to responding swiftly to social changes. Moreover, social scientists are now employing AI techniques to model and simulate complex social phenomena, exploring how social systems might behave under varying conditions and assessing the potential impacts. For example, in studies of online public opinion, during sudden events (such as epidemics or natural disasters), online rumors can spread rapidly and incite social panic. Researchers can utilize big data technologies to monitor the dynamic process of rumor propagation, combined with social network models to analyze key nodes and accelerating factors (such as highly influential users or trending topics), and subsequently develop real-time public opinion monitoring systems that employ sentiment analysis and text mining for early detection of rumors.

4 The Transformation of Social Science Researchers in the Age of Artificial Intelligence

Against the backdrop of rapid advancements in artificial intelligence, traditional social science scholars—such as economists, sociologists, and political scientists—are gradually shifting from conventional academic research pathways toward approaches that increasingly rely on technology and emphasize interdisciplinary integration. They are systematically acquiring digital skills and deeply applying AI tools and methods to enhance their capacity to understand and address complex contemporary social, economic, and political issues.

4.1 Recasting Academic Roles: From Traditional Knowledge Production to Technologically Enabled Knowledge Production

First, there is the pursuit of cross-disciplinary technological collaboration. Traditionally, social science researchers have confined their inquiries to their own disciplinary boundaries, seldom seeking technical partnerships. In the AI era, however, contemporary social scientists are collaborating more closely with computer scientists, data engineers, and AI experts to tackle social issues. For example, they may work with computer scientists to investigate algorithmic bias and its impact on vulnerable groups or join forces with data scientists to analyze large-scale social media datasets in order to examine shifts in public opinion and social behavior. Such collaborations not only break down disciplinary barriers but also present new challenges and opportunities. In this process, researchers must balance traditional social science methods with technical tools—maintaining acute insights into social phenomena while also developing technical analytical capabilities to address increasingly complex sociotechnical issues. Through these efforts, social science scholars can achieve a more comprehensive understanding of the interplay between technology and society.

Second, AI is being leveraged to uncover tacit knowledge. Traditionally, knowledge production has been a human-centric endeavor that synthesizes new practical experiences with existing academic theories, viewpoints, and propositions—thereby endowing established academic systems with fresh conceptual, temporal, and cultural significance and constructing new intellectual frameworks and discourses [34]. Yet this mode of knowledge production has its limitations, as captured by traditional adages that imply the ineffable nature of certain insights. In the AI era, technological augmentation has expanded the capabilities and methodologies available to social science researchers as knowledge producers. Breakthroughs in AI have enabled the extraction of tacit knowledge hidden within massive datasets [35]. For instance, researchers at the Massachusetts Institute of Technology analyzed approximately 126,000 pieces of news dissemination data from Twitter spanning 2006–2017 using AI models to reveal patterns in the spread of fake news. They found that fake news propagates faster and more widely than authentic news, a phenomenon especially pronounced in political reporting. By analyzing the sentiment, structure, and transmission pathways of news via AI models, the research team offered new theories of social communication and strategies for governance [36].

Third, scholars must discern the "knowledge illusion." While the widespread adoption of AI facilitates easier access to and analysis of information—thereby enabling the rapid synthesis of knowledge—it also poses new challenges to researchers' critical thinking and creativity. Given the current technical logic underlying large language models, the "knowledge illusion" generated by AI appears intractable [37]. In the course of exploration and innovation, AI systems may very well "speak nonsense in a serious tone." Social science researchers must, therefore, assume the responsibility of rigorously verifying and scrutinizing AI-generated content, especially when addressing complex social issues and ethical concerns. Researchers need to thoroughly understand the operational principles and limitations of AI to ensure the validity and reliability of their work. Moreover, an overreliance on AI in knowledge generation may lead to homogenization in research, as scholars' outputs converge in content and approach. Consequently, social science researchers face the challenge of maintaining originality and creativity in an AI-augmented environment. This requires not only proficiency with technical tools but also the infusion of humanistic insight and social sensitivity to foster genuine innovation.

4.2 *Increased Interaction Between Academic Research and Social Reality*

First, the research outputs of social science scholars are increasingly focused on their societal impact rather than remaining confined solely to academic discourse. This shift ensures that research not only provides theoretical support to academia but also directly influences policy-making and social practice. For instance, studies on algorithmic discrimination, privacy protection, and the impact of AI on the labor market have furnished critical theoretical foundations for governmental regulatory policies [38]. Simultaneously, these studies have exerted a profound influence on public opinion. Clearly, social science researchers are gradually assuming key roles in technological transformations, offering new knowledge frameworks that help both the public and policymakers better understand and address the challenges posed by AI.

Second, the role of social science researchers is transitioning from that of traditional scholars to that of public intellectuals and policy advisors. An increasing number of academics are actively engaging in public policy debates through media, think tanks, and governmental consultations, thereby providing crucial support for society in confronting the complex challenges brought about by AI technologies [39]. For example, on issues such as AI-induced unemployment, algorithmic transparency, and platform monopolies, many social science scholars have offered valuable insights and recommendations. This shift indicates that social science researchers are no longer confined to the ivory tower of academia; rather, they are emerging as pivotal agents of social change, reflecting the academic community's heightened commitment to social responsibility.

Moreover, the growing interaction between academia and society has spurred the development of new research directions and methodologies. In the future, as AI technology continues to evolve, the interplay between academic research and societal practice will become increasingly frequent. Social science researchers will need to further expand their domains to adapt to the novel challenges brought about by technological change. Establishing closer mechanisms for collaboration between academia and society will facilitate more effective knowledge dissemination and policy feedback, thereby promoting sustainable social development. In the AI era, social science researchers are gradually transitioning from traditional academic roles to explorers at the intersection of technology and society, emerging as a crucial force in understanding and guiding the societal impacts of technological change. This transformation not only reflects the profound influence of AI and digital technologies on social science but also provides new perspectives for gauging the evolution of both social science and AI technology.

5 Trends and Outlook

The Royal Society posits that the rapid advancement of artificial intelligence (AI) technology may usher society into a new turning point. AI is not only providing novel tools for scientific inquiry but is also transforming the very structure of scientific development, presenting both unprecedented opportunities and significant challenges. Social science, which investigates human behavior, socio-economic development, and the relationships among individuals as well as between humans and nature, now faces profound shifts. AI is revolutionizing human production and lifestyles, fundamentally altering the interactions between people and technology, as well as between people and nature, and even challenging the ethical and moral foundations of human progress. Consequently, AI is set to have a comprehensive and far-reaching impact on the social sciences. Looking forward, the influence of AI on social science is likely to manifest in at least the following trends:

First, AI cannot completely replace or eliminate social science. Some industry forecasts have predicted that, with the scaling up of parameters in generative AI, the training process will soon exhaust the corpus of human knowledge, precipitating the rapid emergence of a general AI that surpasses human intelligence. Nevertheless, social science represents a crucial crystallization of human knowledge, and its practice embodies uniquely human intellectual capabilities. The primary purpose of creating AI has been to liberate and assist humanity, not to supplant or obliterate it, including the complete replacement of human intelligence. The human brain comprises an enormous number of neurons with highly complex biological mechanisms underlying creative thought. While AI may outperform humans in areas such as computation and logical reasoning, at least in the near term it remains incapable of fully simulating or replicating the creative cognitive processes of the human brain through mathematically modeled electronic neural networks. Moreover, the data used to train AI consist solely of the explicit knowledge generated throughout human development; a vast reservoir of tacit knowledge—knowledge that cannot be fully articulated—remains inaccessible to AI tools, even though such tacit knowledge may prove crucial for human development and future socio-economic progress [40].

Second, AI will accelerate stratification among social science researchers. Although current AI systems cannot entirely replace social scientists, they are rapidly emerging as vital new research tools across various social science disciplines. AI facilitates not only the discovery of new research questions and the standardization of research paradigms but also significantly enhances research efficiency. Consequently, expertise in AI technology will increasingly be inseparable from exceptional research skills. Just as AI may create an "intelligence gap" between different regions or demographic groups, a similar gap is likely to arise among social science researchers. Those who can adopt and effectively leverage advanced AI tools early on will likely experience accelerated professional growth and produce higher-quality research outcomes, thereby catalyzing the formation of new AI-driven research clusters across diverse disciplines.

Third, AI will accelerate interdisciplinary integration. Advances in AI technology are not only reshaping economic and social development but are also introducing new research topics and objects to the social sciences. This dynamic is driving the evolution and innovation of disciplines around new technologies, new models, and new business forms. Moreover, as a general-purpose technology, AI provides robust research tool support across all areas of the social sciences. The recent emergence of fields such as AI economics, computational sociology, and computational linguistics exemplifies the growing interdisciplinary integration among the social sciences, natural sciences, and engineering disciplines, as well as within social science itself. This trend toward convergence is expected to become even more pronounced in the future.

Fourth, data-driven research will emerge as a critical paradigm in the social sciences. Traditional research methodologies in the social sciences have primarily relied on qualitative analysis, quantitative analysis, and simulation studies. However, the advent of big data allows researchers to uncover the inherent connections and regularities underlying societal complexity through sufficiently large datasets. The development of large language models is poised to transform the epistemological and methodological foundations of social science research. Approaches that have traditionally focused on problem-based or event-based studies will increasingly be grounded in data-driven methods. These data-driven large models not only offer fresh perspectives for social science research but also open entirely new avenues for exploring the deeper, structural issues that underpin society.

In summary, the advent of AI is set to reshape the trajectory of social science research in multifaceted ways—by reinforcing the indispensability of human intellectual inquiry, deepening disciplinary stratification, fostering interdisciplinary collaboration, and catalyzing a shift toward data-driven methodologies.

References

1. Kurzweil R. The Singularity Is Nearer: When We Merge with AI[M]. New York:Random House, 2024.
2. Duan L. A Brief Analysis of the Impact of ChatGPT-like Technologies on China [J]. Internet Weekly, 2023, (08): 16–20.
3. Li G. Intelligent Research (AI4R): The Fifth Research Paradigm [J]. Journal of the Chinese Academy of Sciences, 2024, 39(01): 1–9.
4. Zhang B. Review and Outlook on the Development of Artificial Intelligence [R]. Beijing: Tsinghua University "Renwen Tsinghua Forum" Speech, April 2024.
5. Rosenblatt F. The perceptron: a theory of statistical separability in cognitive systems[M]. Washington: United States Department of Commerce, Office of Technical Services, 1958.
6. Guo Y. On the History, Current Status, and Future Development Strategies of Artificial Intelligence [J]. People's Forum: Academic Frontiers, 2021(23): 41–53.
7. Yao Q. Custody and Governance of Training Data for ChatGPT-like Large Models [J]. China Finance, 2023(6): 51–53.
8. Hong Y, Wang S. Recent Developments in Artificial Intelligence and Their Impact on the Research Paradigm in Economics [J]. Journal of the Chinese Academy of Sciences, 2023, 38(3): 353–357.

9. Ya W, Zhou Y, Chen Luyi. Identification and Analysis of Technological Innovation Paths in China's Artificial Intelligence Industry: A Patent Analysis Approach [J]. Science and Technology Management Research, 2019, 39(10): 210–216.
10. Calo R. Robotics and the Lessons of Cyberlaw[J]. California Law Review, 2015, 103(3): 513–563.
11. Dong X, McIntyre S H. The Second Machine Age: Work, Progress, and Prosperity in a Time of Brilliant Technologies[J]. Quantitative Finance, 2014, 14(11): 1895–1896.
12. McAfee A, Brynjolfsson E, Davenport T, et al. Big data: The management revolution[J]. Harvard Business Review, 2012, 90: 61–67.
13. Almasri F. Exploring the Impact of Artificial Intelligence in Teaching and Learning of Science: A Systematic Review of Empirical Research[J]. Research in Science Education, 2024, 54(5): 977–997.
14. Xu M, David J M, Kim S H. The Fourth Industrial Revolution: Opportunities and Challenges[J]. International Journal of Financial Research, 2018, 9(2): 90–95.
15. Guo Z, Fang A. Artificial Intelligence Promoting High-Quality Economic Development: Mechanisms, Issues, and Countermeasures [J]. Guangxi Social Sciences, 2021(8): 8–17.
16. Frank M R, Autor D, Bessen J E, et al. Toward understanding the impact of artificial intelligence on labor[J]. Proceedings of the National Academy of Sciences of the United States of America, 2019, 116(14): 6531–6539.
17. Acquisti A, Brandimarte L, Loewenstein G. Privacy and human behavior in the age of information[J]. Science, 2015, 347(6221): 509–514.
18. West S M. Data Capitalism: Redefining the Logics of Surveillance and Privacy[J]. Business and Society, 2019, 58(1): 20–41.
19. Kuhn T S. The Structure of Scientific Revolutions [M]. Chicago: University Of Chicago Press, 1996.
20. Cai Y, Wan X. A New Research Paradigm for Social Science in the Big Data Era [N]. China Social Sciences Daily, 2019-11-06(3).
21. Yang Y. Changes and Constants in Social Science Research in the Age of Artificial Intelligence [J]. People's Forum: Academic Frontiers, 2024(4): 96–105.
22. Sun M. Exploring New Frontiers in Computational Social Science Research [N]. China Social Sciences News, 2022-06-15(1).
23. Hong Y, Wang S. How Does Big Data Change the Research Paradigm in Economics? [J]. Management World, 2021, 37(10): 40–55.
24. Yan S, Yu G. Intelligent Methods as the "Fifth Paradigm": A New Species of Research Paradigm in the Age of Artificial Intelligence [J]. Academic Exploration, 2024(1): 34–43.
25. Li Q. When Artificial Intelligence Meets Computational Social Science[J]. People's Forum: Academic Frontiers, 2019(20): 6–12.
26. Guan Q. Intelligent Sociology: A New Direction for Sociological Research in the Age of Intelligence [J]. Journal of South China University of Technology (Social Science Edition), 2022, 24(03): 1–9.
27. Luo W, Huo W, Zhao Y, et al. AI-Driven Organizational and Human Resource Management Transformation: Practical Insights and Research Directions [J]. China Human Resource Development, 2022, 39(01): 4–16.
28. Xie R. From Luddites and Neo-Luddites to Digital Workers: An Exploration of Constructivist Approaches in Technological Critique [J]. Dialectics of Nature Communications, 2022, 44(4): 80–89.
29. Li Z, Duan L, Lv B. Design Principles for Practical Ethical Guidelines in the Construction of Intelligent Systems [J]. China Public Administration, 2022(6): 41–48.
30. Wang S. ChatGPT and the Governance of Academic Misconduct: Challenges and Responses [J]. Science and Technology Progress and Policy, 2023, 40(23): 103–110..
31. Duan L, Lv B. Is There a "Reverse Diamond Paradox" in Online Shopping? [J]. Management Review, 2022, 34(9): 134–146.
32. Liu J, Zhang Y. Digital Segregation and Neighborhood Differentiation: The "Balkanization" of Community Information Spaces and Its Governance [J]. Dongyue Forum, 2023, 44(9): 67–77+191.

33. Jiang X. The Shift Toward Complexity in Social Science: Based on Edgar Morin's Complexity Theory [J]. Journal of Systems Science, 2018, 26(3): 20–24+42.
34. Sun Z. Modern Civilization of the Chinese Nation and China's Indigenous Philosophical Knowledge System [J]. Social Sciences in China, 2023, (08): 22–27.
35. Wang W. Tacit Knowledge [M]. Beijing: CITIC Publishing House, 2019.
36. Vosoughi S, Roy D, Aral S. The spread of true and false news online[J]. science, 2018, 359(6380): 1146–1151.
37. Chen W, Luo Y. An Exploration of the "Knowledge Illusion" in Generative Artificial Intelligence and Its Risk Governance [J]. Journal of the Shanghai Academy of Socialism, 2024, (04): 38–51.
38. Liu C. Manifestations, Causes, and Governance Strategies of Algorithmic Discrimination [J]. People's Forum, 2022, (02): 64–68.
39. Xia Z, Huo G. Digital Transformation of Think Tanks and the Emergence of Digital Think Tanks [J]. Think Tank Theory and Practice, 2024, 9(02): 22–28.
40. Long T, Pang Q. Social Science Research in the Age of Artificial Intelligence [N]. China Social Sciences News, 2023-12-22(5).

Xushu Peng the director and researcher of the Research Office of Informatization and Network Economy at the Institute of Quantitative Economics and Technological Economics of the Chinese Academy of Social Sciences, is also a professor and doctoral supervisor at the University of the Chinese Academy of Social Sciences. His main research directions are digital technology innovation and the digital economy.

Litao Duan the associate researcher at the Research Office of Informatization and Network Economy of the Institute of Quantitative Economics and Technological Economics at the Chinese Academy of Social Sciences. His main research directions are platform economy, informatization and information technology economy, and the digital economy.

Yuanhao Ban the assistant researcher at the Research Office of Informatization and Network Economy of the Institute of Quantitative Economics and Technological Economics at the Chinese Academy of Social Sciences. His main research directions are the digital economy, network economy, and international trade.

MIX
Papier aus verantwortungsvollen Quellen
Paper from responsible sources
FSC® C105338

If you have any concerns about our products,
you can contact us on
ProductSafety@springernature.com

In case Publisher is established outside the EU,
the EU authorized representative is:
Springer Nature Customer Service Center GmbH
Europaplatz 3, 69115 Heidelberg, Germany

Printed by Libri Plureos GmbH
in Hamburg, Germany